112 Advances in Polymer Science

Theories and Mechanism of Phase Transitions, Heterophase Polymerizations, Homopolymerization, Addition Polymerization

With contributions by
K. Binder, D. Hunkeler et al., J. P. Kennedy,
I. Majoros, A. Nagy, V. V. Shamanin

With 110 Figures and 33 Tables

Springer-Verlag Berlin Heidelberg GmbH

ISBN 978-3-662-14901-0 ISBN 978-3-540-47989-5 (eBook)
DOI 10.1007/978-3-540-47989-5

Originally published by Springer-Verlag Berlin Heidelberg New York in 1994
Softcover reprint of the hardcover 1st edition 1994

Library of Congress Catalog Card Number 61-642

Typesetting: Macmillan India Ltd., Bangalore-25
SPIN: 10126159 02/3020 - 5 4 3 2 1 0 - Printed on acid-free paper

Editors

Table of Contents

Conventional and Living Carbocationic Polymerizations United. I. A Comprehensive Model and New Diagnostic Method to Probe the Mechanism of Homopolymerizations

I. Majoros, A. Nagy and J.P. Kennedy
Institute of Polymer Science, The University of Akron, Akron, OH 44325-3909, USA

A closed-loop comprehensive model uniting impurity-induced and purposely-added initiator-induced isobutylene (IB) and styrene (St) polymerizations was developed. Both impurity-induced and purposely-induced olefin polymerizations can be both conventional or living, and the reaction conditions will determine whether the prevailing mechanism will be conventional or living. The model was used to elucidate the detailed mechanism of olefin polymerizations and to provide guidance toward preparative advances. The heart of the model is the Winstein ionicity spectrum which in its simplest form consists of three fundamental entities connected by two equilibria: a) a dormant species (in fact the initiator) which can be either a protic impurity or a purposefully added cationogen ("HX" or RX); b) a polarized covalent dipole $(\sim \overset{\delta\oplus}{C} --- \overset{\delta\ominus}{X} MtX_n)$ which arises from the dormant species under the influence of excess Friedel-Crafts acid coinitiator (MtX_n) and is the source of living chains, and c) a dissociated ionic species ($\sim C^{\oplus} + MtX_{n+1}^{\ominus}$) which yields conventional nonliving (i.e., transfer-dominated) chains. Under conventional conditions in charges containing a stoichiometric excess of MtX_n relative to the initiator (be this protic impurity or purposely-added cationogen), the contribution of the highly reactive ionic species to the polymerization rate is overwhelming and it is difficult to find evidence for the presence of the much less reactive dipole which is responsible for the relatively slow living propagation. Upon the addition of an electron pair donor (ED), hereafter electron donor, to a conventional charge, the ED and the excess MtX_n instantaneously form a complex $MtX_n \cdot ED$, which in turn interacts with the carbocation ($\sim C^{\oplus}$) and thus yields an additional (fourth) species $\sim \overset{\delta\oplus}{C} --- [\overset{\delta\ominus}{M}tX_n \cdot ED] \ldots MtX_{n+1}^{\ominus}$ in the Winstein ionicity spectrum which becomes another source of relatively slowly propagating living chains. The new model led to a comprehensive diagnostic method which in turn led to new insight into mechanistic details of olefin polymerizations in respect of rapid/slow initiation, monomolecular/bimolecular chain transfer, impurity-induced/purposely-induced initiation, rates of ion generation/cationation, and to an examination of the constancy of various rate constants. All the elementary events, namely initiation (comprising of ion generation and cationation), propagation, monomolecular and biomolecular chain transfer and reversible (quasiliving) termination, are controlled by their individual Winstein ionicity spectra, the characteristics of which determine the rates, conversions, product molecular weights and molecular weight distributions (MWD). Specifically, the effects of ED (triethyl amine, TEA) and Friedel-Crafts acid ($TiCl_4$) concentrations, medium polarity (mixtures of $CH_2Cl_2/n\text{-}C_6H_{14}$), temperature (from -20 to $-82\,°C$), and reagent addition sequence (aging) on IB and St polymerization rates, and polyisobutylene (PIB) and polystyrene (PSt) mol. wts. and MWDs were determined quantitatively and analyzed. The model shows the way toward the synthesis of controlled mol. wt. polyolefins of narrow MWD ($M_w/M_n < 1.1$) by living polymerization.

Advances in Polymer Science, Vol. 112

1 Introduction

Since the appearance of the major review on the living carbocationic polymerization of olefins [1], a large body of pertinent additional data have been generated relative to this subject [2–17]. The significance of these data prompts us to combine this recently-acquired information with earlier data and to integrate all kinds of cationic olefin polymerizations into a comprehensive mechanism, be these induced by means of a purposely-added initiator or by an impurity, both of which can lead to conventional (presence of chain transfer and/or termination) or living (absence of chain transfer and irreversible termination) polymerizations.

Rigorous definitions of conventional and living (more precisely quasiliving) carbocationic polymerizations have been developed [1, 18–21]. Very briefly, in conventional carbocationic polymerization chain transfer to monomer and irreversible termination are operational, whereas in living carbocationic polymerizations chain transfer is for all practical purposes absent and termination is present but rapidly reversible (hence the term quasiliving) [18]. It is emphasized that all known living carbocationic polymerizations (including alkyl vinyl ether polymerizations) are in fact quasiliving and, accordingly, have been classified as QL_{RO} systems [1, 18, 22, 23].

This treatise comprises three major parts: 1) the presentation of a novel *comprehensive closed-loop* model of impurity-initiated and purposely-initiated carbocationic polymerizations both of which can be conventional or living, which together with the corresponding kinetic equations derived, provides a valuable roadmap for the at first sight complicated-looking mechanism of these processes. This is the first time that impurity-initiated and purposely-initiated polymerization systems have been combined and their mechanisms discussed as a whole. The model is closed-loop because it accounts for *all* possible routes to the ultimate products arising from *all* starting materials (initiator, coinitiator, monomer) by *all* the elementary steps which can occur during a carbocationic polymerization, i.e., initiation (comprising of ion-generation and cationation), propagation, chain transfer to monomer (both zero and 1st order), and rapidly reversible termination; 2) a detailed quantitative examination of a large amount of data in terms of the new model. Specifically, we study IB and St polymerizations initiated by the 2-chloro-2,4,4-trimethylpentane (TMPCl)/$TiCl_4$ and "H_2O"/$TiCl_4$ combinations in the absence and presence of an electron donor (triethyl amine, TEA), and examine the effect of the concentrations of these chemicals, the polarity of the medium (various mixtures of CH_2Cl_2/n-C_6H_{14}), temperatures in the range from -20 to $-82\,°C$, and reagent addition sequence on the rates, conversions, molecular weights and molecular weights distributions; 3) practical consequences of the model and concluding thoughts.

2 The Model and Mechanistic Considerations

2.1 *The Comprehensive Closed-Loop Model*

Scheme 1 shows the model and the corresponding table the definition of the symbols. This model is *comprehensive* as it encompasses two formally different but fundamentally similar mechanisms: polymerizations induced by a purposely-added initiator (RX) shown on the left side of Scheme 1 and polymerizations induced by adventitious protic impurities "HX" on the right side of Scheme 1. The two sides of Scheme 1 are connected by two routes due to chain transfer to monomer; these will be discussed below. The model is valid for Friedel–Crafts acid (MtX_n) coinitiated polymerizations, including all kinds of conventional and living olefin and alkyl vinyl ether polymerizations.

The model is *closed-loop* as it accounts for *all* the recoverable products formed from *all* the starting materials (i.e., from the initiators "HX" or RX, the coinitiator MtX_n, and the monomer M).

The model is based on the proposition that all carbocationic polymerizations of olefins, alkyl vinyl ethers, etc., involve a spectrum of species with differing ionicities connected by equilibria formally expressed by the Winstein ionicity spectrum [1]; the Winstein spectrum starts with a covalent species and progresses through increasingly polarized and ionized species, to fully ionized solvated ("free") ion pairs:

$$\sim C - X + MtX_n \rightleftharpoons \sim C^{\delta\oplus} \text{- - -} \overset{\delta\ominus}{X}MtX_n \rightleftharpoons \text{- - - -} \sim C^{\oplus} + MtX_{n+1}^{\ominus} \, .$$

For the sake of simplicity the large number of possible entities in the Winstein spectrum can be reduced to three representative species [1, 24]:

$$\underset{D}{\sim C - X} + MtX_n \rightleftharpoons \underset{L}{\sim C^{\delta\oplus} \text{- - -} \overset{\delta\ominus}{X}MtX_n} \rightleftharpoons \underset{C^{\oplus}}{\sim C^{\oplus}} + MtX_{n+1}^{\ominus}$$

where D signifies the dormant cationogen, L stands for a subspectrum of polarized (stretched or activated, more-covalent-than-ionic) dipole intermediates leading to living polymerization, and $C^{\oplus}$ summarizes a further subspectrum of ionized entities starting with contact ion pairs through solvent separated ion pairs to fully solvated "free" ion pairs. Earlier reviews [6, 25–28] extensively discussed these matters together with some of the consequences of the Winstein ionicity spectrum in carbocationic polymerization [1, 24].

First we will consider the events indicated on the right side of the scheme, the part which concerns polymerizations induced by ubiquitous protic impurities

"HX". The quotation marks around "HX" indicate that this entity originates from impurities; this symbolism will be used in the text to emphasize the adventitious nature of the species, however, for simplicity sake it was omitted in the scheme. The top set of equilibria (the first Winstein spectrum) summarizes the various species arising from "HX". "HX" stands for any protic initiating impurity, e.g., H_2O (X=OH), HCl (X=Cl) that may form by hydrolysis of the Friedel-Crafts acid: $H_2O + MtX_n \rightarrow H^{\oplus} + MtX_nOH^{\ominus} \rightarrow HX + MtX_{n-1}OH \rightarrow$ etc. It is recognized but neglected in the discussion which follows that the hydroxylated counteranion $MtX_nOH^{\ominus}$ formed *in situ* is somewhat different from the counteranion $MtX_{n+1}^{\ominus}$ that arises from perhalogenated Friedel-Crafts acids (e.g., $TiCl_4$) considered below; in view of the low concentration of, and similarity between $MtX_nOH^{\ominus}$ and $MtX_{n+1}^{\ominus}$, the effect of this simplification is deemed negligible.

The last entity in the first line on the right shows the species $H^{\delta\oplus}$--- $[\overset{\delta\oplus}{M}tX_n \cdot ED] \ldots MtX_{n+1}^{\ominus}$ which arises upon the addition of an ED to a protic initiating system "HX" + MtX_n. As indicated by the straight and curved equilibria signs respectively, this species may form by adding the $MtX_n \cdot ED$ complex to the free ion pair $H^{\oplus} + MtX_{n+1}^{\ominus}$ or directly from "HX" + MtX_n. The $MtX_n \cdot ED$ arises, most likely instantaneously, upon the addition of an ED to the charge which contains a stoichiometric excess of MtX_n.

The reader is reminded that in essentially all carbocationic polymerization induced by Friedel-Crafts acids (MtX_n) described to date, the MtX_n is always in stoichiometric excess relative to that of initiator or protic impurities, and that the $[MtX_n]/[RX + \text{"HX"}]$ ratio is usually in the five to 20 range. Some of the consequences of this circumstance have been discussed (1).

The first set of three vertical arrows characterized by $k_{H^{\oplus},L}$, $k_{H^{\oplus},C^{\oplus}}$, and $k_{H^{\oplus},L(E)}$, respectively, indicate protonation of monomer M by the representative three protic species. These protonations yield three (representative) entities able to cationate incoming M.

The second set of three vertical arrows characterized by $k_{cH^{\oplus},L}$, $k_{cH^{\oplus},C^{\oplus}}$ and $k_{cH^{\oplus},L(E)}$, indicate the three cationation processes involving the living, nonliving free cationic, and living species obtained upon ED addition, respectively.

The products of these three events are three (representative) propagating species shown in the third or last row on the right side of the model. The loops characterized by $k_{pH^{\oplus},L}$, $k_{pH^{\oplus},C^{\oplus}}$ and $k_{pH^{\oplus},L(E)}$, indicate propagation by the living, nonliving free cationic, and living ED-mediated living species; the loops convey the repetitive nature of propagation, that the products of these steps are in the first approximation indistinguishable from the starting species.

The two large split loops originating from $HM_n^{\oplus} + MtX_{n+1}^{\ominus}$ and ascending respectively to $H^{\oplus} + MtX_{n+1}^{\ominus}$ and $HM_n^{\oplus} + MtX^{n+1}$, indicate the two possible chain transfer steps (the subscripts n, for example in $HM_n^{\oplus}$ and MtX_n, are integers 1, 2, . . . , n, and express the degree of polymerization (as in $\overline{DP}_n$) when they are subscripts to M, or the number of halogens connected to the metal in the Friedel-Crafts acid when they are subscripts to MtX); the longer of the two ascending loops characterized by the rate constant $k_{tr,{}^1M}$ indicates monomolecular (zero order in monomer) chain transfer, in the course of which the propagating chain end $HM_n^{\oplus}$ spontaneously expells a proton which reenters the scheme by the species $H^{\oplus} + MtX_{n+1}^{\ominus}$ in the first Winstein spectrum and simultaneously leaves behind a deprotonated product (for example, an olefin-ended product in the case of isobutylene) indicated by $\sim HM_n^{=}$. The shorter ascending loop characterized by $k_{tr,{}^2M}$ indicates chain transfer by the bimolecular (1st order in monomer) process. By this route the growing chain end loses proton directly to monomer (hence the $+M$ on the loop) and reenters the mechanism at $HM_n^{\oplus} + MtX_{n+1}^{\ominus}$ in the second Winstein spectrum. The product of this event is also a deprotonated polymer whose exit of the mechanism is indicated by $\sim HM_n^{=}$. Obviously, these two chain transfer steps represent two possible routes to recoverable product.

We note that both ascending loops are joined by chain transfer events originating at the left-side of the scheme (see the arrows joining at the bottom of the center of Scheme 1). This symbolism indicates that if chain transfer to monomer occurs in polymerizations initiated by purposely added initiator, the expelled $H^{\oplus}$s enter the mechanism at the same sites as the $H^{\oplus}$s expelled from growing ends formed from "HX." After all, a proton is a proton, irrespective where it comes from. *Thus by these dual channels the two polymerizations initiated on the one hand by RX and on the other hand by "HX" are integrated and combined into a comprehensive unit.*

The left half of the scheme, concerning polymerizations induced by RX, is conceptually identical (and therefore the symbolism parallel) to the right half of the model, the only difference being that on the left the headgroup of the polymer R arises from RX whereas on the right from "HX". At the risk of being somewhat redundant, these polymerizations are induced by $RX + MtX_n$ combinations which yield the first Winstein ionicity spectrum on the left. The two or (in the presence of ED) three representative initial species cationate incoming monomer by the process characterized respectively by $k_{cR^{\oplus},L}$, $k_{cR^{\oplus},C^{\oplus}}$ and $k_{cR^{\oplus},L(E)}$ to yield the second Winstein spectrum interconnecting the various propagating species. The smaller loops indicate the two or (in the presence of ED) three propagation routes characterized, respectively by the rate constants $k_{pR^{\oplus},L}$, $k_{pR^{\oplus},C^{\oplus}}$, and $k_{pR^{\oplus},L(E)}$. If chain transfer to monomer is operational, it will originate from $RM_n^{\oplus}$ and the expelled $H^{\oplus}$ reenters the mechanism by the routes indicated by the split ascending arrows; the significance of these twin ascending loops was pointed out in the preceding paragraph.

Thus the model consists of a series of intimately interconnected Winstein spectra. The individual Winstein spectra show only four representative species of the many possible. The total number of Winstein equilibria in a polymerization of this sort will depend on the number average degree of polymerization ($\overline{DP}_n$). For example, the trimerization of a monomer M induced either by "HX" or RX in the absence of chain transfer, will include four interconnected Winstein spectra: one Winstein spectrum for ion generation, followed by one for cationation of the first monomer unit, and two additional Winstein spectra for the two propagating steps. In the presence of chain transfer the number of Winstein spectra will multiply as suggested by the model.

The scheme is *complete* or *closed-loop* since all input (RX, M) ends up in product (RM-X, and $RM^{=}$ if chain transfer to M is operational), and is *comprehensive* since it includes product formed by adventitious impurity ("HX") *and* by purposely added initiator (RX).

Initiation is visualized to occur in two consecutive steps: ion-generation and cationation [1]. Initiation in impurity-induced polymerization consists of ion-generation from "HX" shown in the first Winstein spectrum on the right followed by protonation of M by any one of these species. Similarly, for purposely-induced initiation by RX, the first line on the left depicts ion-generation, which is followed by cationation of M. The chemistries of cationation and propagation are necessarily very similar [1].

The overall rate of polymerization may be determined by initiation, that is by ion-generation *or* cationation. Under ion-generation, in this context, we understand all the species formed from "HX" or RX in the Winstein ionicity $\overset{\delta\ominus}{X}MtX_n$ or $R^{\delta\oplus}$- - - $\overset{\delta\ominus}{X}MtX_n$, leading to living polymerization. According to two previous efforts aimed at elucidating which of the two steps, ion-generation cationation, was rate determining, cationation was found to be the slow step [29, 30]. In line with this information, cationation was considered rate determining in the present study (see later).

Termination is reversible (quasiliving polymerization) [31–33] and polymerization proceeds as long as M is available; irreversible termination can be "forced" (i.e., brought about by the addition of nucleophile, say quenching with methanol). In either event, the only product of termination is HM_nX or RM_nX. During forced termination, the addition of the excess nucleophile rapidly reacts with the excess MtX_n in the charge and thus displaces the mobile equilibria to the left. This explains the experimental fact that essentially the only products obtained by forced termination of cationic polymerizations are polymers carrying a terminal halogen HM_nX or RM_nX. For a further discussion of quenching, see [34].

According to the model, conventional polymerizations (i.e., polymerizations proceeding in the presence of chain transfer to monomer) involve ionic species, whereas living polymerization involve nonionic dipoles, or in the presence of $MtX_n \cdot ED$ the $MtX_n \cdot ED$-complexed species. The rate of free ion induced conventional polymerization is very high relative to that of the dipole induced living process because the rate constants $k_{H^\oplus,C^\oplus}$, $k_{cH^\oplus,C^\oplus}$, and in particular $k_{pH^\oplus,C^\oplus}$ are large. Indeed, $k_{pH^\oplus,C^\oplus}$ must be very large since the concentration of free ionic species is estimated to be exceedingly small (i.e., $\sim 10^{-5}$–10^{-8} mol/L [31]). *Because of this very large difference in the overall k_ps* (that is $k_{pH^\oplus,C^\oplus}$ and $k_{pH^\oplus,L}$), *it is very difficult to find evidence for living polymerization in an "HX"-induced system and the product formed by the ionic species totally overwhelms that formed by the living process arising from the nonionic dipole.*

In systems in which initiation is due to both "HX" and RX, the products usually exhibit a relatively broad (sometimes even bimodal) MWD (see later) because initiation by "HX" is faster than by RX. It is difficult to interpret the exact course of MWD broadening because this phenomenon may be due to at least two effects: relatively slower initiation than propagation in systems containing one type of propagating species, or two or more relatively slowly equilibriating propagating species (that is when the rates of equilibration between the propagating species are lower than those of propagation).

When the objective is living polymerization and ED is added to the system, the Winstein spectra will shift to the right toward $MtX_n \cdot ED$-complexed dipoles and propagation will proceed by the loops characterized by $k_{pH^{\oplus},L(E)}$, or $k_{pR^{\oplus},L(E)}$. Not too surprisingly, polymerizations induced by a purposely added initiator RX in the presence of ED will tend to yield three growing species (readily demonstrated by GPC, see Sect. 4.2.1.4.1). The first species, usually yielding the highest molecular weight product (at least at low or intermediate conversions), is due to "HX" and is proposed to be formed by very rapid propagation characterized by the $k_{pH^{\oplus},C^{\oplus}}$ loop. The second species is most likely due to somewhat slower initiation by RX and is proposed to grow by rapid propagation via the $k_{pR^{\oplus},C^{\oplus}}$ loop. Finally the third by GPC discernible species is proposed to arise by relatively slowest initiation by RX and to grow relatively slowly but to yield living polymer via the loop characterized by $k_{pR^{\oplus},L(E)}$. The product arising from "HX" may be prevented from forming by the use of a proton trap [35] and under well-chosen conditions uniform very narrow molecular weight distribution product ($\bar{M}_w/\bar{M}_n < 1.1$) can be obtained (see later).

According to the model, living polymerizations are obtained in the virtual absence of ions and one way to achieve this is by the addition of an ED to the charges [36]. Under these conditions, the formation of the $MtX_n \cdot ED$ complex will pull the Winstein spectra to the far right. Experimentally the MtX_n is usually added to charges containing the initiator, ED, and M; it is assumed that the complexation $MtX_n + ED \rightarrow MtX_n \cdot ED$ is extremely rapid and that the $MtX_n \cdot ED$ complex instantaneously interacts with the free cations before conventional polymerization can occur. The often broad ($\bar{M}_w/\bar{M}_n \sim 1.5$), skewed, and sometimes bimodal MWDs frequently observed in ED-mediated living polymerizations suggest that the second of these assumptions (i.e., the $MtX_n \cdot ED$ complex very rapidly interacts with free cations) may not be fully valid. Experiments in which MtX_n and ED were premixed (i.e., the charges were "aged" before M addition), gave very narrow MWDs (see Sections 4.1.1.4.5 and 4.2.1.4.5 which substantiates the validity and usefulness of the model. Carbocation stabilization by EDs was questioned by Faust et al. [4, 10, 37].

It is worth mentioning, although this subject is not treated further, that living carbocationic polymerization can also be obtained in the absence of EDs by the use of common anions added to the charges in the form of common anion salts or simple salts solubilized by crown ethers [38–41]. The model is valid for these systems as well: in these instances, obviously, the last species in the Winstein spectra (the last entity in the horizontal lines) is absent, and living polymerization is attained by the common anions pushing the equilibria toward the left, toward the $\overset{\delta\oplus}{RM}$ - - - $\overset{\delta\ominus}{X}MtX_n$ dipole. In living polymerization mediated by common anions the living route is by the loop characterized by $k_{pH^{\oplus},L}$ or $k_{pR^{\oplus},L}$. The rate constants $k_{pH^{\oplus},L}$ and $k_{pH^{\oplus},L(E)}$, or $k_{pR^{\oplus},L}$ and $k_{pR^{\oplus},L(E)}$ are most likely very similar.

Living IB polymerization may also proceed in the absence of added ED; for example, by using the *p*-dicumyl-methyl ether/$TiCl_4$ initiating system in $CH_3Cl/n\text{-}C_6H_{14}$ (40/60 v/v) at $-80\,°C$ [1, 42]. The living nature of this polymerization was demonstrated by linear $\bar{M}_n$ vs W_p plots up to $\bar{M}_n = 126\,000$ starting at the origin. According to these data [42], however, the $\bar{M}_n$s are about 10% higher than the theoretical values. This discrepancy may be due to either ~10% of the initiator not getting incorporated into the polymer, or that the sample was not uniform but bimodal and only the major peak reflecting the growth of the living species was taken into consideration when the $\bar{M}_n$s were calculated.

2.2 Mechanistic Considerations

2.2.1 Definition of Elementary Steps

2.2.1.1 Initiation

Initiation consists of ion-generation and cationation [1, 43]. Ion-generation may be due to adventitious impurities "HX" or to purposely added initiator RX. For simplicity, we consider ion-generation only in the forward direction:

$$\text{"HX"} + MtX_n \xrightarrow{k_{H^{\oplus}ig}} H^{\oplus}MtX_{n+1}^{\ominus} \tag{1}$$

and

$$RX + MtX_n \xrightarrow{k_{R^{\oplus}ig}} R^{\oplus}MtX_{n+1}^{\ominus}\,, \tag{2}$$

where $H^{\oplus}MtX_{n+1}^{\ominus}$ and $R^{\oplus}MtX_{n+1}^{\ominus}$ summarize all the species that arise from "HX" and RX, respectively, upon the addition of the MtX_n coinitiator. $k_{H^{\oplus}ig}$ and $k_{R^{\oplus}ig}$ summarize *all* the ion-generation rate constants that yield *all* the species from "HX" and RX in the absence or presence of electron donor (ED). Cationation is the irreversible first-order-in-monomer addition of monomer (M) to $H^{\oplus}MtX_{n+1}^{\ominus}$ or $R^{\oplus}MtX_{n+1}^{\ominus}$:

$$H^{\oplus}MtX_{n+1}^{\ominus} + M \xrightarrow{k_{H^{\oplus}}} HM^{\oplus}MtX_{n+1}^{\ominus} \xrightarrow[+M]{k_{cH^{\oplus}}} HM_2^{\oplus}MtX_{n+1}^{\ominus} \tag{3}$$

and

$$R^{\oplus}MtX_{n+1}^{\ominus} + M \xrightarrow{k_{cR^{\oplus}}} RM^{\oplus}MtX_{n+1}^{\ominus}\,, \tag{4}$$

where $k_{H^\oplus}$, $k_{cH^\oplus}$ and $k_{cR^\oplus}$ summarize, respectively, the six rate constants ($k_{H^\oplus L}$, $k_{H^\oplus C^\oplus}$, $k_{H^\oplus L(E)}$, $k_{cH^\oplus L}$, $k_{cH^\oplus C^\oplus}$, $k_{cH^\oplus L(E)}$) on the right plus the three rate constants ($k_{cR^\oplus L}$, $k_{cR^\oplus C^\oplus}$, $k_{cR^\oplus L(E)}$) on the left of Scheme 1 leading to the propagating species. (See vertical arrows in Scheme 1.)

2.2.1.2 Propagation

Propagation is the irreversible repetitive first order addition of monomer to the growing chain [1, 43]. The kinetics of propagation have been treated in detail [1]. According to our model, propagation includes species of different ionicities in the absence or presence of electron donor.

$$RM_n^{\oplus}MtX_{n+1}^{\ominus} + M \xrightarrow{k_p} RM_{n+1}^{\oplus}MtX_{n+1}^{\ominus} \quad (5)$$

where $RM_n^{\oplus}MtX_{n+1}^{\ominus}$ and $RM_{n+1}^{\oplus}MtX_{n+1}^{\ominus}$ stand for all species capable of propagation shown in the bottom lines of Scheme 1, and k_p summarizes *all* rate constants of propagation in the Winstein spectra (a minimum of four in the absence of EDs, chain transfer and/or uncontrolled initiation, and a maximum of six in the presence of these processes). After a few propagation steps ($n > 2$) the two sets of k_ps on the left and the right of Scheme 1 become identical because the influence of the different head-groups ("R" or "H") on the polymerization rate becomes negligible.

2.2.1.3 Chain Transfer

2.2.1.3.1 Chain Transfer to Initiator ("Inifering")

As shown in Scheme 1a (which supplements Scheme 1 and helps to visualize the inifering process), chain transfer to initiator [1, 43, 44] involves a growing chain plus unionized initiator (RX):

$$HM_n^{\oplus}MtX_{n+1}^{\ominus} + RX \xrightarrow{k_{tr,I}} R^{\oplus}MtX_{n+1}^{\ominus} + HM_nX \quad (6)$$

or

$$RM_n^{\oplus}MtX_{n+1}^{\ominus} + RX \xrightarrow{k_{tr,I}} R^{\oplus}MtX_{n+1}^{\ominus} + RM_nX. \quad (7)$$

This irreversible process affects the polymerization only if the rate of ionization of RX (i.e., ion-generation) is slow relative to that of propagation; if initiation is instantaneous or very rapid the effect of chain transfer to initiator is negligible [1, 24, 45]. If the rate of ionization of "HX" is high relative to that of propagation, chain transfer to "HX" may be neglected. Since HM_nX and RM_nX (see Scheme 1a) can be reactivated by the quasiliving equilibrium (see Sect. 2.1, and Scheme 1), chain transfer to initiator is by no means a termination reaction [see also 43].

Purposely-Induced Pzn

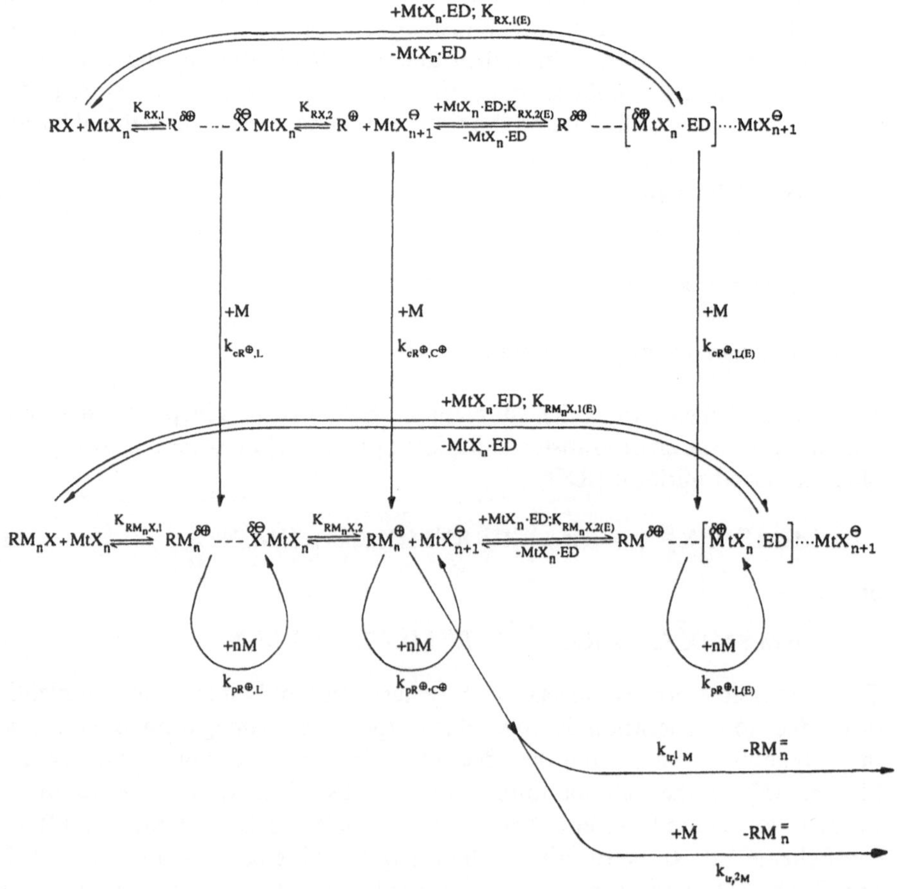

Scheme 1. A comprehensive closed-loop mechanism

Impurity-Induced Pzn

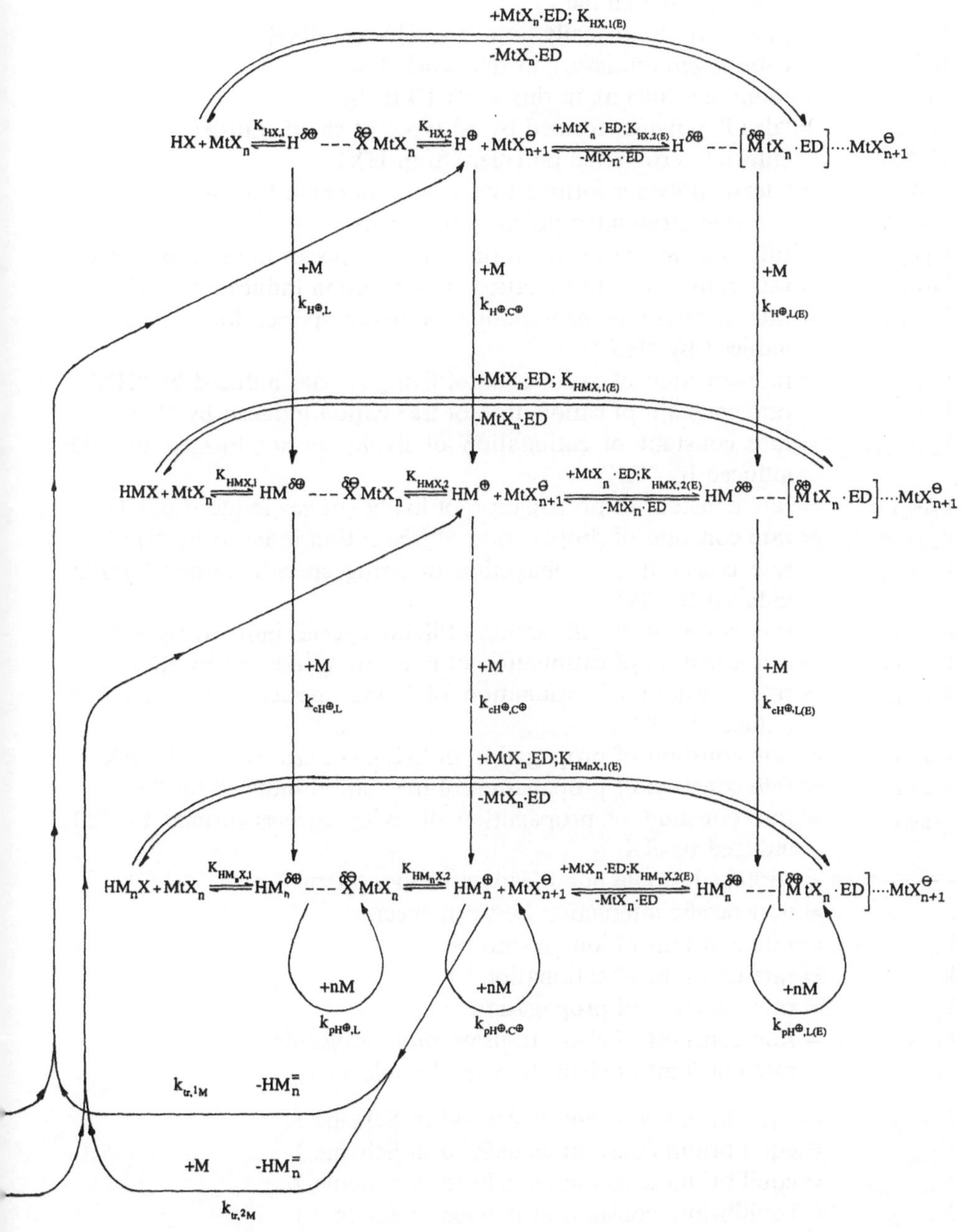

of carbocationic polymerizations

Symbols and Abbreviations used in Scheme 1

MtX_n	= Friedel-Crafts (or Lewis) acid; in this work $TiCl_4$
ED	= electron pair donor, hereafter electron donor; in this work triethyl amine
HX	= protic species (usually impurity "HX" or "H_2O")
RX	= cationogen (initiator); in this work TMPCl
M	= monomer (olefin); in this work IB or St
$HM_n^=$	= "dead" polymer formed by all types of chain transfer
HM_nX	= chlorine-terminated polymer (from HX)
$RM_n^=$	= "dead" polymer formed by all types of chain transfer
RM_nX	= chlorine-terminated polymer (DP = n)
$k_{H^\oplus,L}$	= rate constant of protonation of living species induced by "HX"
$k_{H^\oplus,C^\oplus}$	= rate constant of protonation of free cation induced by "HX"
$k_{H^\oplus,L(E)}$	= rate constant of protonation of living species formed by ED induced by "HX"
$k_{cH^\oplus,L}$	= rate constant of cationation of living species induced by "HX"
$k_{cH^\oplus,C^\oplus}$	= rate constant of cationation of free cation induced by "HX"
$k_{cH^\oplus,L(E)}$	= rate constant of cationation of living species formed by ED induced by "HX"
$k_{pH^\oplus,L}$	= rate constant of propagation of living species induced by "HX"
$k_{pH^\oplus,C^\oplus}$	= rate constant of propagation of free cation induced by "HX"
$k_{pH^\oplus,L(E)}$	= rate constant of propagation of living species formed by ED induced by "HX"
$k_{cR^\oplus,L}$	= rate constant of cationation of living species induced by RX
$k_{cR^\oplus,C^\oplus}$	= rate constant of cationation of free cation induced by RX
$k_{cR^\oplus,L(E)}$	= rate constant of cationation of living species formed by ED induced by RX
$k_{pR^\oplus,L}$	= rate constant of propagation of living species induced by RX
$k_{pR^\oplus,C^\oplus}$	= rate constant of propagation of free cation induced by RX
$k_{pR^\oplus,L(E)}$	= rate constant of propagation of living species formed by ED induced by RX
- - - - -	= stretched dipole bond leading to living Pzn
. . . . . .	= nonspecific interaction between species
k_{ig}	= rate constant of ion generation
k_c	= rate constant of cationation
k_p	= rate constant of propagation
$k_{tr,{}^1M}$	= rate constant of chain transfer, monomolecular
$k_{tr,{}^2M}$	= rate constant of chain transfer, bimolecular
$K_{HX,1}$	= equilibrium constant as defined in Scheme 1
$K_{HX,2}$	= equilibrium constant as defined in Scheme 1
$K_{HX,1(E)}$	= equilibrium constant as defined in Scheme 1
$K_{HX,2(E)}$	= equilibrium constant as defined in Scheme 1
$K_{RX,1}$	= equilibrium constant as defined in Scheme 1
$K_{RX,2}$	= equilibrium constant as defined in Scheme 1
$K_{RX,1(E)}$	= equilibrium constant as defined in Scheme 1

$K_{RX,2(E)}$ = equilibrium constant as defined in Scheme 1
$K_{RMnX,1}$ = equilibrium constant as defined in Scheme 1
$K_{RMnX,2}$ = equilibrium constant as defined in Scheme 1
$K_{RMnX,1(E)}$ = equilibrium constant as defined in Scheme 1
$K_{RMnX,2(E)}$ = equilibrium constant as defined in Scheme 1
$K_{HMX,1}$ = equilibrium constant as defined in Scheme 1
$K_{HMX,2}$ = equilibrium constant as defined in Scheme 1
$K_{HMX,1(E)}$ = equilibrium constant as defined in Scheme 1
$K_{HMX,2(E)}$ = equilibrium constant as defined in Scheme 1
$K_{HM_nX,1}$ = equilibrium constant as defined in Scheme 1
$K_{HM_nX,2}$ = equilibrium constant as defined in Scheme 1
$K_{HM_nX,1(E)}$ = equilibrium constant as defined in Scheme 1
$K_{HM_nX,2(E)}$ = equilibrium constant as defined in Scheme 1

Equilibrium constants characterizing the four equilibria each in the five Winstein spectra in Scheme 1.

Left Side of Scheme 1

1. row:

$$K_{RX,1} = \frac{k_{RX,1}}{k_{RX,-1}}; \quad K_{RX,2} = \frac{k_{RX,2}}{k_{RX,-2}}$$

$$K_{RX,1(E)} = \frac{k_{RX,1(E)}}{k_{RX,-1(E)}}; \quad K_{RX,2(E)} = \frac{k_{RX,2(E)}}{k_{RX,-2(E)}}$$

2. row:

$$K_{RM_nX,1} = \frac{k_{RM_nX,1}}{k_{RM_nX,-1}}; \quad K_{RM_nX,2} = \frac{k_{RM_nX,2}}{k_{RM_nX,-2}}$$

$$K_{RM_nX,1(E)} = \frac{k_{RM_nX,1(E)}}{k_{RM_nX,-1(E)}}; \quad K_{RM_nX,2(E)} = \frac{k_{RM_nX,2(E)}}{k_{RM_nX,-2(E)}}$$

Right side of Scheme 1

1. row:

$$K_{HX,1} = \frac{k_{HX,1}}{k_{HX,-1}}; \quad K_{HX,2} = \frac{k_{HX,2}}{k_{HX,-2}}$$

$$K_{HX,1(E)} = \frac{k_{HX,1(E)}}{k_{HX,-1(E)}}; \quad K_{HX,2(E)} = \frac{k_{HX,2(E)}}{k_{HX,-2(E)}}$$

2. row:

$$K_{HMX,1} = \frac{k_{HMX,1}}{k_{HMX,-1}}; \quad K_{HMX,2} = \frac{k_{HMX,2}}{k_{HMX,-2}}$$

$$K_{HMX,1(E)} = \frac{k_{HMX,1(E)}}{k_{HMX,-1(E)}}; \quad K_{HMX,2(E)} = \frac{k_{HMX,2(E)}}{k_{HMX,-2(E)}}$$

3. row:

$$K_{HM_nX,1} = \frac{k_{HM_nX,1}}{k_{HM_nX,-1}}; \quad K_{HM_nX,2} = \frac{k_{HM_nX,2}}{k_{HM_nX,-2}}$$

$$K_{HM_nX,1(E)} = \frac{k_{HM_nX,1(E)}}{k_{HM_nX,-1(E)}}; \quad K_{HM_nX,2(E)} = \frac{k_{HM_nX,2(E)}}{k_{HM_nX,-2(E)}}$$

2.2.1.3.2 Chain Transfer to Monomer

Chain transfer to monomer can be either monomolecular and zero order in monomer:

$$HM_n^{\oplus} \xrightarrow{k_{tr,\,^1M}} HM_n^{=} + H^{\oplus} \xrightarrow[k_{H^{\oplus}C^{\oplus}}]{+\,M} HM^{\oplus} \tag{8a}$$

or

$$RM_n^{\oplus} \xrightarrow{k_{tr,\,^1M}} RM_n^{=} + H^{\oplus} \xrightarrow[k_{H^{\oplus}C^{\oplus}}]{+\,M} HM^{\oplus} \tag{8b}$$

where $k_{tr,\,^1M} < k_{H^{\oplus}C^{\oplus}}$, or bimolecular and first order in monomer:

$$HM_n^{\oplus} + M \xrightarrow{k_{tr,\,^2M}} HM_n^{=} + HM^{\oplus} \tag{9a}$$

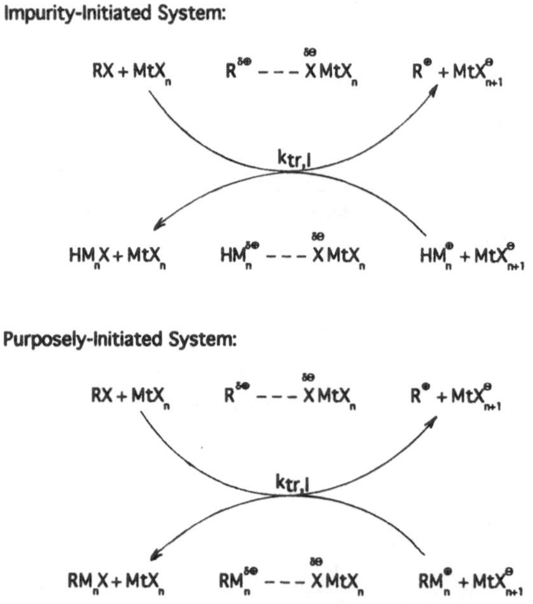

Scheme 1a. Visualization of Chain Transfer to Initiator

or

$$RM_n^{\oplus} + M \xrightarrow{k_{tr,\,^2M}} RM_n^{=} + HM^{\oplus} \tag{9b}$$

where $k_{tr,\,^1M}$ and $k_{tr,\,^2M}$ are the zero order and first order chain transfer rate constants, respectively. The difference between the chain transfer rate constants for the growing polymers carrying R or H head-groups is considered insignificant (see Scheme 1, bottom rows).

2.2.1.4 Termination

Irreversible termination is in principle absent in carbocationic polymerization coinitiated by perhalogenated Friedel-Crafts acids MtX_n. Reversible termination, however, exists in both impurity-induced or purposely-induced processes [43, 46, 47]. Indeed, because of the recognition that irreversible termination does not occur in MtX_n-coinitiated carbocationic polymerizations, we have developed the concept of quasiliving polymerizations the focal point of which is reversible termination [1]. The fundamentals of reversible termination and its consequences have been treated recently in great detail [18, 48, 49], and the reader is referred to these sources for further discussion on this subject.

As indicated by the bottom rows (Winstein spectra) in Scheme 1, the propagating ability of any of the three (or four) growing species may be temporarily lost by pushing the equilibria all the way to the left. However, the species RM_nX and/or HM_nX are not dead, they are merely dormant, and can be resurrected by the excess Friedel-Crafts acid in the charge (i.e., termination is reversible). The rates of resurrection are controlled by the equilibrium constants $K_{RM_nX,1}$ or $K_{HM_nX,1}$. If these constants are small resurrection will be slow and the system will appear "dead"; in other words, irreversibly terminated. This may be the case, for example, in one of the oldest conventional cationic polymerization system, the "H_2O"/$AlCl_3$/IB/CH_3Cl/n-C_6H_{14} system [50]. (Resurrection may be slow in this case because of the very low solubility of $AlCl_3$ and consequently low $AlCl_3$ concentration in the medium. In other systems resurrection is rapid because of the large excess of readily soluble Friedel-Crafts acids in the media.)

2.2.2 A new Diagnostic Method to Probe Comprehensively the Mechanism of MtX_n-Coinitiated Carbocationic Polymerization

The purpose of this phase of our studies was to develop a diagnostic methodology for the study of the mechanism of MtX_n-coinitiated carbocationic polymerizations outlined in the comprehensive model (Scheme 1). This goal was achieved by computer simulation of various effects determining the number of polymer chains (N). The results can be presented by four easily obtainable plots which have the diagnostic power to probe the mechanism of polymerization (See also Sect. 2.3):

1) total number of polymer chains (N) vs conversion (C), $N = f(C)$;
2) Molecular weight vs conversion, $\bar{M}_n = f(C)$;
3) rate as a function of time, $-\ln(1 - C) = f(t)$;
4) conversion as a function of time, $C = f(t)$.

The detailed analysis of various scenarios resulted in increased insight into the overall polymerization mechanisms outlined in Scheme 1.

Table 1. The 28 fundamental scenarios

Symbols	Presence or absence of events that determine the number of polymer chains (N): Initiation, Fast	Initiation, Slow: Ion gener. rate det.	Initiation, Slow: Cation-ation rate det.	Initiator is "HX"	Chain transfer: Zero order	Chain transfer: First order	Mathematical description of individual scenarios
A1	1	0	0	0	0	0	$N = I_0$ $-\frac{dM}{dt} = k_p \cdot I_0 \cdot M$
A2a	1	0	0	0	1	0	$\frac{dN_{tr}}{dt} = k_{tr,1M} \cdot I_0 \quad t = 0;$ $N_{total} = I_0 \quad M = M_0$ $-\frac{dM}{dt} = k_p \cdot I_0 \cdot M \quad N_{total} = I_0 + N_{tr}$
A2b	1	0	0	0	0	1	$\frac{dN_{tr}}{dt} = k_{tr,2M} \cdot I_0 \cdot M \quad t = 0;$ $N_{total} = I_0 \; M = M_0$ $-\frac{dM}{dt} = k_p \cdot I_0 \cdot M + \underline{k_{tr,2M} \cdot I_0 \cdot M}$ $N_{total} = I_0 + N_{tr}$
A2c	1	0	0	0	1	1	$\frac{dN_{tr}}{dt} = k_{tr,1M} \cdot I_0 + k_{tr,2M} \cdot I_0 \cdot M \quad t = 0;$ $N_{total} = I_0 \quad M = M_0$ $-\frac{dM}{dt} = k_p \cdot I_0 \cdot M + \underline{k_{tr,2M} \cdot I_0 \cdot M}$ $N_{total} = I_0 + N_{tr}$
A3a	0	1	0	0	0	0	$\frac{dN_i}{dt} = k_{ig}(I_0 - N_i) \cdot LA;$ $t = 0; \quad N_i = 0 \quad M = M_0$ $-\frac{dM}{dt} = k_p \cdot N_i \cdot M \quad N_{total} = N_i$
A3b	0	0	1	0	0	0	$\frac{dN_i}{dt} = k_c(I_0 - N_i) \cdot M$ $t = 0; \quad N_i = 0 \quad M = M_0$ $-\frac{dM}{dt} = k_p \cdot N_i \cdot M + \underline{k_c(I_0 - N_i)M};$ $N_{total} = I_i$

Resulting expressions			
N_{total} vs conv. (formula) $10^3\ N_{total}$ vs conv. (equation)	$\bar{M}_n$ vs conv. (formula) $10^{-3}\ \bar{M}_n$ vs conv. (equation)	$-\ln(1-\text{Conv.})$ vs time formula equation	Conv. vs time formula equation
$= I_0$	$= W_p \cdot C/I_0$	$= k_p \cdot I_0 \cdot t$	$= 1 - e^{-k_p \cdot I_0 \cdot t}$
$= 3$	$= 100C$	$= 0.3t$	$= 1 - e^{-0.3t}$
$= I_0 - c^{I} \ln(1-C)$	$= W_p \cdot C/(I_0 - c^{I} \ln(1-C)$	Same as A1	Same as A1
$= 3 - 0.2\ln(1-C)$	$= 300C/(3 - 0.2\ln(1-C))$		
$= I_0 - c^{II} \cdot M_0 \cdot C$	$= W_p \cdot C/(I_0 + c^{II} \cdot M_0 \cdot C)$	Same as A1	Same as A1
$= 3 + 0.6 \cdot C$	$= 300C/(3 + 0.6C)$		
$= I_0 - c^{I} \ln(1-C) + c^{II} \cdot M_0 \cdot C$	$= W_p \cdot C/(I_0 - c^{I} \ln(1-C)$ $+ c^{II} \cdot M_0 \cdot C$	Same as A1	Same as A1
$= 3 - 0.2\ln(1-C) + 0.6 \cdot C$	$= 300C/(3 - 0.2\ln(1-C) + 0.6C)$		
$C = 1 - e^{\frac{k_p}{k_{ig} LA}\left(I_0 \ln\left(\frac{I_0 - N_i}{I_0}\right) + N_i\right)}$ implicit	$= W_p \cdot C/N_i$	$= k_p \cdot I_0 \left[t + \frac{1}{k_{ig} \cdot LA}(e^{-k_{ig} \cdot LA \cdot t} - 1)\right]$	
$10^3 \cdot N_i = 3 \cdot (1 - e^{-5.18 \cdot C^{0.821}})$	$= 100C/(1 - e^{-5.18 \cdot C^{0.821}})$	$= 0.3 \cdot \left[t + 1.22(e^{-0.82t} - 1)\right]$ $= 1 - e^{-0.3[t + 1.22(e^{-2.82t} - 1)]}$	
$C = -\frac{k_p}{k_c M_0}\left(N_i + I_0 \ln\left(\frac{I_0 - N_i}{I_0}\right)\right)$ implicit	$= W_p \cdot C/N_i$	$= k_p \cdot I_0 \left[t + \frac{1}{k_c \cdot \bar{M}}(e^{-k_c \cdot \bar{M} \cdot t} - 1)\right]$	
$10^3 N_i = (1 - e^{-7.53C^{0.612}})$	$= 100C/(1 - e^{-7.53 \cdot C^{0.612}})$	$= 0.3\left[t + 0.42(e^{-2.4t} - 1)\right]$ $= 1 - e^{-0.3[t + 0.42(e^{-2.4t} - 1)]}$	

Table 1. (Contd.)

Symbols	Presence or absence of events that determine the number of polymer chains (N)						Mathematical description of individual scenarios
	Initiation			Initiator is "HX"	Chain transfer		
	Fast	Slow					
		Ion gener. rate det.	Cationation rate det.		Zero order	First order	
A4aa	0	1	0	0	1	0	$\frac{dN_i}{dt} = k_{ig}(I_0 - N_i)\cdot LA;$ $\frac{dN_{tr}}{dt} = k_{tr,{}^1M}\cdot N_i;$ $N_{total} = N_{tr} + N_i$ $-\frac{dM}{dt} = k_p\cdot N_i\cdot M$ $t = 0;\quad N_{tr} = N_i = 0;\quad M = M_0$
A4ab	0	1	0	0	0	1	$\frac{dN_i}{dt} = k_{ig}(I_0 - N_i)\cdot LA;$ $\frac{dN_{tr}}{dt} = k_{tr,{}^2M}\cdot N_i\cdot M;$ $N_{total} = N_{tr} + N_i$ $-\frac{dM}{dt} = k_p\cdot N_i\cdot M + \underline{k_{tr,{}^2M}\cdot N_i\cdot M}$ $t = 0;\quad N_{tr} = N_i = 0;\quad M = M_0$
A4ac	0	1	0	0	1	1	$\frac{dN_i}{dt} = k_{ig}\cdot(I_0 - N_i)LA;$ $\frac{dN_{tr}}{dt} = k_{tr,{}^1M}\cdot N_i + k_{tr,{}^2M}N_i\cdot M$ $N_{total} = N_{tr} + N_i$ $-\frac{dM}{dt} = k_p\cdot N_i\cdot M + k_{tr,{}^2M}\cdot N_i\cdot M$ $t = 0;\quad N_{tr} = N_i = 0;\quad M = M_0$
A4ba	0	0	1	0	1	0	$\frac{dN_i}{dt} = k_c\cdot(I_0 - N_i)\cdot M;$ $\frac{dN_{tr}}{dt} = k_{tr,{}^1M}\cdot N_i \quad N_{total} = N_{tr} + N_i$ $-\frac{dM}{dt} = k_p\cdot N_i\cdot M + \underline{k_c(I_0 - N_i)M}$
A4bb	0	0	1	0	0	1	$\frac{dN_i}{dt} = k_c(I_0 - N_i)M;$ $\frac{dN_{tr}}{dt} = k_{tr,{}^2M}\cdot N_i\cdot M \quad N_{total} = N_{tr} + N_i$ $-\frac{dM}{dt} = k_p\cdot N_i\cdot M + k_c(I_0 - N_i)M$ $+ \underline{k_{tr,{}^2M}\cdot N_i\cdot M}$

Resulting expressions			
N_{total} vs conv. (formula) $10^3\ N_{total}$ vs conv. (equation)	$\bar{M}_n$ vs conv. (formula) $10^{-3}\ \bar{M}_n$ vs conv. (equation)	$-\ln(1-\text{Conv.})$ vs time formula equation	Conv. vs time formula equation
$= N_i - c^{I}\ln(1-C)$	$= W_p \cdot C/N_{total}$	Same as A3a	Same as A3a
$= 3(1-e^{-5.18 \cdot C^{0.821}}) - 0.2\ln(1-C)$	$= 300C/(3(1-e^{-5.18 \cdot C^{0.821}}) - 0.2\ln(1-c))$		
$= N_i + c^{II} \cdot M_0 \cdot C$	$= W_p \cdot C/N_{total}$	Same as A3a	Same as A3a
$= 3(1-e^{-5.18 \cdot C^{0.821}}) + 0.6 \cdot C$	$= 300C/(3(1-e^{-5.18 \cdot C^{0.821}}) + 0.6C)$		
$= N_i - c^{I}\ln(1-C) + c^{II} \cdot M_0 \cdot C$	$= W_p \cdot C/N_{total}$	Same as A3a	Sane as A3a
$= 3(1-e^{-5.18 \cdot C^{0.821}})$ $- 0.2\ln(1-C) + 0.6 \cdot C$	$= 300C/(3(1-e^{-5.18 \cdot C^{0.821}})$ $- 0.2\ln(1-C) + 0.6C)$		
$= N_i - c^{I}\ln(1-C)$	$= W_p \cdot C/N_{total}$	Same as A3b	Same as A3b
$= 3(1-e^{-7.53 \cdot C^{0.612}}) - 0.2\ln(1-C)$	$= 300C/(3(1-e^{-7.53 \cdot C^{0.612}})$ $- 0.21\ln(1-C))$		
$= N_i + c^{II} \cdot M_0 \cdot C$	$= W_p \cdot C/N_{total}$	Same as A3b	Same as A3b
$= 3(1-e^{-7.53 \cdot C^{0.612}}) + 0.6 \cdot C$	$= 300C/(3(1-e^{-7.53 \cdot C^{0.612}}) + 0.6C)$		

Table 1. (Contd.)

Symbols	Presence or absence of events that determine the number of polymer chains (N): Initiation – Fast	Initiation – Slow: Ion gener. rate det.	Initiation – Slow: Cation-ation rate det.	Initiator is "HX"	Chain transfer: Zero order	Chain transfer: First order	Mathematical description of individual scenarios
A4bc	0	0	1	0	1	1	$\frac{dN_i}{dt} = k_c(I_0 - N_i)M$; $\frac{dN_{tr}}{dt} = k_{tr,1M} \cdot N_i \cdot k_{tr,2M} \cdot M$ $N_{total} = N_{tr} + N_i$ $-\frac{dM}{dt} = k_p \cdot N_i \cdot M + \underline{k_c \cdot (I_0 - N_i)M} + \underline{k_{tr,2M} \cdot N_i \cdot M}$
B1	0	0	0	1	0	0	$N_{total} = N_{H_2O}$ $-\frac{dM}{dt} = k_p \cdot N_{H_2O} \cdot M$
B2a	0	0	0	1	1	0	$\frac{dN_{tr}}{dt} = k_{tr,1M} \cdot N_{H_2O}$ $N_{total} = N_{H_2O} + N_{tr}$ $-\frac{dM}{dt} = k_p \cdot N_{H_2O} \cdot M$
B2b	0	0	0	1	0	1	$\frac{dN_{tr}}{dt} = k_{tr,2M} \cdot N_{H_2O} \cdot M$ $N_{total} = N_{H_2O} + N_{tr}$ $-\frac{dM}{dt} = k_p \cdot N_{H_2O} \cdot M + \underline{k_{tr,2M} \cdot N_{H_2O} \cdot M}$
B2c	0	0	0	1	1	1	$\frac{dN_{tr}}{dt} = k_{tr,1M} \cdot N_{H_2O} + k_{tr,2M} \cdot N_{H_2O} \cdot M$; $N_{total} = N_{H_2O} + N_{tr}$ $-\frac{dM}{dt} = k_p \cdot N_{H_2O} \cdot M + \underline{k_{tr,2M} \cdot N_{H_2O} \cdot M}$
C1	1	0	0	1	0	0	$N_{total} = I_0 + N_{H_2O}$ $-\frac{dM}{dt} = k_p(I_0 + N_{H_2O}) \cdot M$
C2a	1	0	0	1	1	0	$\frac{dN_{tr}}{dt} = k_{tr,1M} \cdot (I_0 + N_{H_2O})$; $N_{total} = I_0 + N_{H_2O} + N_{tr}$ $-\frac{dM}{dt} = k_p(I_0 + N_{H_2O}) \cdot M$

Resulting expressions			
N_{total} vs conv. (formula) $10^3\ N_{total}$ vs conv. (equation)	$\bar{M}_n$ vs conv. (formula) $10^{-3}\ \bar{M}_n$ vs conv. (equation)	$-\ln(1-\text{Conv.})$ vs time formula equation	Conv. vs time formula equation
$= N_i - c^{I}\ln(1-C) + c^{II}\cdot M_0\cdot C$ $= 3(1-e^{-7.53\cdot C^{0.612}}) - 0.2\ln(1-C) + 0.6\cdot C$	$= W_p\cdot C/N_{total}$ $= 300C/(3(1-e^{-7.53\cdot C^{0.612}}) - 0.2\ln(1-C) + 0.6C)$	Same as A3b	Same as A3b
$= N_{H_2O}$ $= 1$	$= W_p\cdot C/\cdot N_{H_2O}$ $= 300C$	$= k_p\cdot N_{H_2O}\cdot t$ $= 0.1t$	$= 1 - e^{-0.1t}$
$= N_{H_2O} - c^{I}\ln(1-C)$ $= 1 - 0.2\ln(1-C)$	$= W_p\cdot C/N_{total}$ $= 300C/(1 - 0.2\ln(1-C)$	Same as B1	Same as B1
$= N_{H_2O} + c^{II}\cdot M_0\cdot C$ $= 1 + 0.6\cdot C$	$= W_p\cdot C/N_{total}$ $= 300C/(1 + 0.6C)$	Same as B1	Same as B1
$= N_{H_2O} - c^{I}\ln(1-C) + c^{II}\cdot M_0\cdot C$ $= 1 - 0.2\ln(1-C) + 0.6\cdot C$	$= W_p\cdot C/N_{total}$ $= 300C/(1 - 0.2\ln(1-C) + 0.6C)$	Same as B1	Same as B1
$= I_0 + N_{H_2O}$ $= 4$	$= W_p\cdot C/N_{total}$ $= 75C$	$= k_p (I_0 + N_{H_2O})\cdot t$ $= 0.4\cdot t$	$= 1 - e^{-0.4t}$
$= I_0 + N_{H_2O} - c^{I}\ln(1-C)$ $= 4 - 0.2\ln(1-C)$	$= W_p\cdot C/N_{total}$ $= 300\ C/(4 - 0.2\ln(1-C))$	Same as C1	Same as C1

Table 1. (Contd.)

Symbols	Presence or absence of events that determine the number of polymer chains (N)						Mathematical description of individual scenarios
	Initiation			Initiator is "HX"	Chain transfer		
	Fast	Slow					
		Ion gener. rate det.	Cation-ation rate det.		Zero order	First order	
C2b	1	0	0	1	0	1	$\frac{dN_{tr}}{dt}=k_{tr,{}^{2}M}\cdot(I_0+N_{H_2O})\cdot M$ $N_{total}=I_0+N_{H_2O}+N_{tr}$ $-\frac{dM}{dt}=k_p(I_0+N_{H_2O})\cdot M$ $\underline{+k_{tr,{}^{2}M}(I_0+N_{H_2O})\cdot M}$
C2c	1	0	0	1	1	1	$\frac{dN_{tr}}{dt}=k_{tr,{}^{1}M}\cdot(I_0+N_{H_2O})$ $+k_{tr,{}^{2}M}(I_0+N_{H_2O})\cdot M$ $N_{total}=N_{H_2O}+N_{tr}$ $-\frac{dM}{dt}=k_p\cdot(I_0+N_{H_2O})\cdot M$ $\underline{+k_{tr,{}^{2}M}(I_0+N_{H_2O})\cdot M}$
C3a	0	1	0	1	0	0	$\frac{dN_i}{dt}=k_{ig}(I_0-N_i)\cdot LA;$ $N_{total}=N_i+N_{H_2O}$ $-\frac{dM}{dt}=k_p(N_i+N_{H_2O})\cdot M$
C3b	0	0	1	1	0	0	$\frac{dN_i}{dt}=k_c(I_0-N_i)\cdot M$ $N_{total}=N_i+N_{H_2O}$ $-\frac{dM}{dt}=k_p(N_i+N_{H_2O})\cdot M$

Resulting expressions			
N_{total} vs conv. (formula) $10^3\ N_{total}$ vs conv. (equation)	$\bar{M}_n$ vs conv. (formula) $10^{-3}\ \bar{M}_n$ vs conv. (equation)	$-\ln(1-\text{Conv.})$ vs time formula equation	Conv. vs time formula equation
$=I_0+N_{H_2O}+c^{II}\cdot M_0\cdot C$ $=4+0.6\cdot C$	$=W_p\cdot C/N_{total}$ $=300\cdot C/(4+0.6C)$	Same as C1	Same as C1
$=I_0+N_{H_2O}-c^{I}\ln(1-C)$ $+c^{II}\cdot M_0\cdot C$ $=4-0.2\ln(1-C)+0.6\cdot C$	$=W_p\cdot C/N_{total}$ $=300C/(4-0.2\ln(1-C)$ $+0.6C)$	Same as C1	Same as C1
$C=1-e^{\frac{k_p}{k_{ig}LA}\left[\left(\ln\left(\frac{I_0-N_i}{I_0}\right)\right)\cdot (N_{H_2O}+I_0)+N_i\right]}$ implicit $10^3\ N_{tot.}=1+3(1-e^{-4.486\cdot C^{0.993}})$	$=W_p\cdot C/(N_{H_2O}+N_i)$ $=300\ C/(1+3(1-e^{4.486\cdot C^{0.993}}))$	$=k_p\cdot\left[(N_{H_2O}+I_0)\cdot t+\frac{I_0}{k_{ig}\cdot LA}(e^{-k_{ig}LAt}-1)\right]$ $=100\left[0.004t+0.00366(e^{-0.82t}-1)\right]$ $=1-e^{-\left(0.4t+0.366(e^{-0.82t}-1)\right)}$	
$C=\frac{-k_p}{k_c\cdot M_0}\left[N_i+(I_0+N_{H_2O})\cdot \ln\left(\frac{I_0-N_i}{I_0}\right)\right]$ implicit $10^3\ N_{tot.}=1+3(1-e^{-6.94\cdot C^{0.73}})$	$=W_p\cdot C/(N_{H_2O}+N_i)$ $=300\ C/(1+3(1-e^{-6.94\cdot C^{0.733}}))$	$=k_p\cdot\left[(N_{H_2O}+I_0)\cdot t+\frac{I_0}{k_c\cdot \bar{M}}(e^{-k_c\bar{M}\cdot t}-1)\right]$ $=100\left[0.004t+.00125(e^{-2.4t}-1)\right]$ $=I-e^{-\left(0.4t+0.125(e^{-2.4t}-1)\right)}$	

Table 1. (Contd.)

Symbols	Presence or absence of events that determine the number of polymer chains (N)						Mathematical description of individual scenarios
	Initiation			Initiator is "HX"	Chain transfer		
	Fast	Slow					
		Ion gener. rate det.	Cation-ation rate det.		Zero order	First order	
C4aa	0	1	0	1	1	0	$\frac{dN_i}{dt}=k_{ig}(I_0-N_i)\cdot LA;$ $\frac{dN_{tr}}{dt}=k_{tr,{}^1M}(N_i+N_{H_2O})$ $N_{total}=N_i+N_{H_2O}+N_{tr}$ $-\frac{dM}{dt}=k_p(N_i+N_{H_2O})\cdot M$
C4ab	0	1	0	1	0	1	$\frac{dN_i}{dt}=k_{ig}(I_0-N_i)\cdot LA;$ $\frac{dN_{tr}}{dt}=k_{tr,{}^2M}(N_i+N_{H_2O})\cdot M$ $N_{total}=N_i+N_{H_2O}+N_{tr}$ $-\frac{dM}{dt}=k_p(N_i+N_{H_2O})\cdot M$ $\underline{+k_{tr,{}^2M}(N_i+N_{H_2O})\cdot M}$
C4ac	0	1	0	1	1	1	$\frac{dN_i}{dt}=k_{ig}(I_0-N_i)\cdot LA;$ $\frac{dN_{tr}}{dt}=k_{tr,{}^1M}(N_i+N_{H_2O})$ $+k_{tr,{}^2M}(N_i+N_{H_2O})\cdot M$ $-\frac{dM}{dt}=k_p(N_i+N_{H_2O})\cdot M$ $\underline{+k_{tr,{}^2M}(N_i+N_{H_2O})M;}$ $N_{total}=N_i+N_{H_2O}+N_{tr}$
C4ba	0	0	1	1	1	0	$\frac{dN_i}{dt}=k_c(I_0-N_i)\cdot M;$ $\frac{dN_{tr}}{dt}=k_{tr,{}^1M}\cdot(N_i+N_{H_2O})$ $N_{total}=N_i+N_{H_2O}$ $-\frac{dM}{dt}=k_p(N_i+N_{H_2O})\cdot M$ $\underline{+k_c(I_0-N_i)\cdot M}$

Resulting expressions			
N_{total} vs conv. (formula) $10^3\ N_{total}$ vs conv. (equation)	$\bar{M}_n$ vs conv. (formula) $10^{-3}\ \bar{M}_n$ vs conv. (equation)	− ln(1 − Conv.) vs time formula equation	Conv. vs time formula equation
$= N_{H_2O} + N_i - c^{I} \ln(1-C)$ $= (1+3)(1-e^{-4.486 \cdot C^{0.993}})$ $-0.2 \ln(1-C)$	$= W_p \cdot C/(N_{H_2O} + N_i - c^{I} \ln(1-C))$ $= 300\, C \Big/ \begin{pmatrix} 1+3(1-e^{-4.486 \cdot C^{0.993}}) \\ -0.2 \ln(1-C) \end{pmatrix}$	Same as C3a	Same as C3a
$= N_{H_2O} + N_i - c^{II} M_0 \cdot C$ $= 1 + 3(1-e^{-4.486 \cdot C^{0.993}}) + 0.6C$	$= W_p \cdot C/(N_{H_2O} + N_i - c^{II} M_0 \cdot C)$ $= 300\, C/(1+3(1-e^{-4.486 \cdot C^{0.993}})$ $+0.6\, C)$	Same as C3a	Same as C3a
$= N_{H_2O} + N_i - c^{I} \ln(1-C) + c^{II} \cdot M_0 \cdot C$ $= 1 + 3(1-e^{-4.486 \cdot C^{0.993}})$ $-0.2 \ln(1-C) + 0.6 \cdot C$	$= W_p \cdot C \Big/ \begin{pmatrix} N_{H_2O} + N_i - c^{I} \ln(1-C) \\ + c^{II} M_0 \cdot 3C \end{pmatrix}$ $= 300C \Big/ \begin{pmatrix} 1+3(1-e^{-4.486 \cdot C^{0.993}}) - \\ 0.2 \ln(1-C) + 0.6C \end{pmatrix}$	Same as C3a	Same as C3a
$= N_{H_2O} + N_i - c^{I} \ln(1-C)$ $= 1 + 3(1-e^{6.94 \cdot C^{0.73}})$ $-0.2 \ln(1-C)$	$= W_p \cdot C/(N_{H_2O} + N_i - c^{I} \ln(1-C))$ $= 300C \Big/ \begin{pmatrix} 1+3(1-e^{-6.94 \cdot C^{0.73}}) \\ -0.2 \ln(1-C) \end{pmatrix}$	Same as C3b	Same as C3b

Table 1. (Contd.)

Symbols	Presence or absence of events that determine the number of polymer chains (N)						Mathematical description of individual scenarios
	Initiation			Initiator is "HX"	Chain transfer		
	Fast	Slow					
		Ion gener. rate det.	Cation-ation rate det.		Zero order	First order	
C4bb	0	0	1	1	0	1	$\frac{dN_i}{dt}=k_c(I_0-N_i)\cdot M;$ $\frac{dN_{tr}}{dt}=k_{tr,^2M}\cdot(N_i+N_{H_2O})\cdot M;$ $N_{total}=N_i+N_{H_2O}+N_{tr}$ $-\frac{dM}{dt}=k_p(N_i+N_{H_2O})\cdot M$ $+\underline{k_c(I_0-N_i)\cdot M}$ $+\underline{k_{tr,^2M}(N_i+N_{H_2O})\cdot M}$
C4bc	0	0	1	1	1	1	$\frac{dN_i}{dt}=k_c(I_0-N_i)\cdot M;$ $\frac{dN_{tr}}{dt}=k_{tr,^1M}\cdot(N_i+N_{H_2O})$ $+k_{tr,^2M}(N_i+N_{H_2O})\cdot M$ $-\frac{dM}{dt}=k_p(N_i+N_{H_2O})\cdot M$ $+\underline{k_c(I_0-N_i)\cdot M+k_{tr,^2M}(N_i+N_{H_2O})\cdot M}$ $N_{total}=N_i+N_{H_2O}+N_{tr}$

Symbols and Abbreviations used in Table 1

N	= concentration of polymer molecules
N_i	= concentration of polymer molecules formed from active initiator
N_{tr}	= concentration of chains produced by chain transfer to monomer
N_{H_2O}	= impurity (water) concentration
N_{total}	= $N = N_i + N_{tr} + N_{H_2O}$
* "HX"	= initiation by adventitious (protic) impurities
I_0	= initiator concentration
M_0	= initial monomer concentration
M	= monomer concentration
$\bar{M}$	= integral average monomer concentration
* LA or MtX_n	= coinitiator concentration
$\bar{M}_n$	= number average molecular weight

Resulting expressions			
N_{total} vs conv. (formula) $10^3\ N_{total}$ vs conv. (equation)	$\bar{M}_n$ vs conv. (formula) $10^{-3}\ \bar{M}_n$ vs conv. (equation)	$-\ln(1-\text{Conv.})$ vs time formula equation	Conv. vs time formula equation
$=N_{H_2O}+N_i+c^{I}\cdot M_0\cdot C$ $=1+3(1-e^{-6.94\cdot C^{0.73}})+0.6\cdot C$	$=W_p\cdot C/(N_{H_2O}+N_i+c^{I}\cdot M_0\cdot C)$ $=300\,C/(1+3(1-e^{-6.94\cdot C^{0.73}})+0.6\cdot C)$	Same as C3b	Same as C3b
$=N_{H_2O}+N_i-c^{I}\ln(1-C)$ $+c^{II}\cdot M_0\cdot C=1+3(1-e^{-6.94\cdot C^{0.73}})$ $-0.2\ln(1-C)+0.6\cdot C$	$=W_p\cdot C\Big/\left(\begin{matrix}N_{H_2O}+N_i-c^{I}\cdot\ln(1-C)\\+c^{II}M_0\cdot C\end{matrix}\right)$ $=300C\Big/\left(\begin{matrix}1+3(1-e^{-6.94\cdot C^{0.73}})\\-0.2\ln(1-C)+0.6\cdot C\end{matrix}\right)$	Same as C3b	Same as C3b

W_p = weight of polymer in unit volume
C = monomer conversion
t = time

* k_{ig} = ion-generation rate constant
* k_c = cationation rate constant
* k_p = overall propagation rate constant
* $k_{tr,\,^1M}$ = rate constant of chain transfer, monomolecular
* $k_{tr,\,^2M}$ = rate constant of chain transfer, bimolecular

c^{I} = zero order chain transfer constant: $c^{I} = k_{tr,\,^1M}/k_p$
c^{II} = first order chain transfer constant: $c^{II} = k_{tr,\,^2M}/k_p$
- - - - = negligible terms

*For more detailed description see also parts 2.1 (Table of Symbols for Scheme 1) and 2.2.1 (Definition of Elementary Steps)

2.2.2.1 The 28 Fundamental Scenarios

Table 1 summarizes and defines the 28 fundamental scenarios considered, their mathematical description, and the expressions derived for the above-mentioned plots. The very large amount of information assembled in the table is subdivided into four main columns: the first column, *Symbols*, shows abbreviations for the scenarios specificed in column two and will be used in the discussion.

The second column, *Presence or Absence of Events that Determine the Number of Polymer Chains* (*N*), comprises 6 sub-columns and shows the 28 scenarios selected for examination. Among the very large number of theoretically possible scenarios we have selected for analysis only those which seemed to be closest to real-life situations. Specifically, we have examined the effects on N of fast and slow initiation (and in the case of slow initiation, the respective effects of slow ion-generation and slow cationation), the effect of initiation by adventitious protic impurity (i.e., "HX" = "H_2O"), and the effects of chain transfer to monomer, specifically both monomolecular or zero-order-in-monomer chain transfer and bimolecular or first-order-in-monomer chain transfer. The presence or absence of these effects is indicated by the symbols, 1 or 0 in column two. The organization of the 28 scenarios is as follows:

The first 12 scenarios (symbol A in column one) represent systems in which the polymerizations are induced by purposely-added initiator RX. The next four scenarios (symbol B in column one) concern systems in which initiation is due to "H_2O" and it was assumed that "H_2O"-induced initiation is very rapid, not rate determining. In the 12 further scenarios (symbol C in column one) initiation occurs both by purposely-induced initiator RX *and* impurity-induced initiator "HX." The first scenario in Table 1 (symbol A1) is the "ideal" system, i.e., a polymerization in which initiation is fast and complicating side-reactions do not occur. The digits 2–4 following the capital letters, and the lower case letters *a–c* following the digits, indicate the *presence* and *nature* of chain transfer and slow initiation that may be present; the digit 2 indicates the presence of chain transfer to monomer and *a*, *b*, or *c* indicates the nature of this side reaction which can be zero-order (*a*), first-order (*b*), or both (*c*); the digit 3 indicates slow initiation and the lower case letters *a* or *b* that follow this number indicate the nature of the rate determining step, ion-generation (*a*) or cationation (*b*); the digit 4 indicates the presence of both chain transfer *and* slow initiation; this digit is always followed by two letters: the first which indicates the cause of slow initiation due to RX (i.e., slow ion-generation *a* or slow cationation *b*), whereas the second which indicates the presence and type of chain transfer *a*, *b*, or *c* (as above).

The third column, *Mathematical Descriptions of Individual Scenarios*, shows the differential equations of N and M together with the initial conditions in terms of t, N_i, N_{tr}, N_{total}, and M (see legend to Table 1).

The fourth column, *Resulting Expressions*, is divided into four subcolumns which contain the N vs Conv., $\bar{M}_n$ vs Conv., $-\ln(1 - \text{Conv.})$ vs Time, and Conv. vs Time formulas, and the numerical equations for presentation of the functions. The following initial conditions were chosen to calculate the numerical equations:

Molecular weight of a hypothetical monomer unit = 100 g/mol
Total polymer weight/total volume = 6 g/20 ml
Initiator concentration $[I_0] = 3.10^{-3}$ mol/l
Impurity (water) concentration $[N_{H_2O}] = 10^{-3}$ mol/l
Chain transfer constant (zero order): $c^{I} = k_{tr,{}^1M}/k_p = 2 \cdot 10^{-4}$ mol/l
Chain transfer constant (first order): $c^{II} = k_{tr,{}^2M}/k_p = 2 \cdot 10^{-4}$
Initial monomer concentration $[M]_0 = 3$ mol/l
Friedel-Crafts acid (MtX_n) concentration $[TiCl_4] = 4.1 \cdot 10^{-2}$ mol/l
Propagation rate constant $k_p = 100$ l/(mol · min)
Ion generation rate constant $k_{ig} = 20$ l/(mol · min)
Cationation rate constant $k_c = 1$ l/(mol · min)
$[N_i]$ = concentration of polymer molecules formed from the initiator
$[N_{tr}]$ = concentration of chains produced by chain transfer to monomer

In scenarios A2b, A2c, and A3b the second terms in the $-$dM/dt expressions can be neglected because they are small relative to the first terms (the negligible terms are underlined by dotted lines). In a few scenarios (A3a, A3b, C3a, C3b) explicit solutions could not be derived and graphical methods were applied to obtain the equations from computer fitted curves.

For the sake of simplicity and because of the large number (28) of fundamental scenarios considered, the rate constants in Scheme 1 are considered to be true constants, although we have established (see parts 2.3 and 3) that in many instances they are in fact not constants (they should preferably be termed rate coefficients) and are strongly influenced by the equilibria: dormant $\rightleftharpoons$ living $\rightleftharpoons$ ionic. By disregarding these equilibria and assuming that the rate coefficients are constant, we were able to derive differential equations with rather simple analytical solutions. The detailed consideration of these multi-part equilibria would have lead to numerical solutions of the differential equations (see e.g., [45a, b]) and thus to an enormous increase in the length of this treatment.

2.2.2.2 Diagnostic Plots

In line with the analysis above, diagnostic plots were computer generated to depict the various scenarios in terms of N = f(C), $\bar{M}_n$ = f(C), $-\ln(1-C)$ = f(time), and C = f(time).

The scenarios in series A concern initiation by purposely added initiator RX. Figures 1–8 (A–D) and 10 (A-D) show the diagnostic plots for the ideal case (instantaneous initiation, scenario A1, dotted lines) together with the effect of the various complicating factors, i.e., instantaneous initiation plus zero order chain transfer to monomer (scenario A2a), instantaneous initiation plus first order chain transfer to monomer (scenario A2b), instantaneous initiation plus both

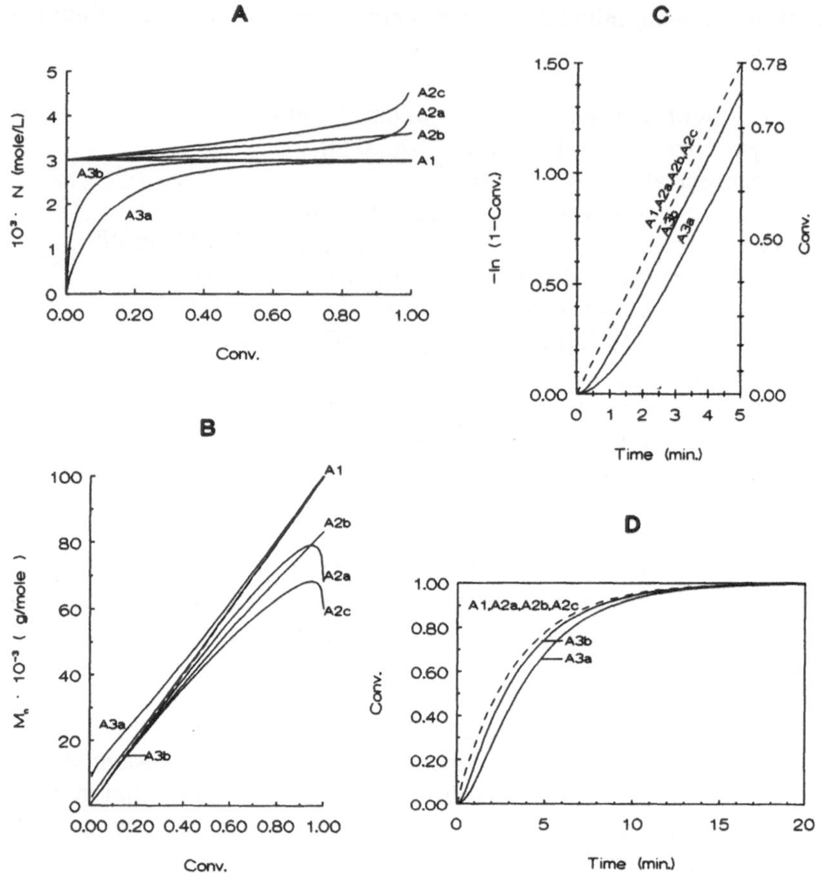

Fig. 1A. Number of polymer chains vs conversion. **B** Number average molecular weight vs conversion. **C** Rate of polymerization vs time. **D** Conversion vs time. Diagnostic plots for instantaneous initiation (A1), with various chain transfers (A2a, A2b, A2c) and for slow initiations (A3a, A3b)

kinds of chain transfer to monomer (scenario A2c), show ion-generation (scenario A3a) and slow cationation (scenario A3b).

As shown in Fig. 1B, slow initiation leads to higher than theoretical mol. wts. at low conversions, but the effect diminishes with increasing conversion. Chain transfer to monomer is difficult to discern at low molecular weights (or conversions), but it becomes more pronounced at higher molecular weights, particularly with zero order chain transfer, as pointed out for example in [1, 20, 51].

Since chain transfer to monomer does not change the number of growing chains, the conversion vs time (Fig. 1D) and rate vs time (Fig. 1C) plots run very close to the theoretical one. The only exception is the scenario with slow initiation (A3a, Fig. 1C) which shows a noticeable downward shift after a brief induction period. The magnitude of this shift depends on the extent of slow initiation.

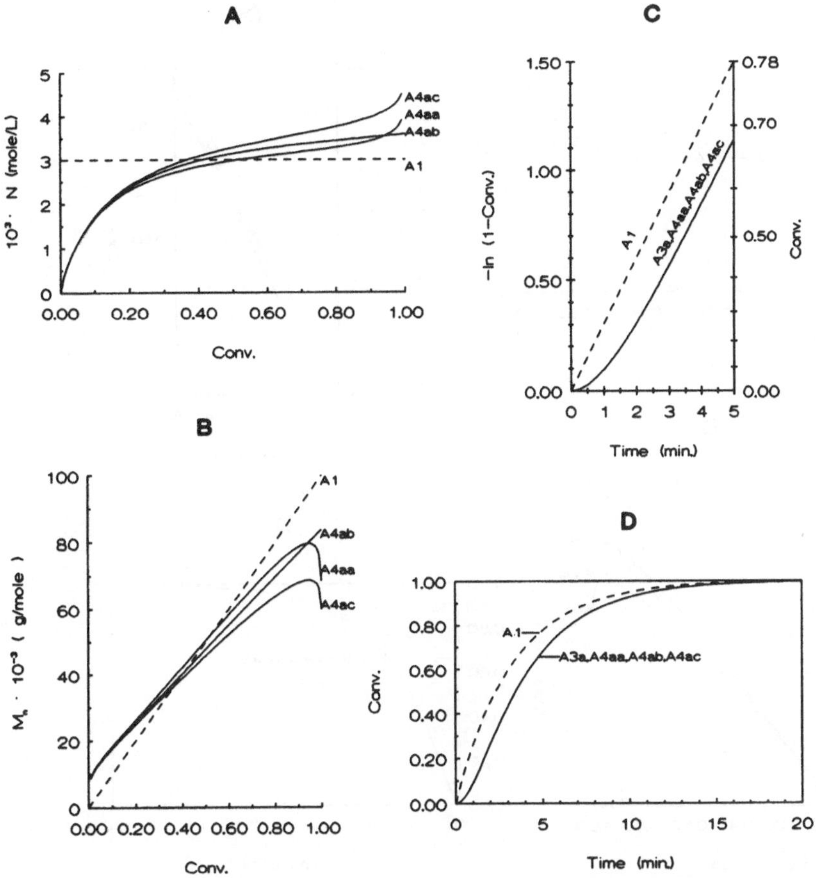

Fig. 2A. Number of polymer chains vs conversion. **B** Number average molecular weight vs conversion. **C** Rate of polymerization vs time. **D** Conversion vs time. Diagnostic plots for slow ion generation plus various chain transfers A4aa, A4ab, A4ac

Figure 2 shows the simultaneous effect of slow ion-generation plus chain transfer(s) to monomer on the diagnostic curves A4aa, A4ab and A4ac. Since these effects show maxima at different conversions (for slow initiation at low conversion, for chain transfer at high conversion) the initial parts of Fig. 2A, B are very similar to the plots shown in Fig. 1A, B (see curve A3a), whereas the latter parts of Fig. 2A, B resemble those of Fig. 1A, B (see curves for scenarios A2a, A2b, A2c). Since chain transfer was assumed to have no measurable effect on the rates, scenarios shown by curves A4aa, A4ab, and A4ac must be identical with that of A3a in Fig. 2C, D. The same explanation holds for scenarios A4ba, A4bb, and A4bc (Fig. 3A–D) where cationation is rate determining.

The scenarios in series B concern initiation only by impurity ("HX"). Slow initiation was not considered because according to the experimental data

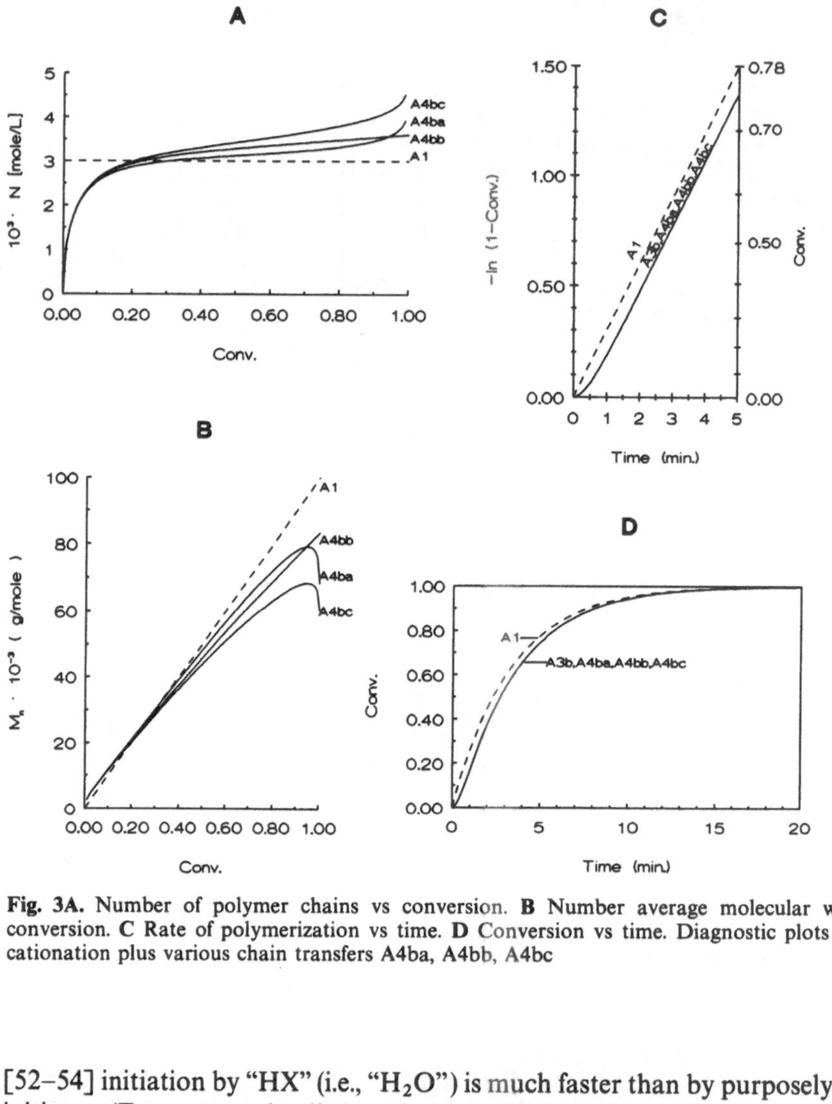

Fig. 3A. Number of polymer chains vs conversion. **B** Number average molecular weight vs conversion. **C** Rate of polymerization vs time. **D** Conversion vs time. Diagnostic plots for slow cationation plus various chain transfers A4ba, A4bb, A4bc

[52–54] initiation by "HX" (i.e., "H_2O") is much faster than by purposely added initiator. (For a more detailed analysis, see Sects. 4.1.2, 4.2.2.).

Figure 4A–D shows the effect of initiation by "H_2O" (B1), and initiation by "H_2O" plus chain transfer to monomer (scenarios B2a, B2b, and B2c). Since $[N_{H_2O}] < [I_0]$, the slope of the line B1 in the $\bar{M}_n$ vs conversion plot is higher than that of A1, but the polymerization rate is lower than that of A1. The presence of chain transfers affect the N vs C and M_n vs C plots just the same as in scenarios A2a, A2b, and A2c (see curves for scenarios B2a, B2b and B2c in Fig. 4A, B) but evidently they do not affect the rate or conversion vs time functions; (cf. curves B1 depicting scenarios of B2a, B2b, and B2c in Fig. 4C, D).

In series C both purposeful-initiation and "H_2O" induced-initiation exist. Figure 5A–D shows the diagnostic plots for instantaneous initiation by "H_2O"

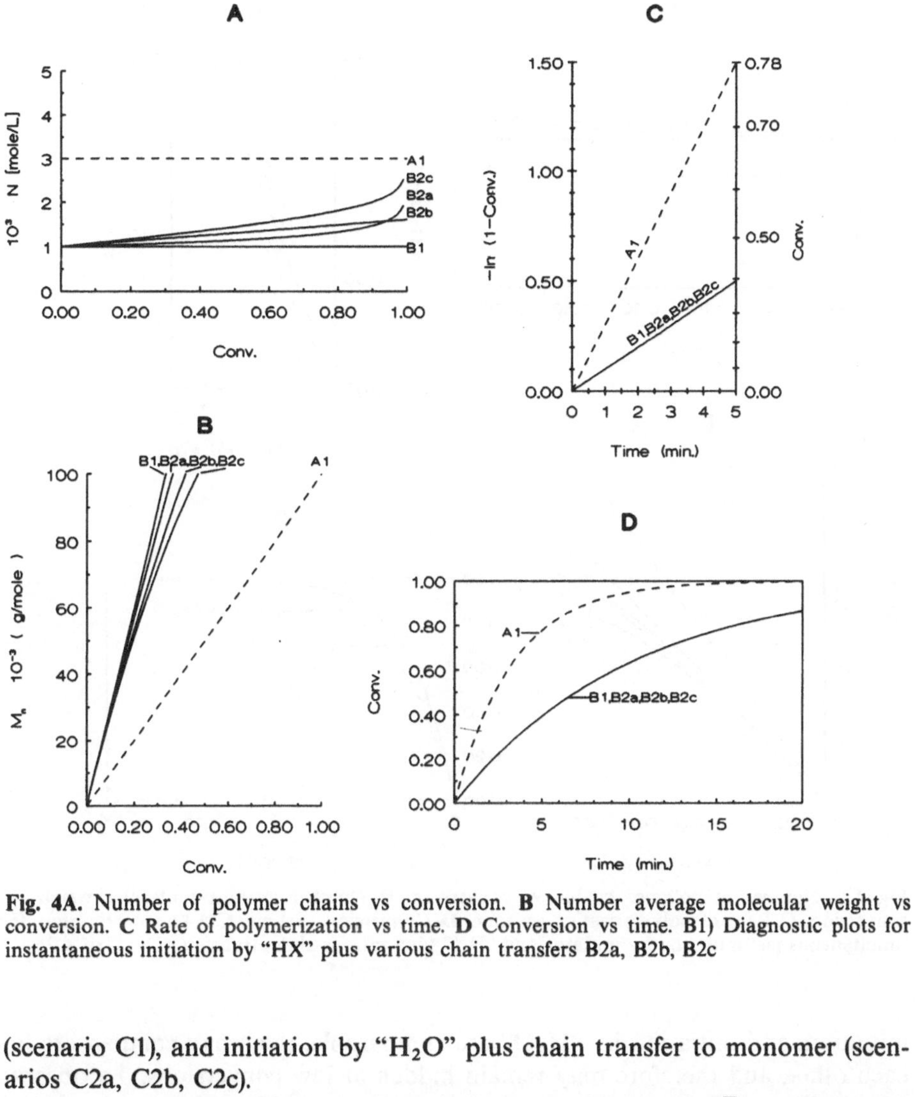

Fig. 4A. Number of polymer chains vs conversion. **B** Number average molecular weight vs conversion. **C** Rate of polymerization vs time. **D** Conversion vs time. **B1**) Diagnostic plots for instantaneous initiation by "HX" plus various chain transfers B2a, B2b, B2c

(scenario C1), and initiation by "H_2O" plus chain transfer to monomer (scenarios C2a, C2b, C2c).

Since in scenario C1, N_{total} is larger than I_0, the diagnostic $\bar{M}_n$ vs C plot will be below the "ideal" scenario of A1 just after the onset of polymerization, and the rate of polymerization will be higher than that of A1 (see Fig. 5C). Chain transfer has the same effect as in scenarios of e.g., B2a, B2b, and B2c (see Fig. 5A–D). If initiation due to added initiator is slow, but initiation by "H_2O" is relatively fast, the diagnostic plots will run below that of A1 in the N vs C, $-\ln(1-C)$ vs t and C vs t plots (Fig. 6A, C, D, scenarios C3a and C3b) but M_n will be higher than theoretical (Fig. 6B, scenarios C3a and C3b) at low conversions. After N_{total} becomes larger than I_0, the curves that run under the theoretical line will cross above it (see Fig. 6A) and those running over the theoretical line will cross under it (see Fig. 6B, C, D). If the effects of slow

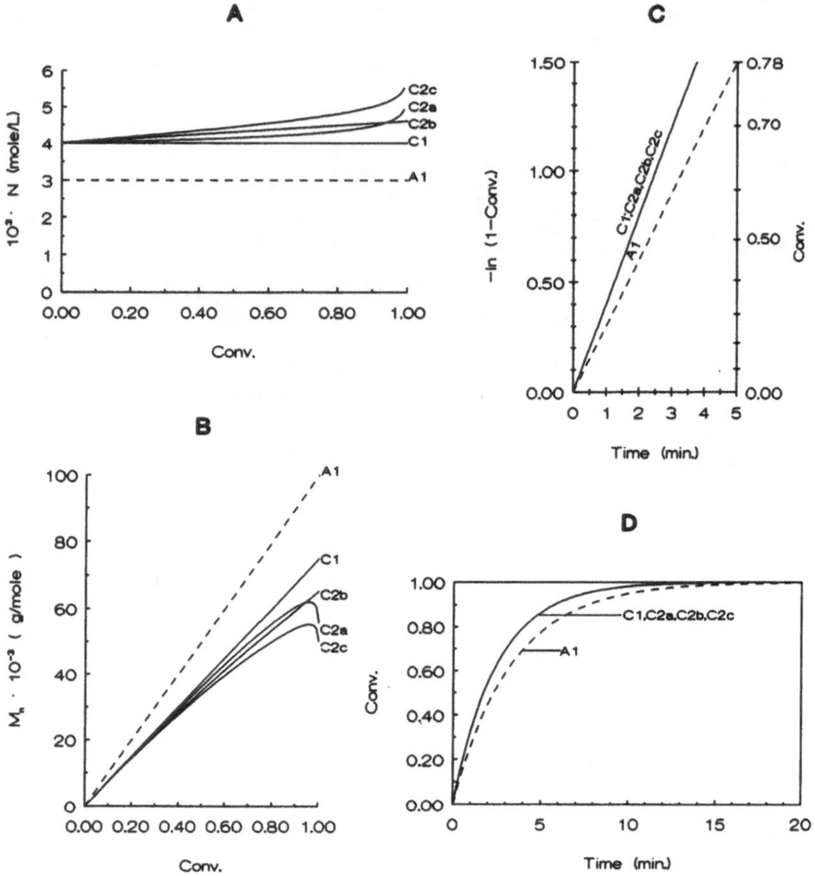

Fig. 5A. Number of polymer chains vs conversion. **B** Number average molecular weight vs conversion. **C** Rate of polymerization vs time. **D** Conversion vs time. C1) Diagnostic plots for simultaneous fast initiation by initiator and "HX" plus various chain transfers C2a, C2b, C2c

initiation, and initiation by "H_2O" are comparable, they may compensate for each other, and therefore may remain hidden at low conversions. To observe nonideal behavior the reaction must be carried out to high conversions and/or molecular weights [20, 45, 51].

If scenarios C3a and C3b are combined with chain transfer to monomer, the diagnostic curves will change as shown in Figs. 7A–D and 8A–D. Chain transfer, similarly to initiation by "H_2O", slightly reduces the effect due to slow initiation at low conversions in the $\bar{M}_n$ vs C plot (see Fig. 7B), while it becomes more pronounced with increasing conversions (see scenarios C4aa, C4ab, and C4ac). Evidently, chain transfer shows no additional effects on the rate vs time, or conversion vs time plots (Fig. 7C, D). The same description holds for plots corresponding to scenarios C4ba, C4bb, and C4bc shown in Fig. 8A–D. The only difference between Figs. 7A–D and 8A–D is in the degree of slow initiation.

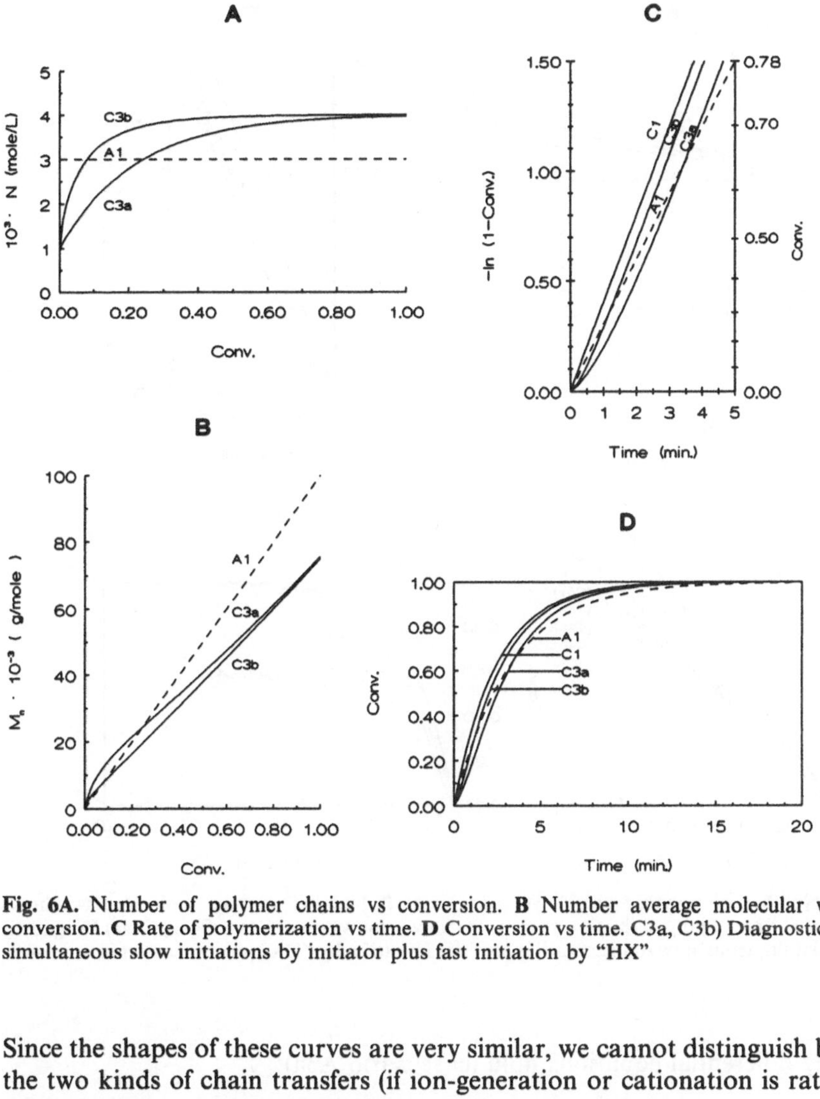

Fig. 6A. Number of polymer chains vs conversion. **B** Number average molecular weight vs conversion. **C** Rate of polymerization vs time. **D** Conversion vs time. C3a, C3b) Diagnostic plots for simultaneous slow initiations by initiator plus fast initiation by "HX"

Since the shapes of these curves are very similar, we cannot distinguish between the two kinds of chain transfers (if ion-generation or cationation is rate determining). (This problem will be further discussed in Sect. 2.3).

Some comments on "infering". The 28 fundamental scenarios do not formally include chain transfer to the initiator ("infering"), that is when the initiator is consumed both by slow initiation and chain transfer [44, 46, 47]. Recently Ivan and Kennedy [24], and Szwarc [45] have treated such a scenario and both groups of authors agreed that infering and ion generation provide the same ion pair (see Eqs. (6) and (7)). If ion-generation is fast, significant infering cannot occur. Hence, in our treatment we consider only scenario A3, in which ion generation determines the rate of initiation and combine it with chain transfer to initiator: when both ion-generation and chain transfer to initiator are slow, the

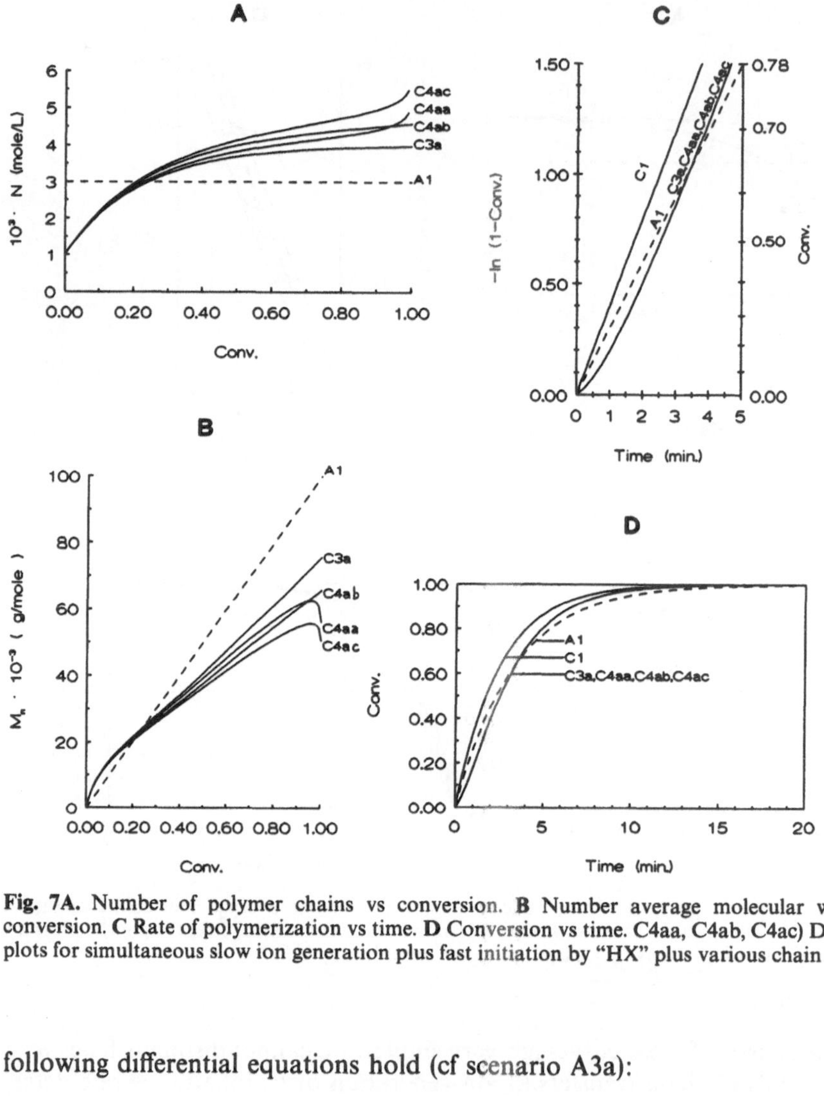

Fig. 7A. Number of polymer chains vs conversion. **B** Number average molecular weight vs conversion. **C** Rate of polymerization vs time. **D** Conversion vs time. C4aa, C4ab, C4ac) Diagnostic plots for simultaneous slow ion generation plus fast initiation by "HX" plus various chain transfers

following differential equations hold (cf scenario A3a):

$$\frac{dN_i}{dt} = k_{ig}(I_0 - N_i) \cdot (MtX_n) + k_{tr,I}(I_0 - N_i) \cdot N_i \tag{10}$$

$$-\frac{dM}{dt} = k_p \cdot N_i \cdot M \,. \tag{11}$$

We can get the active initiator concentration (N_i) (i.e., fraction of initiator incorporated into the polymer) as a function of time by solving Eq. (10):

$$N_i = \frac{I_0(1 - e^{-(I_0 k_{tr,I} + k'_{ig}) \cdot t})}{1 + I_0 k_{tr,I} \cdot e^{-(I_0 k_{tr,I} + k'_{ig}) \cdot t}} \tag{12}$$

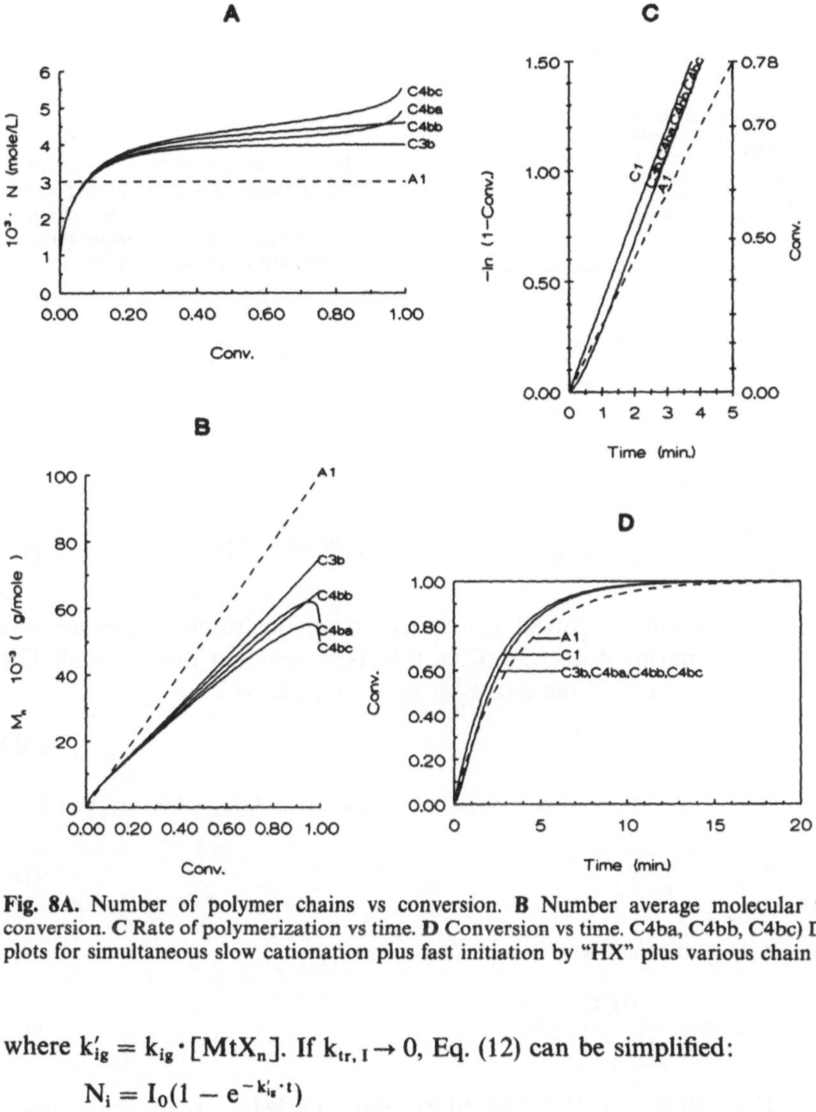

Fig. 8A. Number of polymer chains vs conversion. **B** Number average molecular weight vs conversion. **C** Rate of polymerization vs time. **D** Conversion vs time. C4ba, C4bb, C4bc) Diagnostic plots for simultaneous slow cationation plus fast initiation by "HX" plus various chain transfers

where $k'_{ig} = k_{ig} \cdot [MtX_n]$. If $k_{tr,I} \to 0$, Eq. (12) can be simplified:

$$N_i = I_0(1 - e^{-k'_{ig} \cdot t}) \tag{13}$$

which is identical with the time function of scenario A3a (not shown in Fig. 1A–D).

Figure 9 shows N_i vs time at different inifer constants ($c_{tr,I} = k_{tr,I}/k_p$). As shown, chain transfer to initiator accelerates ionization of initiator. We can get the $N_i = f$(conversion) function, after dividing Eq. (10) by Eq. (11), and separating the variables:

$$\int \frac{N_i \cdot dN_i}{(I_0 - N_i)\left[\dfrac{k'_{ig}}{k_p} + \dfrac{k_{tr,I}}{k_p} \cdot N_i\right]} = -\int \frac{dM}{M} \tag{14}$$

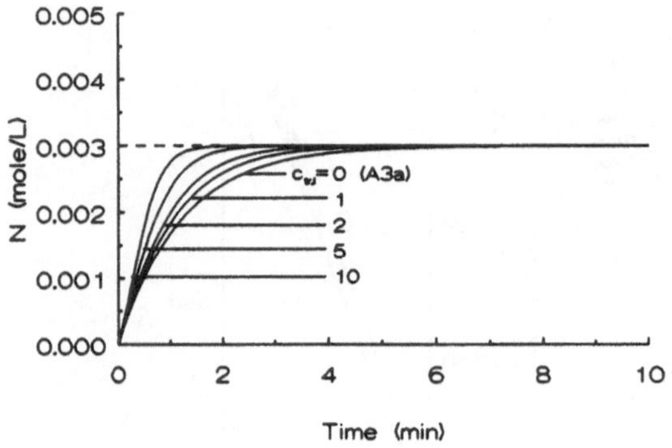

Fig. 9. Diagnostic N vs time plot for simultaneous slow ion generation (scenario A3a) and chain transfer to initiator at various inifer constants: $c_{tr,I} = 0, 1, 2, 5, 10$

and integration yields:

$$1 - \frac{M}{M_0} = C = 1 - e^{\frac{k_p}{I_0 k_{tr,I} + k'_{ig}}\left[I_0 \ln\left(\frac{I_0 - N_i}{I_0}\right) + \frac{k'_{ig}}{k_{tr,I}} \ln\left(1 + \frac{k_{tr,I}}{k'_{ig}} N_i\right)\right]} \tag{15}$$

Since Eq. (15) cannot be solved in explicit form, the same procedure was followed as in scenarios A3a, A3b, C3a, C3b (see above in this section). The equation of the computer fitted curve for $c_{tr,I} = k_{tr,I}/k_p = 2$ is:

$$N_i = I_0 (1 - e^{-7 \cdot C^{0.845}}) \tag{15b}$$

and is shown in Fig. 10A and is compared to scenario A3a, where $c_{tr,I} = 0$.

$$\text{The } \bar{M}_n = \frac{W_p \cdot C}{N_i} \tag{16}$$

equation, or its numerical form:

$$10^{-3} \bar{M}_n = \frac{100C}{1 - e^{-7 \cdot C^{0.845}}} \tag{17}$$

is shown in Fig. 10B ($c_{tr,I} = 2$) compared to scenario A3a ($c_{tr,I} = 0$). As the figure shows, the $\bar{M}_n$ vs C plot is closer to scenario A1, because inifering reduces the effect of slow initiation. By substituting Eq. (12) into Eq. (11), and separating the variables, integration yields the polymerization rate (expressed as $-\ln(1 - C)$) vs time:

$$\ln\left(\frac{M_0}{M}\right) = -\ln(1 - C) = k_p \cdot I_0 \left\{t + \left[\frac{1 + I_0 \cdot k_{tr,I}}{I_0 \cdot k_{tr,I} \cdot (k'_{ig} + I_0 \cdot k_{tr,I})}\right] \left[\ln \frac{1 + I_0 \cdot k_{tr,I} \cdot e^{-(I_0 k_{tr,I} + k'_{ig}) t}}{1 + I_0 \cdot k_{tr,I}}\right]\right\}. \tag{18}$$

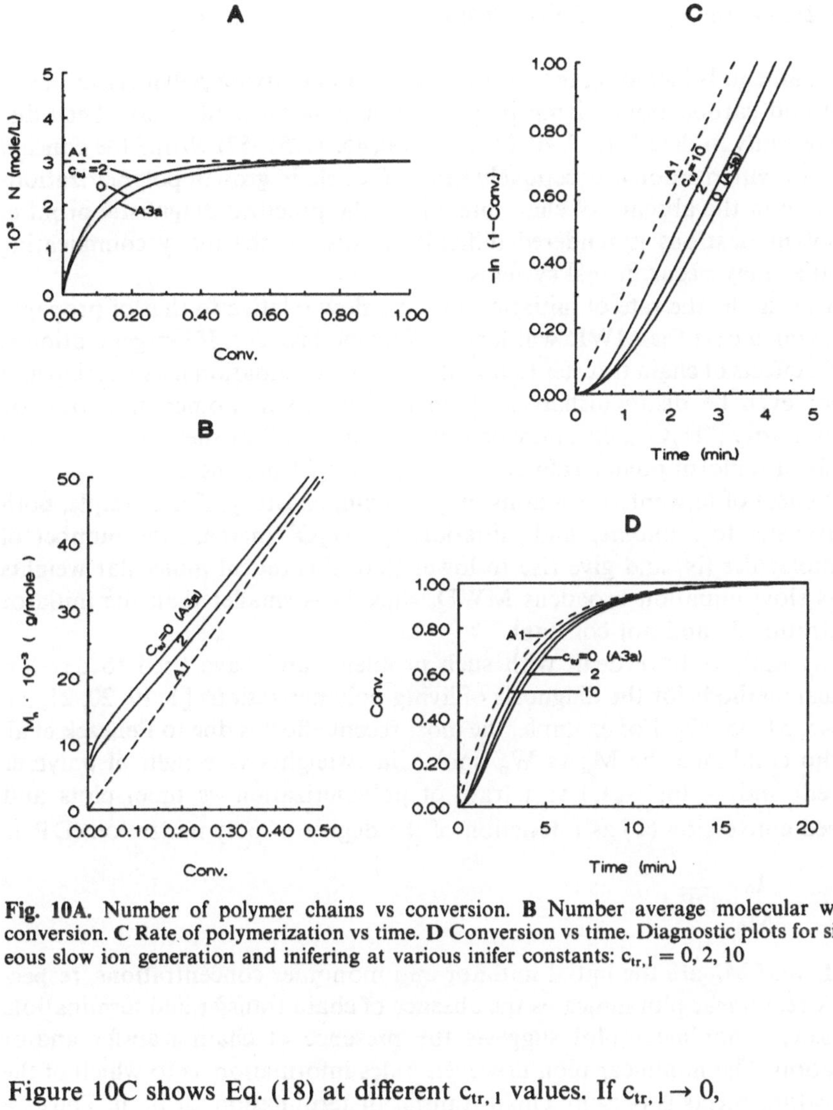

Fig. 10A. Number of polymer chains vs conversion. **B** Number average molecular weight vs conversion. **C** Rate of polymerization vs time. **D** Conversion vs time. Diagnostic plots for simultaneous slow ion generation and inifering at various inifer constants: $c_{tr,I} = 0, 2, 10$

Figure 10C shows Eq. (18) at different $c_{tr,I}$ values. If $c_{tr,I} \rightarrow 0$,

$$- \ln(1 - C) = k_p \cdot I_0 \left[t + \frac{1}{k'_{ig}} (e^{-k'_{ig} \cdot t} - 1) \right] \tag{19}$$

which is identical with the expression for scenario A3a. The same applies for the conversion vs. time curves, Fig. 10D.

As forcefully demonstrated by these basic scenarios, even under simplified circumstances, and with rate constants that are truly constant and do not vary with conditions, the effect of the combinations of the possible reactions may be compensatory and therefore undetectable by the diagnostic plots (see Sect. 2.2.2.3).

2.2.2.3 Diagnosis of Living Polymerizations

The fundamentals including the vigorous definition of living polymerizations in general and carbocationic living polymerizations in particular have been discussed recently in detail (1, 19, 20, 21, 24, 29, 30, 45, 51, 55–57). While the concept of living polymerization is deceptively simple (i.e., chain growth polymerizations proceeding in the absence of chain breaking), the practical diagnostic proof of such polymerizations is rendered difficult because of the many complicating effects that may occur in real systems.

For example, the rate of initiation may be slow relative to that of propagation, in which case the MWD will tend to become broader. If ion-generation is slow, the effects of chain transfer to initiator and slow initiation may overlap and may not even be distinguishable. Chain transfer to monomer, initiation by protic impurity ("H_2O"), etc., may be present which will further complicate the synthesis of uniform predetermined molecular weight polymers.

The effect of unwanted reactions may be compensatory. For example, both chain transfer to monomer and initiation by "H_2O" increase the number of macromolecules (N) and give rise to lower than theoretical molecular weights whereas slow initiation broadens MWD, since N is smaller than the initiator concentration I_0 and not constant.

Many authors have dealt with such problems and have tried to develop analytical methods for the diagnosis of living polymerizations [1, 19, 20, 21, 24, 29, 30, 45, 51, 55–57]. For example, the most recent effort is due to Penczek et al. [21] who combined the $\bar{M}_n$ vs W_p (molecular wieights vs weight of polymer produced) and $-\ln(1-C)$ vs t (rate of polymerization vs time) plots and expressed conversion (C) as a function of the degree of polymerization ($\overline{DP}_n$):

$$C = \frac{I_0}{M_0} \cdot \overline{DP}_n ,$$

where I_0 and M_0 are the initial initiator and monomer concentrations, respectively. A rectilinear plot indicates the absence of chain transfer and termination; conversely, a nonlinear plot suggests the presence of chain transfer and/or termination. The nonlinear plot, however, hides information as to which of the complicating effects is present: chain transfer or termination, or both. Thus we feel that *the simultaneous examination of linear $\bar{M}_n$ vs W_p (or N vs W_p), and rate vs time (or conversion vs time), plots is still the best method to probe the livingness of polymerizations.*

2.3 Methodology to Determine Kinetic Parameters

In Section 2.2.2 we have simulated the effects of various complicating events by computer generated diagnostic plots. In this section we show how to treat experimental data to obtain kinetic parameters in the presence of the following complicating events: 1) slow initiation, 2) slow initiation plus chain transfer to monomer, and 3) slow initiation plus initiation by "H_2O" plus chain transfer to monomer (cf scenarios A3a or b; A4ac or bc; C4ac or bc, in Table 1). As a result of our treatment of these scenarios, all the other scenarios will become self-explanatory.

As mentioned in Sect. 2.2.1.1 slow initiation is controlled either by the rate of ion-generation or cationation (cf. Eqs. (1)–(4). Based on earlier results [58, 59] a convenient diagnostic method has been developed [1, 29] to elucidate the nature of the rate determining step for rapid carbocationic polymerizations and in IMA (incremental monomer addition) experiments. For only one monomer addition, i.e., for AMI (all monomer in or for the first monomer addition in IMA) experiments, the relevant expressions are as follows:

if $k_c \ll k_{ig}$ and $k_p/k_c \gg 1$,

$$-\ln(1 - I_{eff}) - I_{eff} = \frac{k_c}{k_p} \cdot \frac{[M_0]}{[I_0]} \cdot C = Q_1 C \tag{20}$$

and if $k_{ig} \ll k_c$ and $k_p/k_{ig} \gg 1$,

$$-\ln(1 - I_{eff}) - I_{eff} = \frac{k_{ig}}{k_p} \cdot \frac{[MtX_n]}{[I_0]} \cdot \ln(1 - C) = Q_2 \ln(1 - C) \tag{21}$$

$$\text{where } I_{eff} = \frac{[N]}{[I_0]}$$

(cf. scenarios A3a and A3b in Table 1). In Eq. (20) the independent variable is monomer conversion (C), while in Eq. (21) it is $-\ln(1 - C)$. Since at low conversions $-\ln(1 - C) \cong C$ (see Fig. 11), we cannot use the above diagnostic plots because in our fundamental scenarios (see Fig. 1A) and by our measurements (see later), $I_{eff} \gg 90\%$ even below $\sim 50\%$ monomer conversion. However, even at this relatively high conversion, the plots of both Eq. (20) and Eq. (21) are graphically indistinguishable, see Fig. 11B (for better comparison we used the same constants, e.g., $Q_1 = Q_2 = 2.73$ in both cases, that is $k_c = 2.73$ L/(mole·min) for Eq. (20) and $k_c = 100$ L(mole·min) for Eq. (21), the other parameters: k_{ig}, k_p, $[M_0]$, $[I_0]$ and $[MtX_n]$ were the same as in Sect. 2.2.2 (see initial conditions)). Hence, we have to rely on literature data to decide the rate determining step of initiation.

2.3.1 Slow Initiation

According to available evidence in the systems investigated to date [29, 30, 56], the rate of cationation controls the rate of initiation. In line with this information therefore, we assume that upon addition of a stoichiometric excess of MtX_n, all the initiator becomes active for cationation, see Eqs. (1) and (2), and Eqs. (3)

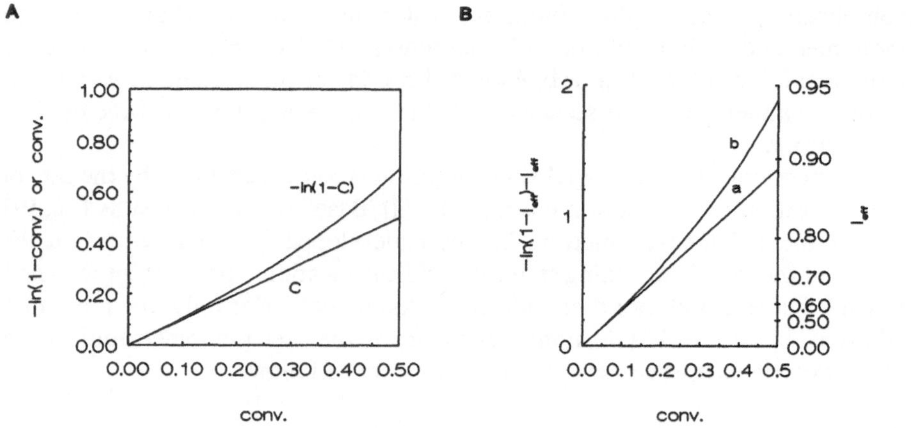

Fig. 11A. The difference between C and $-\ln(1-C)$ up to 50% conversion. **B** The difference between the diagnostic slow initiation plot. See [7] for cationation and ion generation up to 50% conversion. (See text)

and (4). After cationation, the concentration of the active species $[RM_n^\oplus]$, or in general $[RM^\oplus]$, is equal to that of polymer chains [N]:

$$[RM^\oplus] \equiv [N] = \frac{W_p}{\bar{M}_n}\,. \tag{22}$$

Since $k_{ig} > k_c$, the rate of initiation from Eqs. (3) and (4) is:

$$\frac{d[N]}{dt} = k_c[R^\oplus][M] \tag{23}$$

where [M] is monomer concentration and $[R^\oplus]$ the initiating concentration before cationation. In the absence of protic impurity "HX", the purposely-added initiator I_0 is present either as an active initiating cation $R^\oplus$ or is incorporated into a polymer chain N:

$$[I_0] = [R^\oplus] + [N] \tag{24}$$

hence

$$[R^\oplus] = [I_0] - [N] \tag{25}$$

and

$$\frac{d[N]}{dt} = k_c([I_0] - [N])\cdot[M]\,. \tag{26}$$

Consequently, the cationation rate coefficient can be expressed as

$$k_c = \frac{d[N]/dt}{([I_0] - [N])\cdot[M]}\,. \tag{27}$$

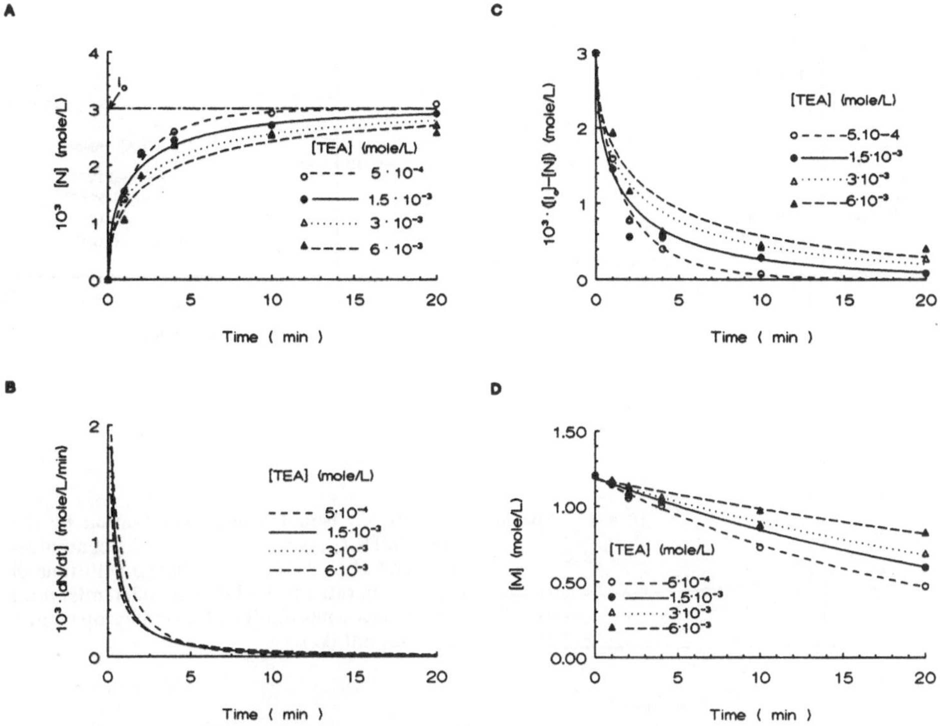

Fig. 12A–D. Isobutylene polymerization by the $TMPCl/TiCl_4$ initiating system at various TEA concentrations and at $-45\,°C$: **A** the change with time of number of macromolecules; **B** its differential; **C** $I_0 - N$; **D** monomer concentration

It is possible to obtain k_c by the following route: we can readily obtain [N] (from W_p and $\bar{M}_n$) vs time (t) plots, and thence obtain d[N]/dt vs t plots. We can also readily obtain $[I_0] - [N]$ vs t, and [M] vs t plots, which is all we need to get k_c by Eq. (27).

Let us illustrate these operations by the data in Figs. 12A–D and 13 A. Figure 12A shows representative [N] vs t plots for the $TMPCl/TiCl_4/IB/TEA/-45\,°C$ system at various TEA concentrations. The experimental data are computer fitted to yield the corresponding d[N]/dt vs t relationship shown in Fig. 12B. Figure 12C shows the $[I_0] - [N]$ vs t plot, and Fig. 12D the [M] vs t relationship. These plots provide the information needed to computer generate the corresponding k_c vs t plots shown in Fig. 13A. More results will be shown in Sect. 4.1.1.1.

In this treatment we have neglected the number of chains produced by chain transfer to monomer; however, it should be kept in mind that the number of polymer chains calculated by Eq. (22) includes those generated by both initiation and chain transfer. Neglecting chains that arose by chain transfer may lead to increased k_c, as if initiation were faster (see Sect. 2.3.2 and 2.3.3).

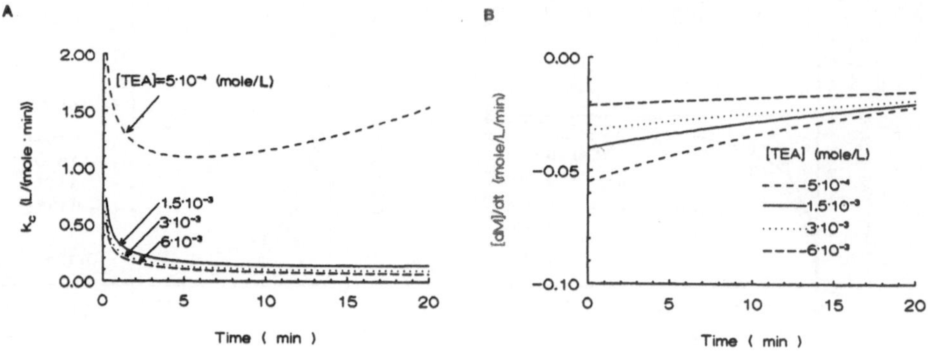

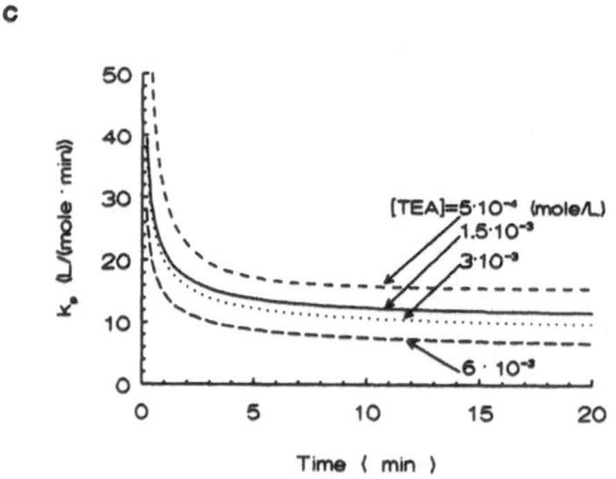

Fig. 13A–C. Isobutylene polymerization by the TMPCl/$TiCl_4$ system at various TEA concentrations and at − 45 °C: **A** the change with time of cationation rate constant (k_c); **B** differential ratio of monomer concentration; **C** overall propagation rate constant (k_p)

The simplest way to express propagation is:

$$RM^{\oplus} + M \xrightarrow{k_p} RMM^{\oplus} + M \xrightarrow{k_p} RMMM^{\oplus} \equiv RM_n^{\oplus} \tag{28}$$

where $RM^{\oplus}$ is the cation arising from RX (or "HX") plus monomer, k_p is the overall rate constant of propagation, and $RMM^{\oplus}$, $RMMM^{\oplus}$, ... $RM_n^{\oplus}$ signify the growing cations. The monomer is consumed by propagation and cationation:

$$-\frac{dM}{dt} = k_p \cdot [N] \cdot [M] + k_c \cdot [R^{\oplus}][M]. \tag{29}$$

By substituting Eq. (23), we obtain:

$$-\frac{dM}{dt} = k_p \cdot [N][M] + \frac{d[N]}{dt} \tag{30}$$

and thus

$$k_p = \frac{-\frac{d[M]}{dt} - \frac{d[N]}{dt}}{[N] \cdot [M]}. \tag{31}$$

To solve for k_p by Eq. (31) we need d[M]/dt. This information was obtained by computer differentiating the experimental points in Fig. 12D. The corresponding d[M]/dt vs time plot is shown in Fig. 13B. The other information needed (i.e., the [N] vs t, [M] vs t, and d[N]/dt vs t plots) has already been presented in Fig. 12A, 12D, and 12B. By using Eq. (31) we can now calculate the overall k_p. Figure 13C shows representative k_p vs t curves. According to this result, k_p is not a true constant and its value is rapidly decreasing at short times until it levels off and becomes a "constant".

According to Scheme 1, the overall k_p includes the rate constants of 6 elementary steps: $k_{pR^\oplus,L}$; $k_{pR^\oplus,C^\oplus}$; $k_{pR^\oplus,L(E)}$; $k_{pH^\oplus,L}$; $k_{pH^\oplus,C^\oplus}$; $k_{pH^\oplus,L(E)}$. If $[R^\oplus,L]$, $[R^\oplus,C^\oplus]$ and $[R^\oplus,L(E)]$ stand for the concentration of all the growing species on the left side of Scheme 1, and $[H^\oplus,L]$, $[H^\oplus,C^\oplus]$ and $[H^\oplus,L(E)]$ for those on the right, and [N] is the total concentration of dormant and growing polymer chains (see Eq. (23)):

$$k_p = \frac{[R^\oplus,L]}{[N]}\cdot k_{pR^\oplus,L} + \frac{[R^\oplus,C^\oplus]}{[N]}\cdot k_{pR^\oplus,C^\oplus} + \frac{[R^\oplus,L(E)]}{[N]}\cdot k_{pR^\oplus,L(E)} + \frac{[H^\oplus,L]}{[N]}\cdot k_{pH^\oplus,L} + \frac{[H^\oplus,C^\oplus]}{[N]}\cdot k_{pH^\oplus,C^\oplus} + \frac{[H^\oplus,L(E)]}{[N]}\cdot k_{pH^\oplus,L(E)} \tag{32}$$

These concentrations are determined by eight equilibria characteried by eight equilibrium rate constants (see Definitions for Scheme 1) related to each other:

$$K_{RM_nX,1}\cdot K_{RM_nX,2} = K_{RM_nX,1(E)}\cdot K_{RM_nX,2(E)} \tag{33}$$

and similarly,

$$K_{HM_nX,1}\cdot K_{HM_nX,2} = K_{HM_nX,1(E)}\cdot K_{HM_nX,2(E)} \tag{34}$$

In this case, however, we neglect initiation by "HX". Thus the last 3 terms of Eqs. (32) and (34) can be neglected. While we do not have sufficient information to determine all the individual propagation rate constants and equilibrium constants, changes in the overall propagation rate constant will indicate changes in the relative contribution of the different growing species to N. The higher an individual propagation rate constant, and the lower the concentration of its corresponding species, the more sensitive will the overall k_p be of a concentration change. Since the propagation rate constant of the ionic species is several orders of magnitude larger than those of the living species [31, 59], a relatively slight change in $[R^\oplus,C^\oplus]$ (or in general $[R^\oplus,C^\oplus] + [H^\oplus,C^\oplus]$) the concentrations of which are low, will lead to a large change in k_p.

At this point it is instructive to examine the message conveyed by the k_p vs t plots in Fig. 13C. Apparently, at the very beginning of the polymerization the k_ps are extremely high indicating a very large contribution of dissociated ionic species to propagation. Very rapidly, however, these nonliving species are converted by the ED into living ones. Evidently, the interaction between the ionic species and $MtX_n \cdot ED$ complexes requires some time. The k_p decreases with increasing ED concentration. The concentration of ionic species was found to be very sensitive to medium polarity, temperature and additives, and their effects have been investigated (see Sect. 4.1.1.1).

By the use of the above "differential" method for the determination of k_c and k_p, we have in fact determined k_c and k_p *as a function of time* (see Eqs. (27) and (31)); thus, these rate coefficients are not the usual rate constants which are commonly obtained by the conventional "integral" method. (The reader is reminded, that in the fundamental scenarios, k_c and k_p were treated as real constants to enable the integration of differential equations.) Our experiments (see later) have proven that k_c and k_p are *not* true rate constants but are functions of time or conversion (see e.g., Figs. 13A and 13C and sect. 4.1.1.1).

2.3.2 Slow Initiation Plus Chain Transfer to Monomer

As mentioned earlier, the effects of slow initiation and chain transfer on the total number of macromolecules N compensate each other. At increasingly higher conversions (or molecular weights), the effect of slow initiation becomes increasingly less important and after all the initiator is used up the increase in N will be due only to chain transfer. This was the starting point for our investigation aimed at the determination of the chain transfer constant c^{I} and c^{II}; the precise calculation of these parameters requires relatively high molecular weight samples [20, 51].

Figure 14A shows experimental N data (see Eq. (22)) plotted against monomer conversion. The additivity of N is valid only for its monomer conversion dependent functions:

$$[N] = [N_{tr}] + [N_i] \tag{35}$$

where N_{tr} and N_i are the polymer molecules formed by chain tranfer to monomer and initiation by the initiator, respectively. The conversion dependent form of N_{tr} is (see, e.g., [1], or scenario A4bc in Table 1):

$$[N_{tr}] = -c^{\mathrm{I}} \ln(1 - C) + c^{\mathrm{II}}[M_0] \cdot C \tag{36}$$

which shows that the number (or concentration) of polymer chains produced by chain transfer to monomer $[N_{tr}]$ depends only on the zero order (c^{I}) and first order (c^{II}) chain transfer constants and initial monomer concentration $[M_0]$. After a sufficiently long reaction time (beyond $\sim 60\%$ conv.) $[N_i] \sim [I_0]$, so Eq. (35) can be written:

$$[N] = [I_0] - c^{\mathrm{I}} \ln(1 - C) + c^{\mathrm{II}}[M_0] \cdot C \tag{37}$$

or

$$\frac{1}{\overline{DP}_n} = \frac{[I_0]}{[M_0] \cdot C} + c^{\mathrm{II}} - c^{\mathrm{I}} \frac{\ln(1 - C)}{[M_0]C} \tag{38}$$

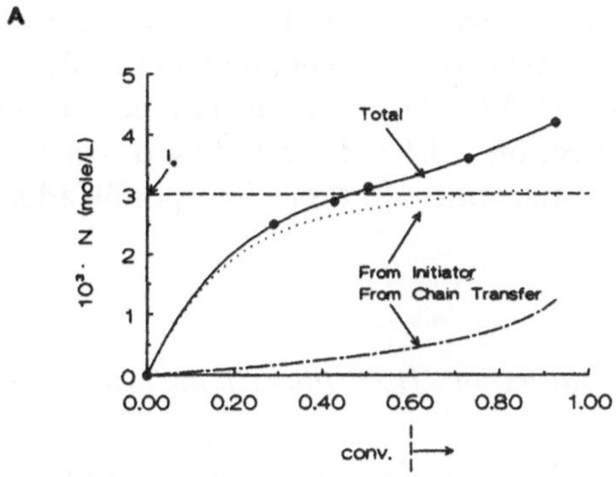

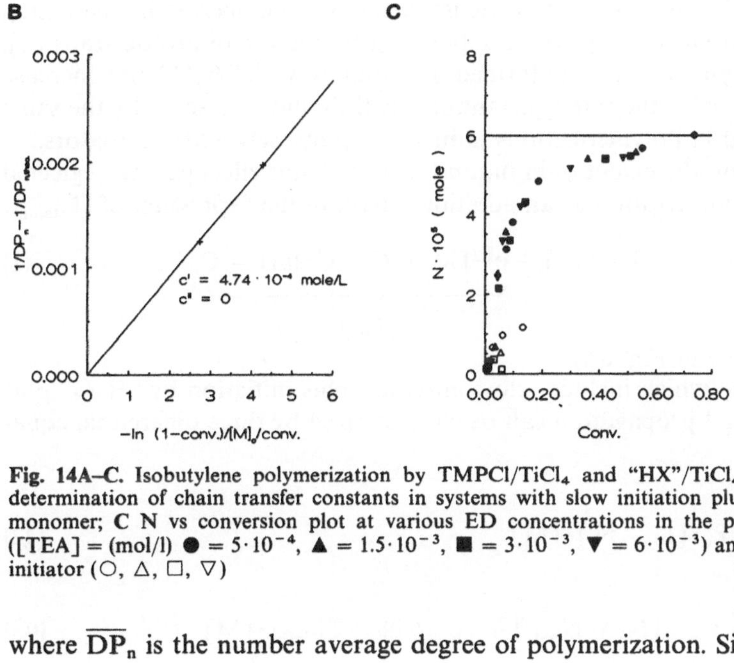

Fig. 14A–C. Isobutylene polymerization by TMPCl/$TiCl_4$ and "HX"/$TiCl_4$ at $-60\,°C$: **A, B** determination of chain transfer constants in systems with slow initiation plus chain transfer to monomer; **C** N vs conversion plot at various ED concentrations in the presence of initiator ([TEA] = (mol/l) ● = $5 \cdot 10^{-4}$, ▲ = $1.5 \cdot 10^{-3}$, ■ = $3 \cdot 10^{-3}$, ▼ = $6 \cdot 10^{-3}$) and in the absence of initiator (○, △, □, ▽)

where $\overline{DP}_n$ is the number average degree of polymerization. Since

$$\frac{[I_0]}{[M_0]\cdot C} = \frac{1}{\overline{DP}_{n,\text{theo}}} \tag{39}$$

thus

$$\frac{1}{\overline{DP}_n} - \frac{1}{\overline{DP}_{n,\text{theo}}} = c^{\text{II}} - c^{\text{I}}\,\frac{\ln(1-C)}{[M_0]C}. \tag{40}$$

The chain transfer constants can be obtained from the intercept and slop of Eq. (40) (see Fig. 14B). In this experiment, only two points were suitable for the determination of c^{I} and c^{II}, causing high uncertainty in their values (see Fig. 14A). According to these results $c^{\text{II}} \sim 0$, and $c^{\text{I}} = 4.74 \cdot 10^{-4}$ mole/L. From

Eqs. (35) and (36), we can now readily calculate the $[N_i] = f(\text{conversion})$ function (see Fig. 14A). Since the conversion = f(time) function is always available, $[N_i] = f(C)$ can easily be converted to $[N_i] = f(\text{time})$ function. From this point the procedure presented in Section 2.3.1 can be followed. (In this treatment instead of [N], $[N_i]$ stands for the concentration of purposely added initiator). See Eqs. (27) and (31).

2.3.3 *Slow Initiation Plus Initiation by "HX" Plus Chain Transfer to Monomer*

Depending on experimental conditions, the concentration of protic impurities, e.g., "H_2O" = $[N_{H_2O}]$, can reach 10^{-4}–10^{-3} mole/L and the exact "H_2O" concentration can be determined by control experiments (see [52–54, 60]). The effect of unwanted initiation by protic impurities can be masked by increasing the initiator concentration [56], or suppressed by the use of proton traps [1]. The number of macromolecules formed by initiation with "H_2O" may increase due to chain transfer, the rate constant of which should be essentially the same as that obtained in polymerizations induced by purposely added initiators.

Depending on the extent of initiaton by "H_2O", this effect can be neglected or taken into consideration as an additional term in the expression of $[N_{total}]$:

$$[N_{total}] = [N_{H_2O}] + [N_i] + \underbrace{c^{II}\cdot[M_0]\cdot C - c^{I}\cdot\ln(1-C)}_{[N_{tr}]} \tag{41}$$

(cf. scenario C4bc in Fig. 8A).

This scenario which includes slow initiation plus initiation by "H_2O" plus chain transfer and propagation can be characterized by three differential equations:

$$\frac{d[N_i]}{dt} = k_c\cdot[I_0 - N_i]\cdot[M] \tag{42}$$

$$\frac{d[N_{tr}]}{dt} = k_{tr,{}^1M}[N_i + N_{H_2O}] + k_{tr,{}^2M}[N_i + N_{H_2O}]\cdot[M] \tag{43}$$

$$-\frac{d[M]}{dt} = k_p[N_i + N_{H_2O}]\cdot[M] + k_c[I_0 - N_i]\cdot[M] + k_{tr,{}^2M}[N_i + N_{H_2O}]\cdot[M] \tag{44}$$

(The last two terms in Eq. (44) are small relative to the first one and can be neglected.) From our experiments we can readily develop $[N_{total}]$ (see Eq. (22)) versus conversion (C) plots. To obtain the $[N_i] = f(\text{conv.})$ function, we need to know the ["H_2O"] and $[N_{tr}] = f(\text{conv.})$ functions (see Eq. (41)), i.e., the results of control experiments must be available carried out under the same conditions characterized by the following differential equations: (cf scenario B2c in Table 1 and Fig. 4A)

$$\frac{d[N_{tr}]}{dt} = k_{tr,{}^1M} \cdot [N_{H_2O}] + k_{tr,{}^2M} \cdot [N_{H_2O}] \cdot [M] \tag{45}$$

$$-\frac{d[M]}{dt} = k_p \cdot [N_{H_2O}] \cdot [M] + k_{tr,{}^2M} \cdot [N_{H_2O}] \cdot [M] \tag{46}$$

(The second term in Eq. (46) can be neglected.)

$$[N_{total}] = [N_{H_2O}] + [N_{tr}]. \tag{47}$$

Dividing Eq. (43) by Eq. (44) and Eq. (45) by Eq. (46) and solving the differential equations, leads to the same results in both cases (cf Eq. (41), i.e.,

$$[N_{tr}] = c^{II}[M_0] \cdot C - c^{I} \ln(1 - C). \tag{48}$$

By subtracting Eq. (47) from Eq. (41), we obtain $[N] = [N_i] = f(\text{conv.})$ which is easily converted to $[N_i] = f(\text{time})$ functions (see also Sect. 2.3.2).

From Eq. (42):

$$k_c = \frac{d[N_i]/dt}{[I_0 - N_i] \cdot [M]}. \tag{49}$$

Thus K_c can be obtained since all terms in Eq. (49) are available by the route described in Sect. 2.3.1.

Based on Eq. (44):

$$k_p = \frac{-\frac{dM}{dt}}{[N_i + N_{H_2O}][M]} \tag{50}$$

To solve for k_p we needed the $[N_i + N_{H_2O}]$ vs t relationship. Since $[N_{H_2O}]$ can be obtained by extrapolating the $[N_{total}]$ vs conv. plot to 0 conversion in the control experiment and the $[N_i]$ vs time plot is already known, we obtain k_p by Eq. (50) by following the route outlined in Sect. 2.3.1.

In this section we have demonstrated how to obtain k_c and k_p under different experimental conditions by using a "differential" method without assuming that these quantities are constants. This is supported by experimental results (see Sects. 4.1.1.1, 4.2.1.1, 4.2.2.2, 4.2.2.3.1).

3 Experimental

The sources of chemicals, their purification, and the synthesis of TMPCl [36, 61] together with polymerization and characterization methods have been described [62].

4 Experimental Substantiation of the Model

Experiments were carried out with IB and St, arguably the two most important representative olefins polymerizable by cationic means. Polymerizations were initiated by the purposely added initiator 2-chloro-2,4,4-trimethylpentane (TMPCl) or by adventitious moisture, "H_2O" (i.e., absence of purposely added initiator), in conjunction with $TiCl_4$ under various conditions specified in the legends to the figures and tables. The raw data are collected in tables shown in the Appendix. This section concerns the presentation of the data and their interpretation in terms of the model developed in Sect. 2.

4.1 Isobutylene

4.1.1 Controlled Initiation by the TMPCl/TiCl₄/IB System

4.1.1.1 The Effect of Reaction Conditions on k_c and k_p

Experiments have been carried out to substantiate the comprehensive closed-loop model described in Sect. 2. The experiments included the determination of k_c and k_p under various conditions including a study of the effect of temperature on these parameters, an examination of the effect of experimental conditions on $\bar{M}_n$ and $\bar{M}_w/\bar{M}_n$ together with an analysis of the GPC traces obtained. The raw data are shown in Tables 2–8 in the Appendix.

4.1.1.1.1 The Effect of [ED] on k_c and k_p at −25, −45, −60, and −82 °C

Figures 13A, C, 15A–D, and 16A, B show the effect of [ED] (i.e., TEA concentration) on k_c and k_p at different temperatures. In general, k_c decreases with increasing [ED] because the reactivity of the initiating cations is reduced. The effect of [ED] on k_c must be very similar to those on the reactivity of growing species characterized by Eq. (32).

In the presence of insufficient [ED] chain transfer may occur; thus with increasing time (conversion), k_c calculated by Eq. (27) may increase, suggesting a higher than true cationation rate. (See. Sects. 2.2.1, 2.3).

The k_p is also decreasing with increasing [ED] indicating the conversion of highly reactive cationic chain ends to less reactive living species by $MtX_n \cdot ED$ complexes. Since the contribution of $k_{pC^\oplus}$ to the overall k_p is decreasing, the latter must also decrease (see Eq. (32)).

4.1.1.1.2 The Effect of Solvent Polarity on k_c and k_p at −80 °C

According to the data shown in Fig. 17A, B, k_c and k_p are increasing with increasing solvent polarity at −80 °C. Evidently, the higher the polarity the higher the solvation of the ionic chain ends which explains the higher overall k_ps.

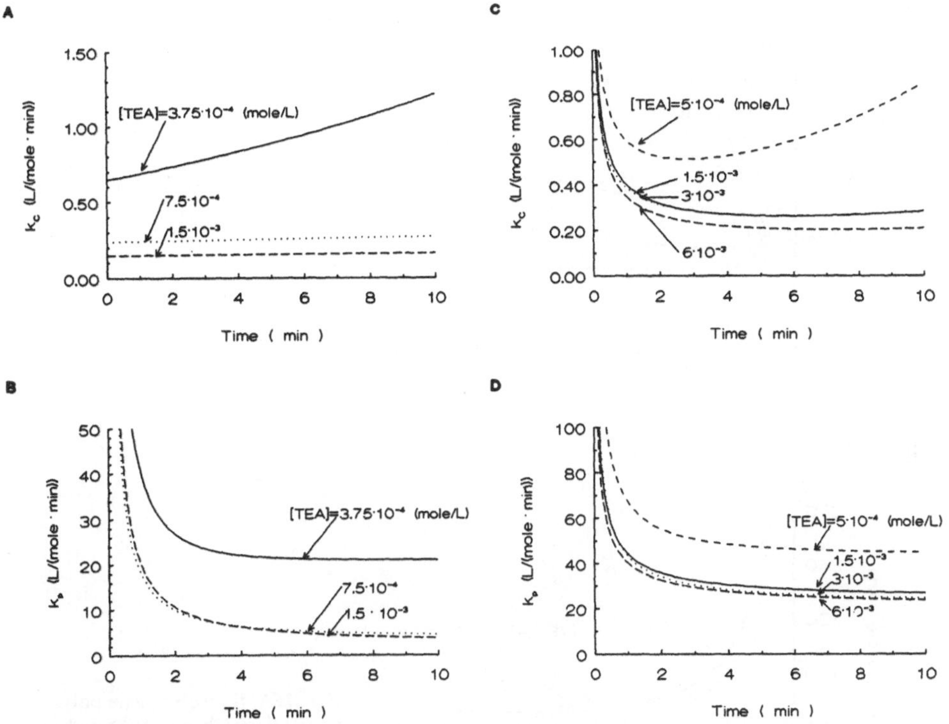

Fig. 15A–D. Isobutylene polymerization: **A, B** change of k_c and k_p with time at various ED concentrations at $-25\,°C$; **C, D** change of k_c and k_p with time at various ED concentrations at $-60\,°C$

4.1.1.1.3 The Effect of [MtX_n] on k_c and k_p at $-80\,°C$

Figures 17C, D, and 18A show k_c and k_p at different [$TiCl_4$]s at $-80\,°C$. Apparently, both k_c and k_p increase with increasing [$TiCl_4$] suggesting that the equilibria governing k_c and k_p are similar and are governed by similar influences.

4.1.1.1.4 The Effect of Aging on k_c and k_p at $-60\,°C$

We postulated that by changing the addition sequence of the ingredients, that is by first mixing the initiator, ED and MtX_n (TMPCl + TEA + $TiCl_4$) and adding the monomer last, the Winstein equilibria shown in the two uppermost rows of Scheme 1 will be more rapidly established, and that the concentration of the undesirable ionic ($MtX_n \cdot$ED-uncomplexed) species will be reduced in proportion to the quantity of $MtX_n \cdot$ED available in the charge (see Sect. 2.1).

Thus the series of experiments have been carried out in which first the solvent-initiator-electron donor and $TiCl_4$ in this order, were premixed, and

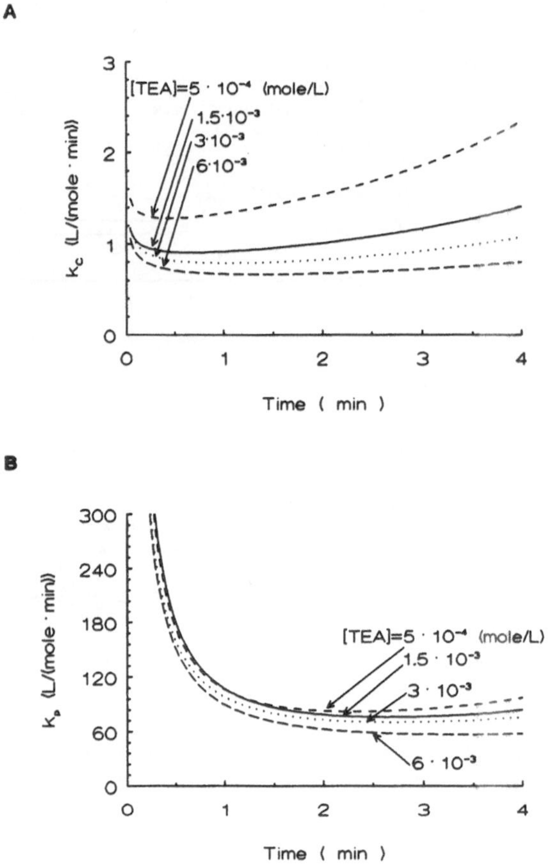

Fig. 16A, B. Isobutylene polymerization. Change of k_c and k_p with time at various ED concentrations at $-82\,°C$

after 4 min of "aging" the monomer was rapidly introduced. The moment of monomer addition was the zero point of the experiment.

Figure 18C shows the change in k_c with time: While in the conventional unaged system k_c drops exponentially from large values to a limit at ~ 0.35, in the aged system the k_c reaches a maximum at ~ 0.52 whence it decreases rapidly to approximately the same limit. As anticipated, according to the plots in Fig. 18D, the k_p versus time profiles of conventional and aged systems are the same within experimental error. Experimental conditions and the raw data are shown in Table 8 in the Appendix.

4.1.1.1.5 The Effect of Temperature on k_c and k_p

A study of the effect of temperature on k_c and k_p gave interesting insight into the nature of these rate constants. Figures 19A–D and 20A–D show the data in the form of log k_c and log k_p vs 1/T plots. Surprisingly, all the plots are linear with

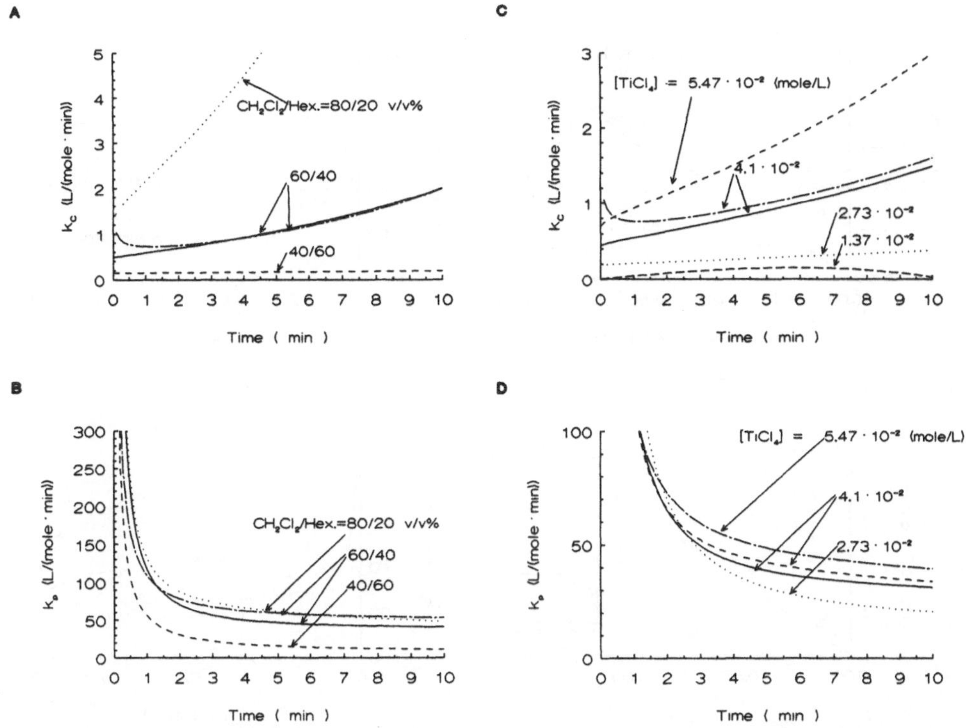

Fig. 17A–D. Isobutylene polymerization at $-80\,°C$: **A, B** change of k_c and k_p with time in different solvent mixtures; **C, D** change of k_c and k_p with time at different MtX_n concentrations

positive slopes. Both k_c and k_p increase with decreasing temperature most likely because of the small but significant increase of the polarity of the medium with decreasing temperatures. Since the contribution of ionic species to the rates of initiation and propagation is conceivably several orders of magnitude larger than that of nonionic species, even a minute increase in the polarity may result in a large increase in the overall rate of polymerization.

4.1.1.2 The Effect of Reaction Conditions on Molecular Weights ($\bar{M}_n$) and Number of Chains (N)

4.1.1.2.1 The Effect of [ED]

Figure 21 shows the $\bar{M}_n$ of PIB (and the number of PIB chains, N, in the insets) as a function of polymer formed W_p at various ED concentrations (expressed by [TMPCl]/[TEA]). The solid lines are "theoretical," i.e., have been calculated by assuming 100% initiation efficiency. At $-25\,°C$ massive chain transfer occurs rather early in the reaction (above ~ 0.5 g or ~ 6000 $\bar{M}_n$) and the addition of

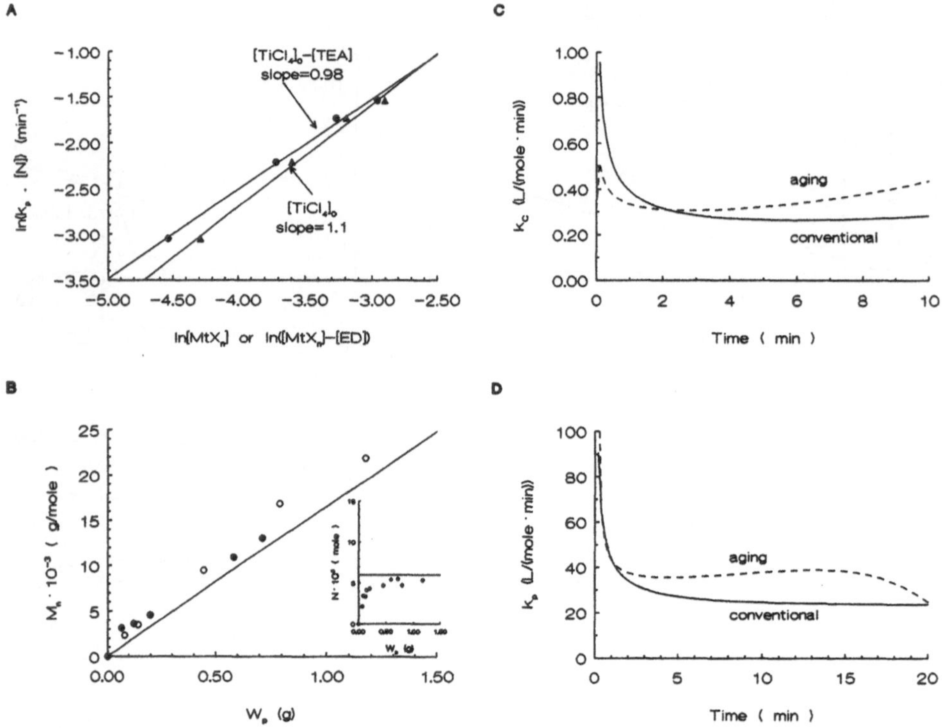

Fig. 18A–D. Isobutylene polymerization: **A** determination of apparent rate order in $TiCl_4$ at $[TEA] = 3 \cdot 10^{-3}$ mol/l at $-80\,°C$; **B** M_n (and N) vs W_p plot in conventional (●) and "aging" (○) experiments at $-60\,°C$; **C, D** change of k_c and k_p with time in conventional and "aging" experiments at $[TEA] = 3 \cdot 10^{-3}$ mol/l at $-60\,°C$ ("aging" time = 4 min)

increasing amounts of TEA does not seem to suppress this undesirable reaction. At $-45\,°C$ the rate of chain transfer is much decreased and the experimental data are close to theoretical up to ~14000 $\bar{M}_n$. At $-60\,°C$ chain transfer is absent up to ~16000 $\bar{M}_n$. At this temperature the rate of initiation is noticeably reduced relative to that of the propagation (notice the initially higher than theoretical $\bar{M}_n$s). At $-82\,°C$ this effect becomes even more pronounced.

According to these data, optimum living IB polymerization and up to $\bar{M}_n$ ~15000 with uniform molecular weight distribution ($\bar{M}_w/\bar{M}_n < 1.2$) PIB can be obtained with the $TMPCl/TiCl_4$ system in CH_2Cl_2/n-C_6H_{14} 60/40 v/v solution at $-45\,°C$ (see also data in Tables 2–5 in the Appendix).

4.1.1.2.2 The Effect of Solvent Polarity at $-80\,°C$

Figure 22A (and the data in Table 6 in the Appendix) illustrate the effect of medium polarity (expressed by the CH_2/Cl_2/*n*-C_6H_{14} ratio from 40/60 to 100/0 v/v) at $[I_0]/[ED] = 1.0$ on the $\bar{M}_n$ (and N) vs W_p profiles. The rate of initiation

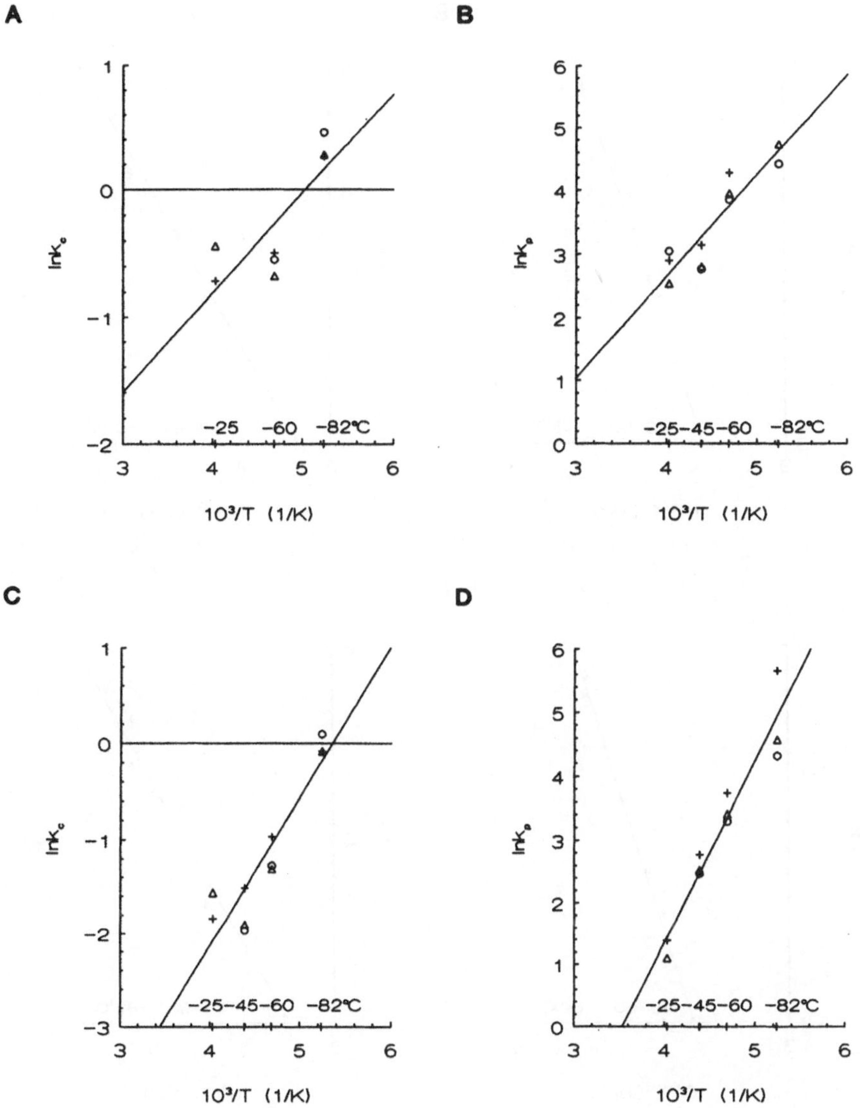

Fig. 19A–D. Isobutylene polymerization: **A, B** the effect of temperature on k_c and k_p [TEA] = $5 \cdot 10^{-4}$ mol/l; **C, D** [TEA] = $1.5 \cdot 10^{-3}$ mol/l (monomer conversion: + = 10%, △ = 30%, ○ = 50%)

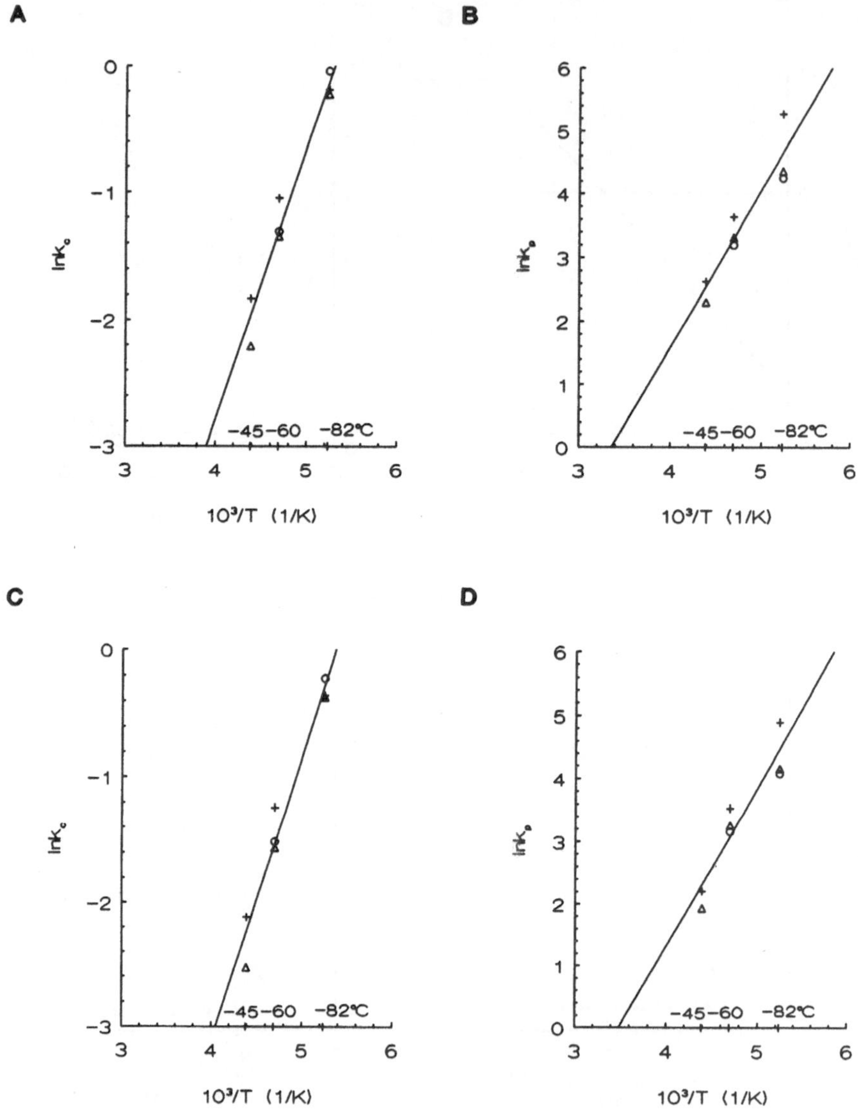

Fig. 20A–D. Isobutylene polymerization. The effect of temperature on k_c and k_p: **A, B** [TEA] = $3 \cdot 10^{-3}$ mol/l; **C, D** [TEA] = $6 \cdot 10^{-3}$ mol/l (monomer conversion: + = 10%, △ = 30%, ○ = 50%)

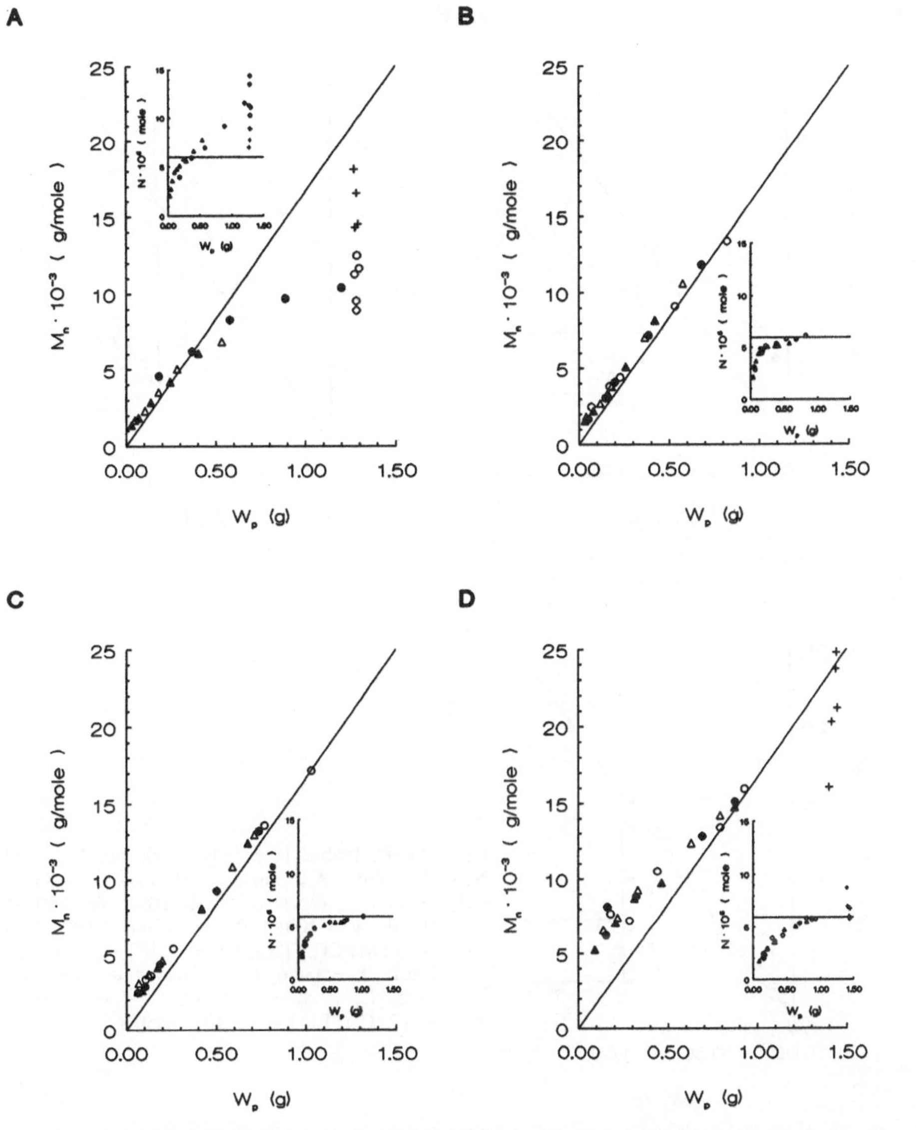

Fig. 21A–D. Isobutylene polymerization. M_n (and N) vs W_p plots at various temperatures and at various ED concentrations expressed by $[I_0]/[TEA]$ ratios: **A** at −25 °C: + = 1:0; ○ = 1:0.042; ● = 1:0.125; △ = 1:0.25; ▲ = 1:0.5; **B** at −45 °C; **C** at −60 °C; **D** at −82 °C: + = 1:0; ○ = 1:0.17; ● = 1:0.5; ▲ 1:1; △ 1:2

A

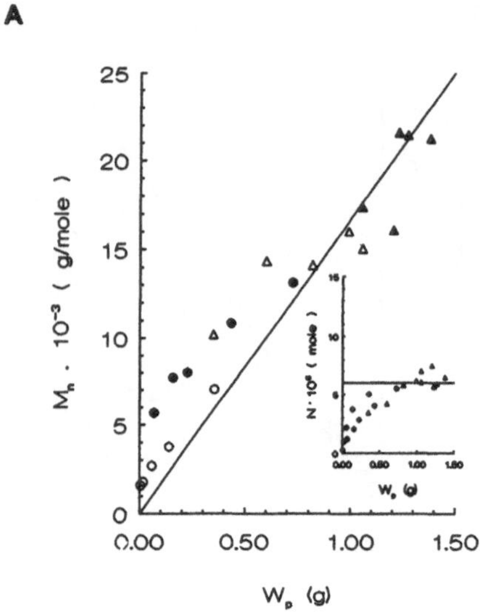

B

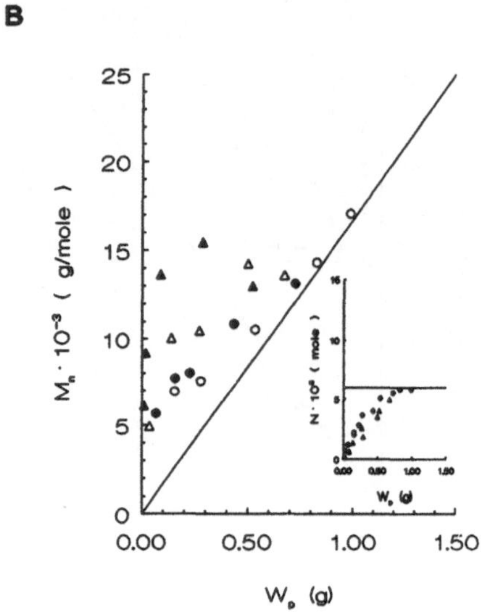

C

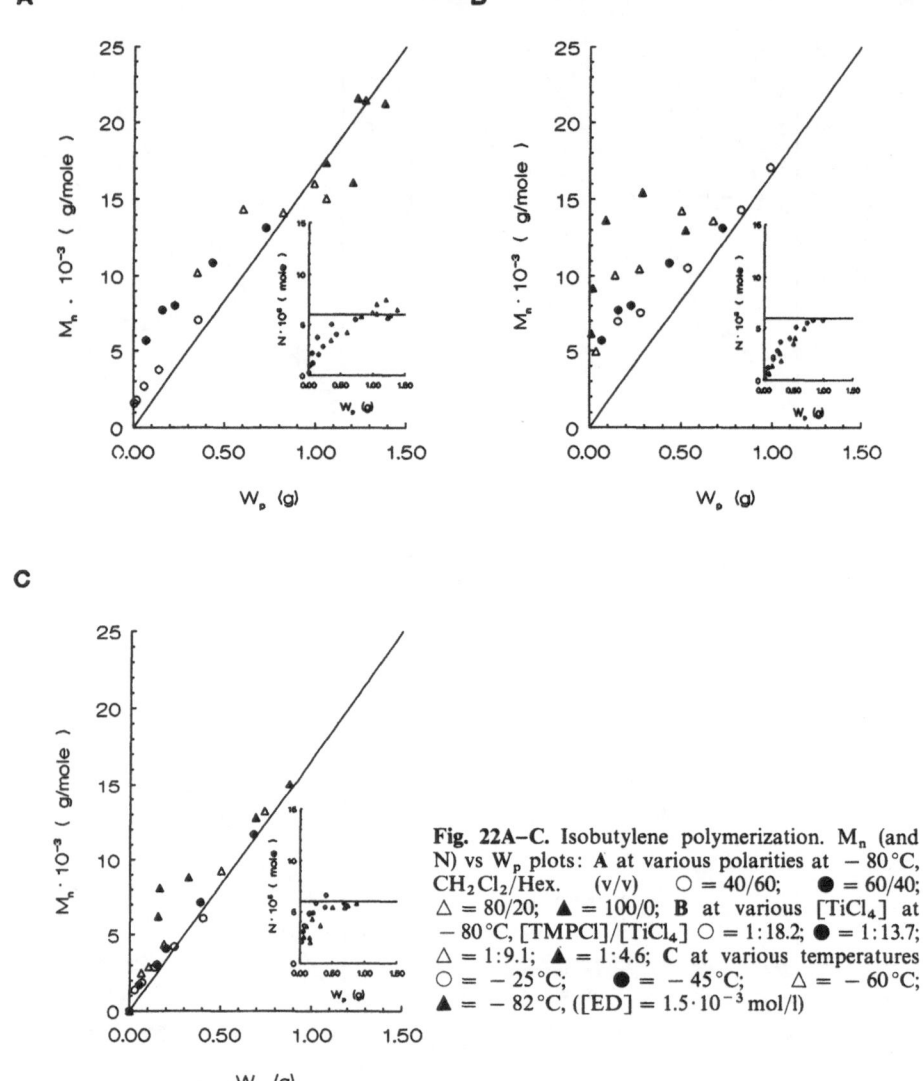

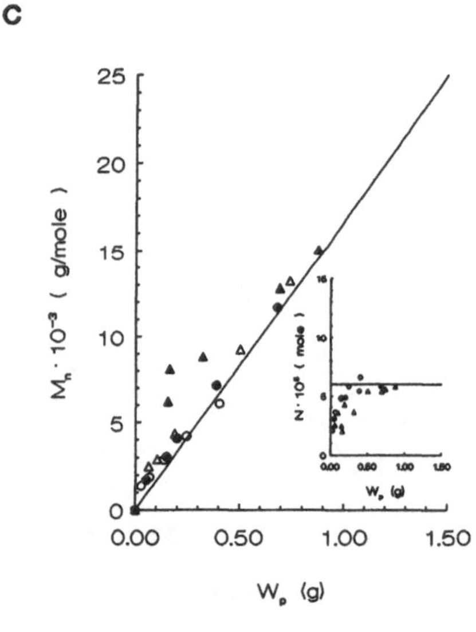

Fig. 22A–C. Isobutylene polymerization. M_n (and N) vs W_p plots: **A** at various polarities at $-80\,°C$, CH_2Cl_2/Hex. (v/v) ○ = 40/60; ● = 60/40; △ = 80/20; ▲ = 100/0; **B** at various $[TiCl_4]$ at $-80\,°C$, $[TMPCl]/[TiCl_4]$ ○ = 1:18.2; ● = 1:13.7; △ = 1:9.1; ▲ = 1:4.6; **C** at various temperatures ○ = $-25\,°C$; ● = $-45\,°C$; △ = $-60\,°C$; ▲ = $-82\,°C$, ($[ED] = 1.5 \cdot 10^{-3}$ mol/l)

relative to that of propagation was found to be slow in every experiment. By the use of neat CH_2Cl_2 the overal rates were too high to be measured and the first samples that could be harvested have already reached $\bar{M}_n \sim 15\,000$.

4.1.1.2.3 The effect of $[MtX_n]$ at $-80\,°C$

Figure 22B (and the data in Table 7 in the Appendix) show the effect of the $[I_0]/[MtX_n]$ ratio on the $\bar{M}_n$ (and N) vs W_p profiles. Evidently, by increasing the

[$TiCl_4$] the relative rate of initiation increases and the system becomes close to theoretical at $\bar{M}_n \approx 15000$.

4.1.1.2.4 *The Effect of Temperature*

Figure 22C shows the effect of temperature in the −25 to −82 °C range. The lower the temperature the lower the relative rate of initiation. At the relatively high temperature of −25 °C, there is evidence for chain transfer (see the datum below the theoretical line). At −45 °C, and −60 °C close to living systems can be obtained with $\bar{M}_n \sim 15000$ and $\bar{M}_w/\bar{M}_n = 1.19$ and 1.23, respectively. (See data in Tables 2–5 in the Appendix).

4.1.1.2.5 *The Effect of Aging on $\bar{M}_n$ versus W_p Plots at −60 °C*

The effect of relatively brief (4 min) aging is quite apparent upon comparing the data shown in Fig. 18B (the raw data are in Tables 4 and 8 in the Appendix). Evidently, aging decreases the extent of slow initiation at low conversions, i.e., the experimental points obtained with the aged system move closer to the solid theoretical line suggesting that a larger number of polymer chains grow at the beginning in the aged system than in the conventional system. With increasing conversions the $\bar{M}_n$s seem to satisfy a straight ascending line (not shown) above the theoretical $\bar{M}_n$ vs W_p plot. The fact that the experimental points run above the theoretical line indicates a constant but lower number of living chains than I_0, and the part of the TMPCl initiator was consumed by side reaction(s) during aging.

4.1.1.3 Apparent Rate Order in Friedel-Crafts Acid

The reaction order of propagation in respect of the Friedel-Crafts acid has been determined. Figure 18A shows the corresponding plot of ln [k_pN] vs ln [$TiCl_4$]. The plot is linear with a slope close to unity indicating first order behavior. The figure also contains a second plot: ln (k_pN) vs ln ([$TiCl_4$] − [TEA]), i.e., the Friedel-Crafts acid concentration minus the electron donor concentration. This plot is also linear with a slope of close to unity suggesting first order behavior.

Contrary to these results, Faust et al. (37) found close to second order behavior in [$TiCl_4$] in similar IB polymerizations initiated by aromatic compounds, e.g., dicumyl chloride. The authors explained their findings by assuming the existence of $TiCl_4$ dimers. The resolution of this intriguing discrepancy between these observations must wait for further investigation.

4.1.1.4 The Effect of Experimental Conditions on Molecular Weights and Molecular Weight Dispersities: GPC Studies

The analysis of the position and shape of GPC traces provides valuable insight into the mechanistic details of polymerizations. To elucidate the nature of undesirable reactions encumbering the living polymerization of IB (i.e., chain transfer, relatively slow initiation or exchange reactions, impurity effects), we have systematically studied the effect of time, temperature, medium polarity, Friedel-Crafts acid and electron donor concentration on the position and shape of GPC traces of PIBs prepared in impurity-initiated and purposely-initiated polymerizations. This section concerns the presentation of the data and information gleaned by the analysis of RI GPC traces.

4.1.1.4.1 The Effect of Time and [ED]

Figure 23A shows a family of RI GPC traces of PIBs obtained in a set of AMI experiments with the TMPCl/$TiCl_4$ initiating system (at TMPCl/$TiCl_4$/TEA = 1:13:7:0.5) at −25 °C as a function of time. At the shortest time or lowest conversion (2 min or 2.2%), the GPC trace indicates the presence of three major products: the highest $\bar{M}_n$ product indicated by the broad peak at $\bar{M}_n \sim 10^{-6}$ is most likely due to rapid initiation by "H_2O" and subsequent rapid nonliving ionic propagation via the $k_{pH^{\oplus}, C^{\oplus}}$ route; (see descending arrow in Scheme 1). The next species ($\bar{M}_n \sim 10^4$) is also nonliving and is proposed to arise somewhat less rapidly from the TMPCl and to propagate by the ionic $k_{pR^{\oplus},C^{\oplus}}$ route. Finally, the lowest molecular weight product with the narrowest molecular weight distribution ($\bar{M}_n \sim 10^3$, $\bar{M}_w/\bar{M}_n = 1.24$) also arises from TMPCl but the growing species is mediated by the $MtX_n \cdot ED$ complex and therefore propogates relatively even more slowly by the $k_{pR^{\oplus}, L(E)}$ route. With increasing time (conversion) the proportion of the latter "living" species increases, its position shifts linearly toward higher molecular weights and its molecular weight dispersity narrows ($\bar{M}_w/\bar{M}_n = 1.06$ at 40 min) (see also data in Table 2 in the Appendix); at the same time, the other two nonliving populations are gradually overtaken and engulfed by the advancing living population. Initiation and propagation by the two nonliving cationic species are fast relative to the $MtX_n \cdot ED$-mediated living entity; however, in time, the nondissociated living $MtX_n \cdot ED$ complex suppresses the concentration of the ionic nonliving species so that at higher conversions the GPC traces show only the living product. The overall MWD of the final product (at 30.8% in Fig. 23A) is determined by the contributions of the two nonliving populations plus the steadily growing living polymer. The relative amounts of the two nonliving species gradually decrease with time and it becomes difficult to distinguish them from the GPC baseline with increasing conversions (see also Sect. 4.2.1.4.1).

The set of data in Fig. 23B indicate the effect of [ED] (expressed by the [TMPCl]/[TEA] ratio) on $\bar{M}_n$ and MWD under the same conditions. In the absence of ED (conventional polymerization) or at low [ED] ([TMPCl]/[TEA] = 1:0.042), polymerizations are rapid and conversions are complete in much less than 10 min, and the GPC traces show broad multimodal distributions suggesting a multiplicity of propagating species. With increasing [TEA] the rate of polymerization is reduced due to the decrease in the concentration of cationic species and a reduction in k_p. At [TMPCl]/[TEA] = 1:0.5 most of the

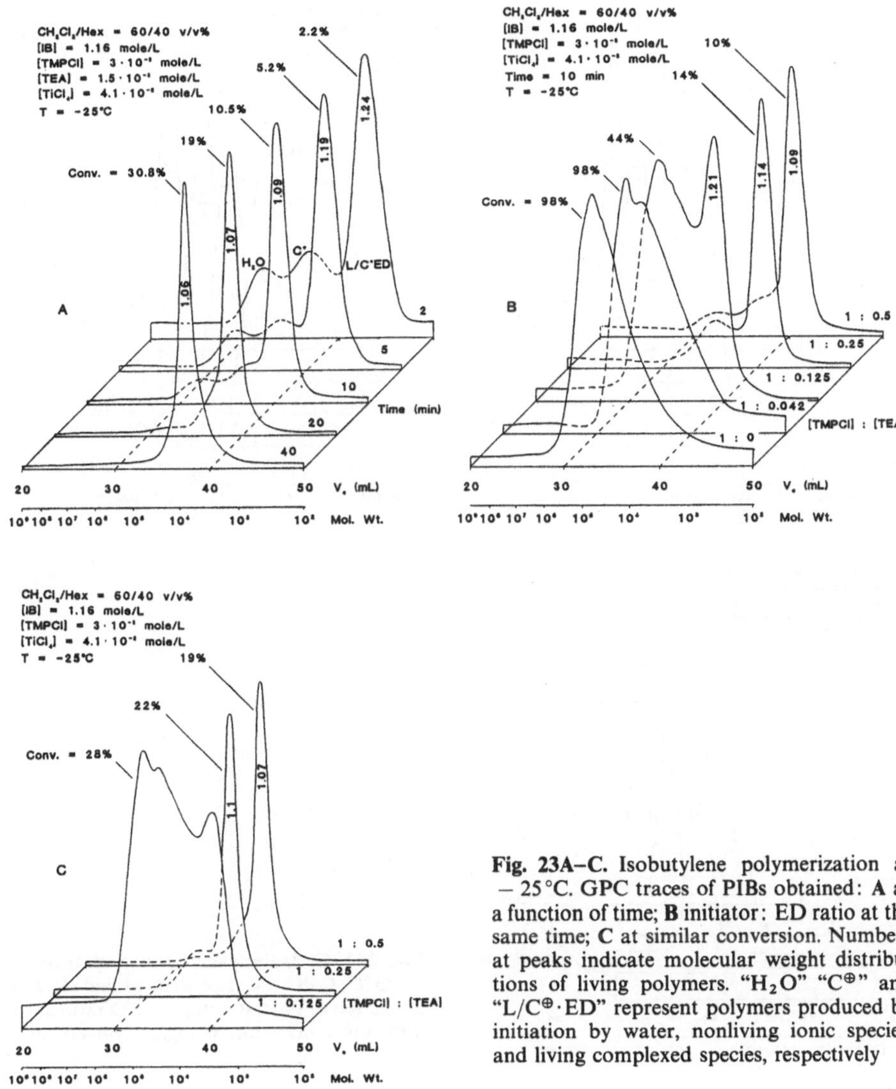

Fig. 23A–C. Isobutylene polymerization at $-25\,°C$. GPC traces of PIBs obtained: **A** as a function of time; **B** initiator: ED ratio at the same time; **C** at similar conversion. Numbers at peaks indicate molecular weight distributions of living polymers. "H_2O" "$C^{\oplus}$" and "$L/C^{\oplus} \cdot ED$" represent polymers produced by initiation by water, nonliving ionic species, and living complexed species, respectively

product is produced by the living species, while the contributions of the nonliving ionic species (GPC peaks at higher molecular weights) are greatly reduced. Under the particular conditions, the conversion reached only 10%, however, the MWD of the predominant species was quite narrow, i.e., $\bar{M}_W/\bar{M}_n = 1.09$.

Figure 23C conveys a similar message: it shows three GPC traces of PIBs obtained at about the same intermediate conversion level (19–28%) at various [ED]s. Thus at low [ED] (i.e., at [TMPCl]/[TEA] = 1:0.125), the GPC trace is multimodal indicating the presence of multiple chain carriers of comparable importance. At the higher [ED] (i.e., at [TMPCl]/[TEA] = 1:0.5), the GPC

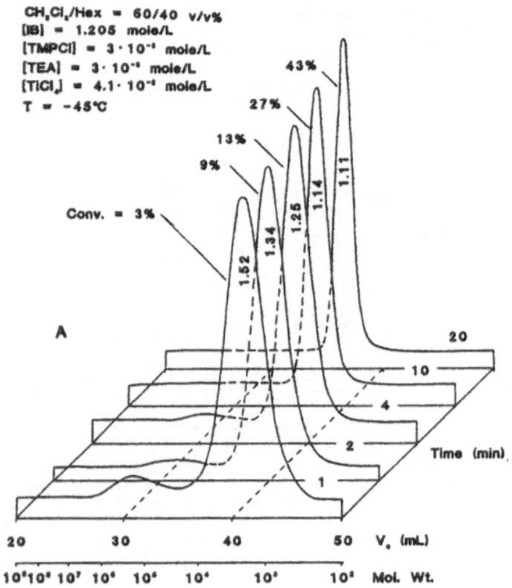

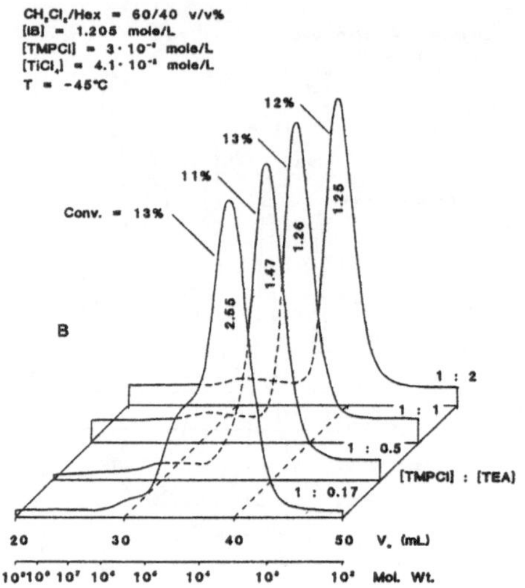

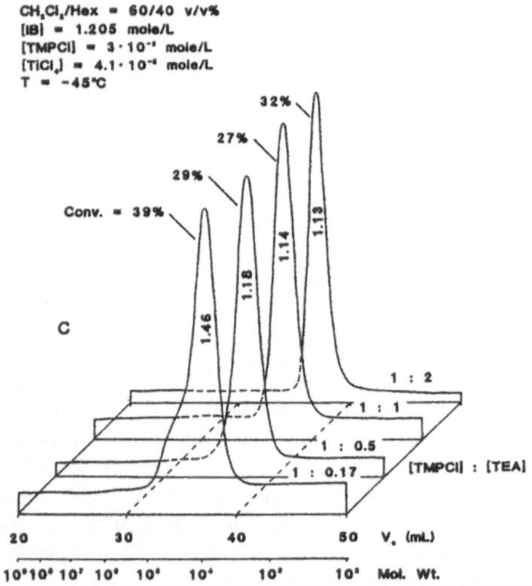

Fig. 24A–C. Isobutylene polymerization at $-45\,^{\circ}C$. GPC traces of PIBs obtained: **A** as a function of time; **B** initiator: ED ratio at the same low conversion; **C** higher conversion

trace exhibits at low molecular weights a major narrow distribution, $\bar{M}_w/\bar{M}_n = 1.07$, indicating the contribution by living species. Evidently with decreasing [ED] the polymerization becomes less and less controlled.

Figure 24A (and the data in Table 3 in the Appendix) show the effect of time (conversion) on the $\bar{M}_n$ and MWD of PIBs prepared at $[I_0]/[ED] = 1$ at $-45\,^{\circ}C$. At low conversion (3%) the GPC trace shows two relatively broad peaks. With increasing conversions (up to 43%) the major peak at low $\bar{M}_n$ becomes progressively narrower while the smaller hump at high $\bar{M}_n$ melts into the base line. Most likely the higher $\bar{M}_n$ species arose by rapid "H_2O" induced polymerization. Simulataneously, the $MtX_n \cdot ED$ complex reacts with the highly reactive cationic chain ends and gives rise to the less reactive living species. Thus the product becomes progressively uniform and the contribution by the initially formed high molecular weight product gradually decreases.

The data collected in Fig. 24B, C reinforce the above conclusions: the most homogeneous product in terms of MWD uniformity (narrowness of GPC traces)

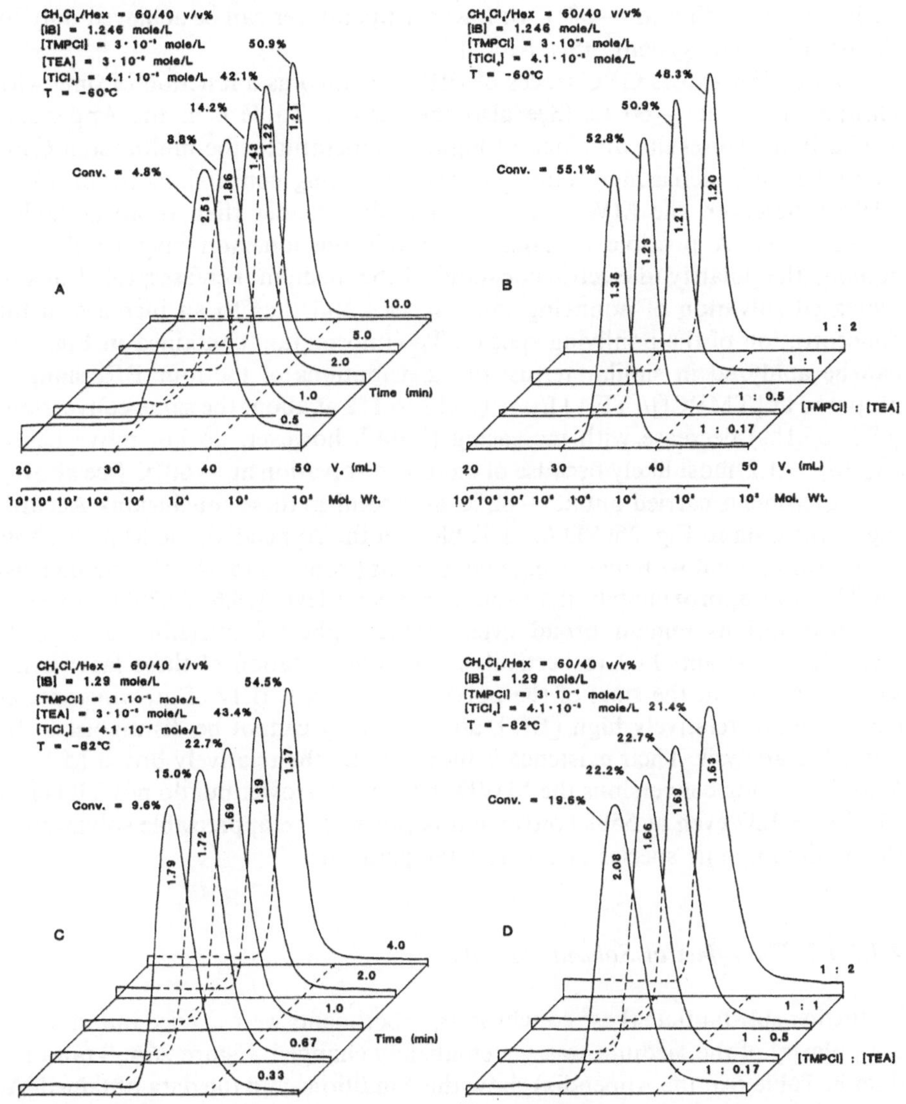

Fig. 25A–D. Isobutylene polymerization. GPC traces of PIBs obtained: **A, C** as a function of time; **B, D** initiator: ED ratio at similar conversion; **A, B** at −60 °C; **C, D** at −82 °C

are obtained at a relatively high [ED] and higher conversions. The product, although definitively better defined, still shows some contribution by the nonliving high molecular weight product in the 11–13% conversion range. In contrast, in the 27–39% conversion range at the same relatively high [TEA], the product is well-defined and its MWD is narrow with $\bar{M}_w/\bar{M}_n = 1.13$.

It is noteworthy that such narrow MWD ($\bar{M}_w/\bar{M}_n \leqq 1.1$) PIBs can be readily prepared at $-25°$ and -45 °C. As will be discussed below, the synthesis of such narrow MWDs at -60 °C or lower temperatures can be achieved only by the use of "aged" systems. (See Sect. 4.1.1.4.5)

Figure 25A shows GPC traces of PIBs prepared as a function of time with $[I_0]/[ED] = 1$ at -60 °C. (See also the data in Table 4 in the Appendix.) Similarly to the results obtained at higher temperatures, the multimodal GPC traces become increasingly uniform with increasing conversions (from 4.8 to 50.9%), however, the MWDs tend to remain broader than those at higher temperatures. A possible explanation of this phenomenon may be that by cooling, the polarity (dielectric constant) of the medium increases; this leads to increased solvation of nonliving ionic species and thus to an increase in the concentration of the nonliving species. The information contained in Fig. 25B can be analyzed in similar terms: the distributions of the four PIB samples prepared at [TMPCl]/[TEA] from 1:0.17 to 1:2 at about the same conversions (48.3–55.1%) decrease with increasing [TEA], however, do not move below $\bar{M}_w/\bar{M}_n = 1.2$, most likely because of increased solvation at -60 °C (see above).

Experiments carried out at -82 °C also confirm these conclusions. According to the data in Fig. 25C, D (and Table 5 in the Appendix), the MWDs show a narrowing trend with increasing conversions (from 9.6 to 54.5%) and increasing [ED] at approximately the same conversion level (19.6–22.7%), however, the distributions remain broad even at the highest conversions and [ED] ($\bar{M}_w/\bar{M}_n = 1.37$ and 1.63 respectively). The interpretation of these facts is the same as above: at the start of the polymerization a variety of growing species exists even at relatively high [ED], although they cannot be distinguished by our GPC analysis. Their existence is indicated by the relatively broad MWDs. With increasing conversions the MWDs become narrower but do not fall below $\bar{M}_w/\bar{M}_n = 1.37$ even at 55% conversion because of the appreciable solvation of the nonliving ionic species at this low temperature.

4.1.1.4.2 The Effect of Solvent Polarity

To further substantiate these conclusions, experiments were carried out in which the polarity of the medium was systematically changed. Figure 26A, B (and the data in Table 6 in the Appendix) show the conditions and the data obtained. At about the same conversion level (24.3–30.2%) the narrowest MWD (most uniform) product was obtained ($\bar{M}_w/\bar{M}_n = 1.32$) in $CH_2Cl_2/n\text{-}C_6H_{14} = 40{:}60$. By increasing the polarity the rate of the polymerization increases rapidly, and it becomes very difficult to stop the reactions at low conversions. In polar media

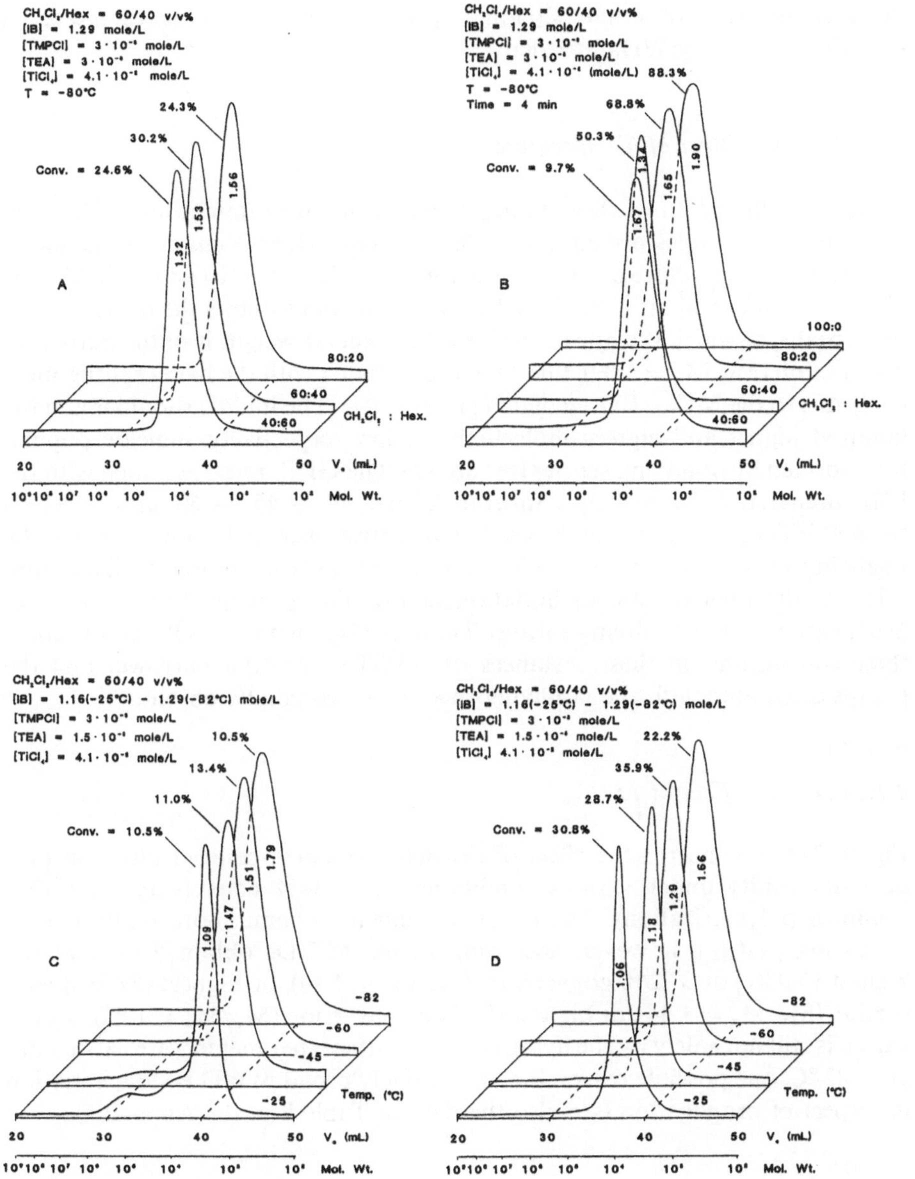

Fig. 26A–D. Isobutylene polymerization. GPC traces of PIBs: **A** as a function of solvent polarity at similar conversions at $-80\,°C$; **B** at the same time at $-80\,°C$; **C** at similar low conversions as a function of temperature; **D** higher conversions as a function of temperature

the concentration of nonliving ionic species is enhanced due to increased solvation and the MWDs broaden.

4.1.1.4.3 The Effect of Temperature

Further confirmation of these thoughts comes upon examination of IR GPC traces of PIB samples obtained at different temperatues. Figure 26C, D shows the effect of temperature at two conversion levels, i.e., at ~ 10 and ~ 30%. The GPC trace obtained at −25 °C at 10.5% conversion exhibits the characteristic three peaks, of which the one at the lowest molecular weight is of the narrowest distribution ($\bar{M}_w/\bar{M}_n = 1.09$); this peak is associated with the living species most likely due to the $MtX_n \cdot$ED-mediated growing site. The middle shoulder and the hump at higher and highest molecular weights, respectively, indicate populations formed by nonliving species (see above). The GPC traces obtained with the PIBs prepared at decreasing temperatures (i.e., at −45, −60, and −82 °C) show MWD broadening and the gradual disappearance of the humps due to the nonliving propagating species. These effects are most likely due to the diminution of the relative rate of initiation and/or the gradual "freezing out" of propagation by the nonliving entities. The data obtained at ~ 30% substantiate these conclusions: in these instances the MWDs are even narrower and the humps associated with the nonliving species are essentially absent.

4.1.1.4.4 The Effect of [MtX_n]

Figure 27A, B concerns the effect of Friedel-Crafts acid concentration on product uniformity under various conditions (at ~ 34% conversions and after 4 min of polymerization). The products tend to become more uniform with increasing [$TiCl_4$], however, according to the MWDs obtained even at the highest [$TiCl_4$] at 37.2% conversion, ($\bar{M}_n/\bar{M}_n = 1.44$), or highest (57.5%) conversion ($\bar{M}_w/\bar{M}_n = 1.44$), or highest (57.5%) conversion ($\bar{M}_w/\bar{M}_n = 1.23$), propagation is still not solely by the living species. As the experiments were carried out at −82 °C, one possible reason for the relatively broad MWD is slow initiation in respect of propagation (see also the data in Table 7 in the Appendix).

4.1.1.4.5 The Effect of "Aging"

As shown in Sect. 4.1.1.4, the addition of $TiCl_4$ to TMPCl/IB in the absence or presence of TEA under the greatest variety of conditions invariably leads to relatively broad molecular weight distribution products or even to multimodal GPC traces. In these experiments the starting point (zero time) of the reactions is the moment of $TiCl_4$ addition to the charges (see Figs. 23–27). Low product uniformity has been assumed to be caused by the time lag between the mixing of

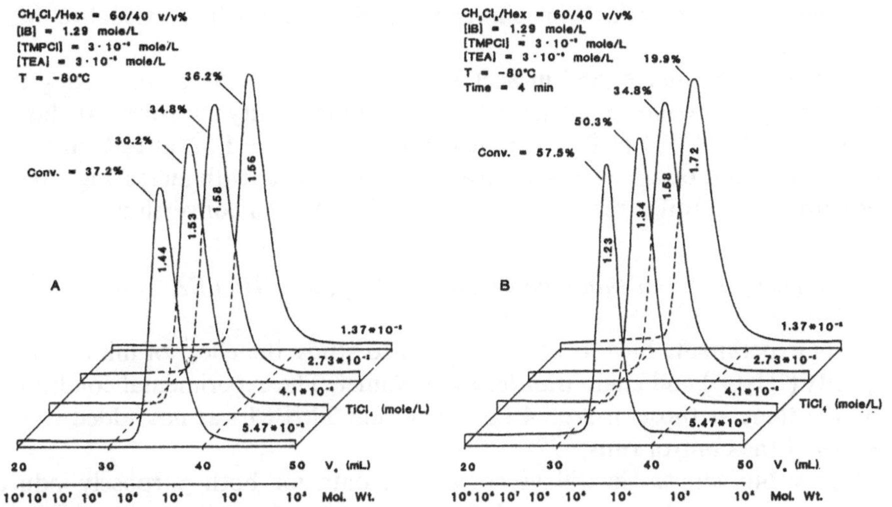

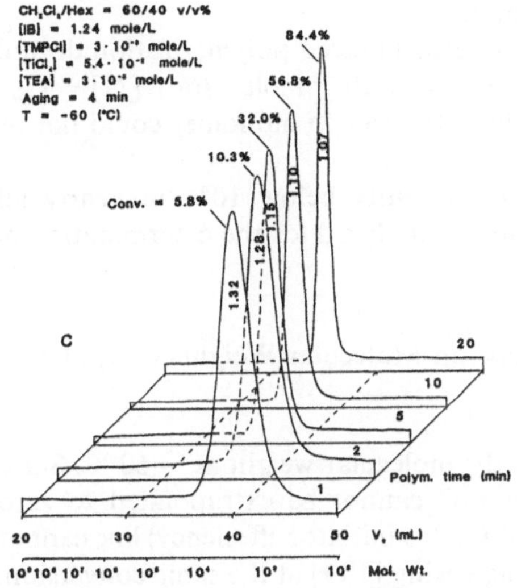

Fig. 27A–C. Isobutylene polymerization. GPC traces of PIBs obtained: **A** as a function of $TiCl_4$ concentration at similar conversions; **B** at the same time at $-80\,°C$; **C** with 4 min aging as a function of polymerization time at $-60\,°C$

the ingredients to produce the first most reactive species, e.g., $H^{\oplus} + MtX^{\ominus}_{n+1}$, leading to nonliving propagation and that of the least reactive species, e.g., $H^{\delta\oplus}\text{---}[\overset{\delta\ominus}{Mt}X_n\cdot ED]\ldots MtX^{\ominus}_{n+1}$, that induces living polymerization. During this lag period, while the latter species is being formed, a multiplicity of chain carriers may be operational and yield broad MWDs. The reaction between "HX" + $TiCl_4 \rightarrow H^{\oplus} + TiCl_5^{\ominus}$ is probably the fastest among the possible routes to active species while it takes some time to convert this highly reactive species to the much less reactive $MtX_n \cdot ED$-complexed species. The reaction RX + $TiCl_4 \rightarrow R^{\oplus} + TiCl_5^{\ominus}$ is probably somewhat slower than "HX" + $TiCl_4 \rightarrow H^{\oplus} + TiCl_5^{\ominus}$ and the complexation of the cations with $MtX_n \cdot ED$ also takes some time. A glance at the Winstein spectra in Scheme 1 helps to visualize these possibilities.

As mentioned above (and in Sect. 4.1.1.1.4 and 4.1.1.2.5), by premixing the ingredients prior to monomer addition, product uniformity improves. As shown in Fig. 27C, the MWD of PIB is quite narrow ($\bar{M}_w/\bar{M}_n = 1.32$) already at 5.8% conversion and it becomes progressively much narrower with increasing monomer conversion until it reaches $\bar{M}_w/\bar{M}_n = 1.07$ at 84.4% conversion.

4.1.2 *Impurity-Induced Initiation by the "H_2O"/$TiCl_4$/IB/TEA System*

Control experiments have been carried out to study the effect of initiation by impurity ("H_2O") and chain transfer to monomer. The experimental conditions were the same as those in Sect. 4.1.1 except that TMPCl was not added to the charges in the control runs.

Figure 14C shows that N vs conversion data for both purposely-induce (filled signs) and impurity-induced (hollow signs) polymerization at different [ED]s. (Some of these data are shown in the insert of Fig. 21C. The raw data are collected in Tables 4 and 9 in the Appendix.)

Contrary to Fig. 14A, the data for initiator-induced polymerization do not exceed the horizontal line at $N = I_0 = 6 \cdot 10^{-5}$ mole (or $[N] = [I_0] = 3 \cdot 10^{-3}$ mole/L), thus the extent of chain transfer to monomer could not be obtained to calculate k_c and k_p.

Since monomer conversions were consistently below 10% in nearly all control experiments, their effect could not be analyzed for the determination of k_c and k_p.

4.1.2.1 The Effect of Reaction Conditions on Molecular Weights ($\bar{M}_n$) and Number of Chains (N)

Figure 28A shows the effect of [ED] on the molecular weight at −60 °C. Since ["H_2O"] is unknown, the data in Fig. 14C cannot be extrapolated to zero monomer conversion, and a "theoretical" (100% initiator effeciency) line cannot be constructed. The $\bar{M}_n$s increase with increasing [ED] at the same conversion, indicating a decreasing number of polymer chains, that is the suppression of chain transfer. Figure 28B shows the effect of medium polarity on the $\bar{M}_n$ (and N) vs W_p plots. The higher the medium polarity, the higher the conversion (that is the overall polymerization rate) and the higher the number of polymer chains due to chain transfer. The raw data are shown in Tables 9 and 10 in the Appendix.

4.1.2.2 The Effects of Reaction Condition on Polymerization Rate

The effects of reaction conditions on the polymerization raté (expressed by $-\ln(1 - C)$ have been investigated. Figure 28C shows the effect of [ED] on the

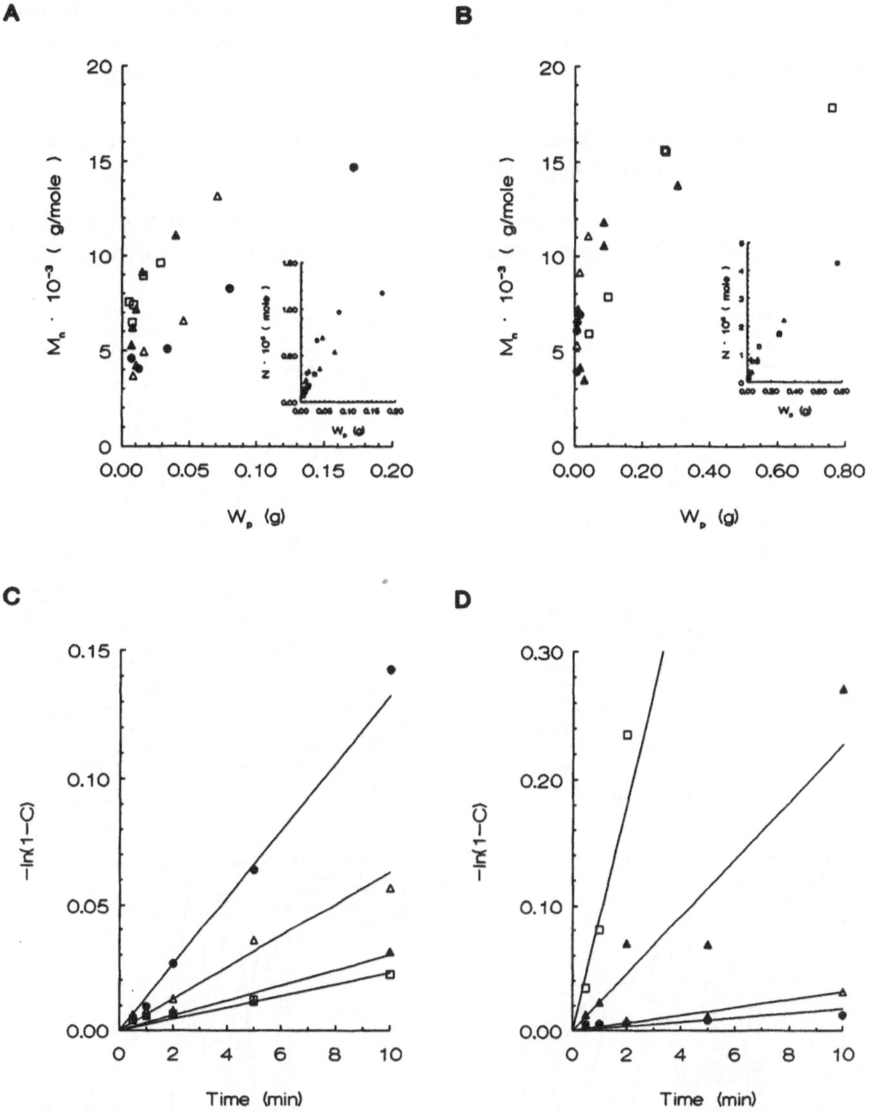

Fig. 28A–D. IB polymerization in the absence of added initiator at $-60\,°C$. M_n (and N) vs W_p plots and rate vs time plots: **A, C** at various ED concentrations: [TEA] (mol/l) ● = $5\cdot10^{-4}$; △ = $1.5\cdot10^{-3}$; ▲ = $3\cdot10^{-3}$; □ = $6\cdot10^{-3}$; **B, D** at various medium polarities: CH_2Cl_2/Hex. (v/v) ● = 40/60; △ = 60/40; ▲ = 80/20; □ = 100/0

rate, i.e., the slope of the $-\ln(1-C)$ vs time plots. The higher the [ED], the lower the slope (i.e., $k_p\cdot[N]$), independent of the extent of chain transfer ([N] = the concentration of growing polymer chains). If initiation by "H_2O" is instantaneous, the decreasing slopes with increasing [ED] indicate decreasing k_p values, that is, the contribution of $k_{pc^\oplus}$ to k_p decreases.

Figure 28D illustrates the effect of polarity on the rate of polymerization. Similar to Fig. 17B, the higher the polarity, the higher the rate (slope of the lines), indicating higher k_ps due to increased solvation of the ionic chain carriers.

4.1.2.3 The Effect of Experimental Conditions on Molecular Weights and Molecular Weight Dispersities: GPC Studies

As mentioned earlier, the position and the shape of GPC traces of products provide valuable information concerning the mechanism of polymerization.

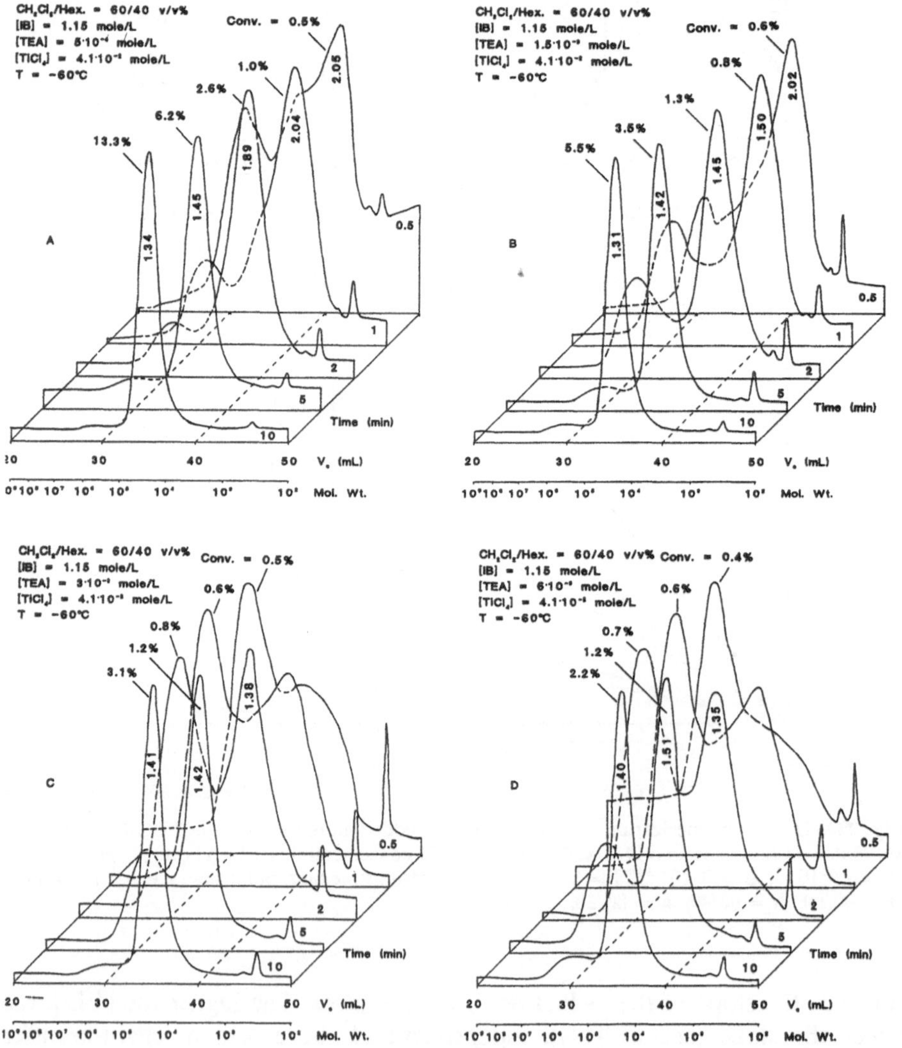

Fig. 29A–D. IB polymerization in the absence of added initiator at − 60 °C. GPC traces of PIBs obtained as a function of time at different ED concentrations; **E** IB polymerization in the absence of added initiator at − 60 °C. GPC RI and UV traces of PIB obtained after 5 min polymerization

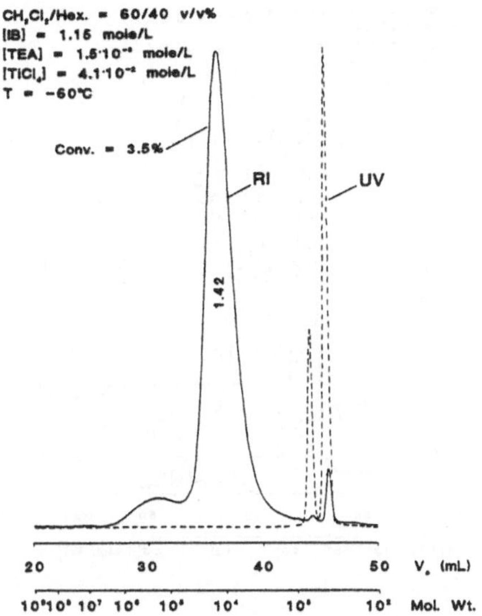

Fig. 29E. (Contd.)

This section concerns a study of the effect of time, [ED], temperature, and medium polarity on the shape and position of GPC traces of PIBs prepared in control experiments. (Tables 9–11 in the Appendix summarize the raw data.)

Figure 29A–D shows a set of GPC traces of PIBs obtained in AMI experiments by the "H_2O"/$TiCl_4$ initiating system at $-60\,°C$ at different [TEA]s ($5 \cdot 10^{-4}$–$6 \cdot 10^{-3}$ mol/l) as a function of time (0.5–10 min). At low conversions (short times), the GPC traces are multimodal, indicating the presence of different kinds of growing species. The peaks at the lowest MWs correspond to 6–8 monomer units. These products show UV activity (see Fig. 29E), which may be due to the olefinic end groups due to chain transfer to monomer. The intermediate peaks tend to narrow with time in all groups of GPC traces. The higher the [ED], the narrower the MWDs at the same or similar conversions because of the increasing contribution of non-dissociated living species to propagation.

The peaks at the highest MWs reflect the presence of ionic species and are suppressed by increasing the conversions and the [ED].

Figure 30A–D represents GPC traces of PIBs obtained in experiments at $-60\,°C$ at different solvent polarities as a function of time. All the traces convey a similar message: the MWD of the main product becomes narrower with increasing conversion. The best solvent composition seems to be a 60/40 vv CH_2Cl_2/n-C_6H_{14} mixture, since a monomodal narrow MWD product is obtained even at $\approx 3\%$ monomer conversion (see Fig. 30B). Figure 31A, B represent experiments carried out at $-60\,°C$ either with a constant solvent composiposition (60/40 v/v CH_2/Cl_2/n-C_6H_{14}) at different [ED], or with a constant [ED] ($3 \cdot 10^{-3}$ mol) at different polarity solvent mixtures.

An effort was made to compare GPC traces of PIBs obtained at different [ED]s at the same low conversion level. As shown by the data in Fig. 31A, by

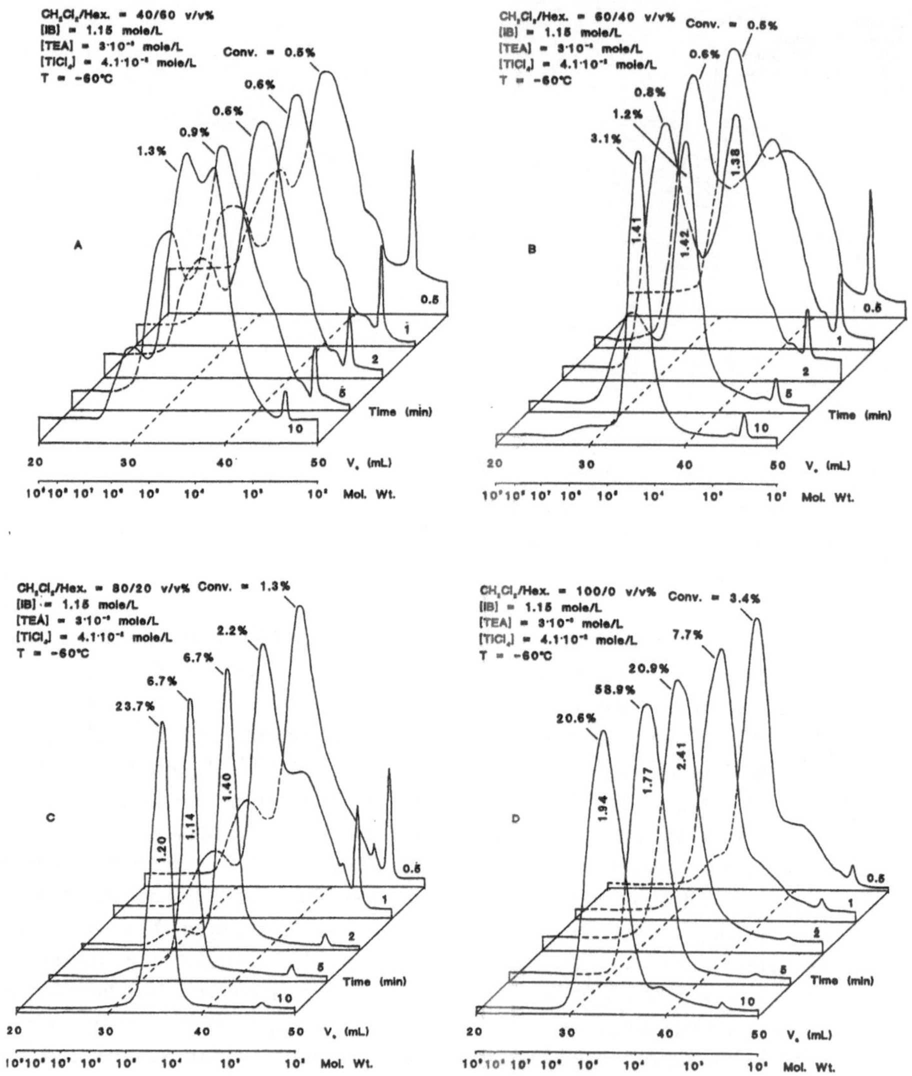

Fig. 30A–D. IB polymerization in the absence of added initiator at $-60\,°C$. GPC traces of PIBs obtained as a function of time at various solvent polarities

working with the same polymerization time (as in purposely-induced experiments, i.e., in the 1–5 min range), the very low conversions did not allow a valid comparison.

Meaningful low conversion results were obtained in various media (i.e., in $CH_2Cl_2/n\text{-}C_6H_{14}$ = 100/0, 80/20, 60/40, and 40/60 v/v) by adjusting the polymerization time (i.e., by carrying out the shortest polymerization with 100% CH_2Cl_2, and the longest run in the less polar 40/60 v/v mixture). As shown by

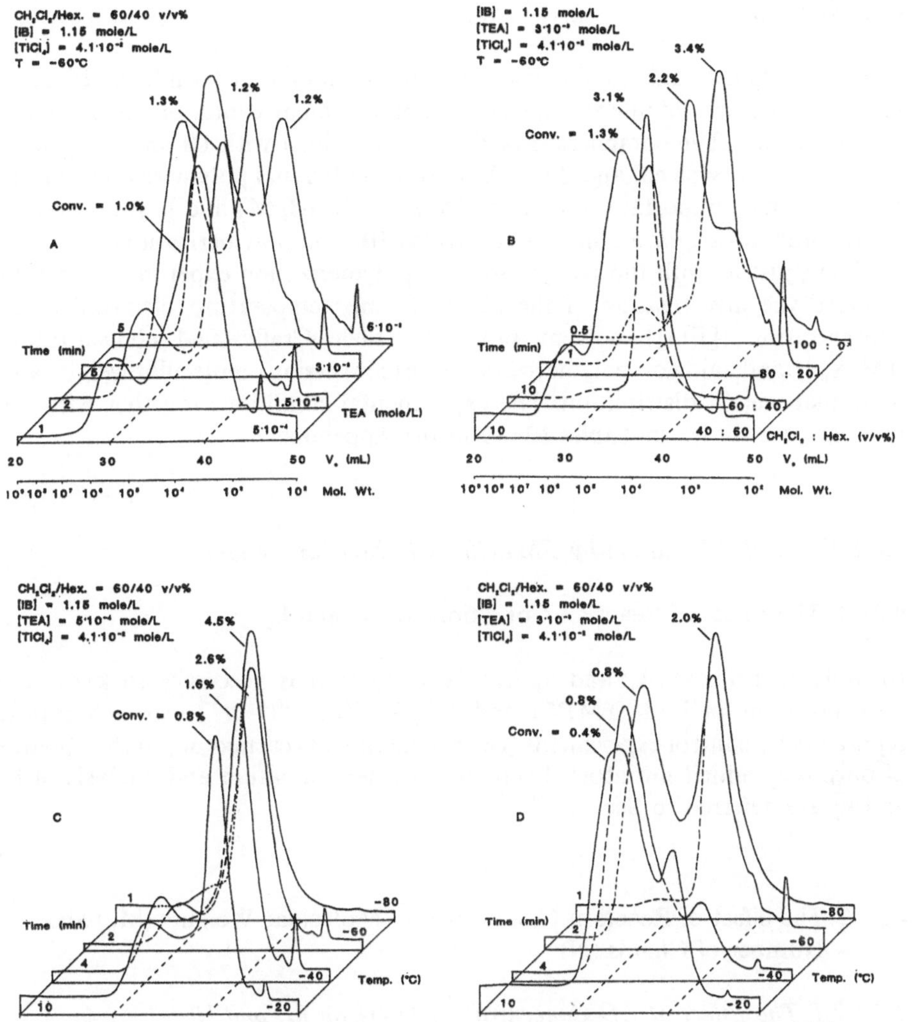

Fig. 31A–D. IB polymerization in the absence of added initiator. GPC traces of PIBs obtained: **A** as a function of ED concentration; **B** solvent polarity; **C** temperture at lower and; **D** higher ED concentration

the GPC traces in Fig. 31B, the narrowest MWD product was obtained, again, with the 60/40 v/v $CH_2Cl_2/n\text{-}C_6H_{14}$ mixture.

Figure 31C, D shows GPC traces of samples prepared as a function of temperature (from $-80\,°C$ to $-20\,°C$) at two different [ED] ($5\cdot10^{-4}$ mol/l, $3\cdot10^{-3}$ mol/l). Since the monomer conversions are very low (0.4%–4.4%), their comparison may be misleading.

4.2 Styrene

After the detailed analysis of purposely-induced and impurity-induced IB polymerizations, it was of interest to examine under similar conditions corresponding St systems. The comparison of the data obtained with IB and St systems promised to provide enhanced insight into the individual polymerization mechanisms of these important monomers [6, 7, 37, 47, 63, 64] and beyond that, to lay the framework for a study of controlled IB/St copolymerizations.

Thus, just as with the IB systems, the polymerization experiments with St included the investigation of the effect of time, temperature, electron donor concentration ([ED]), solvent polarity, Friedel-Crafts acid concentration ($[MtX_n]$), and aging on polymerization rate (conversion), molecular weight, and molecular weight distribution. The experimental conditions together with the raw data are shown in Tables 12–19 in the Appendix.

4.2.1 Controlled Initiation by TMPCl/TiCl$_4$/St/TEA System

4.2.1.1 The Effect of Reaction Conditions on k_c and k_p

In order to calculate k_c and k_p (see Sect. 2.3.3), it is necessary to know the concentrations of both TMPCl and "H_2O". The ["H_2O"] was determined separately by control experiments (i.e., in experiments carried out in the absence of purposely added initiator). Therefore, the determination and analysis of k_c and k_p are deferred to Sect. 4.2.2.

4.2.1.2 The effect of Reaction Conditions on Molecular Weights ($\bar{M}_n$) and Number of Chains (N)

4.2.1.2.1 The Effect of ED Concentration [ED] on $\bar{M}_n$ and N

Figure 32A–D shows the effect of [ED] on PSt $\bar{M}_n$ and N at −20, −40, −60, and −80 °C. All these figures indicate slow initiation ($\bar{M}_n$ higher than theoretical at low conversions), significant chain transfer to monomer $\bar{M}_n$ lower than theoretical at higher conversions), and significant initiation by "H_2O" (see values extrapolated to zero conversion in the insets to the $\bar{M}_n$ vs C plots in Sect. 4.2.2.1). It is difficult to choose optimum conditions for the synthesis of high molecular weight PSts from these experiments because the effects on N of slow initiation on the one hand, and those of chain transfer and initiation by "H_2O" on the other hand, are opposite. Increasing [TEA] seems to suppress chain transfer but the effect is insufficient to obtain well-controlled high MW PSt. (See, also, the raw data in Tables 12–15 in the Appendix.) However, promising results were obtained in two experiments.

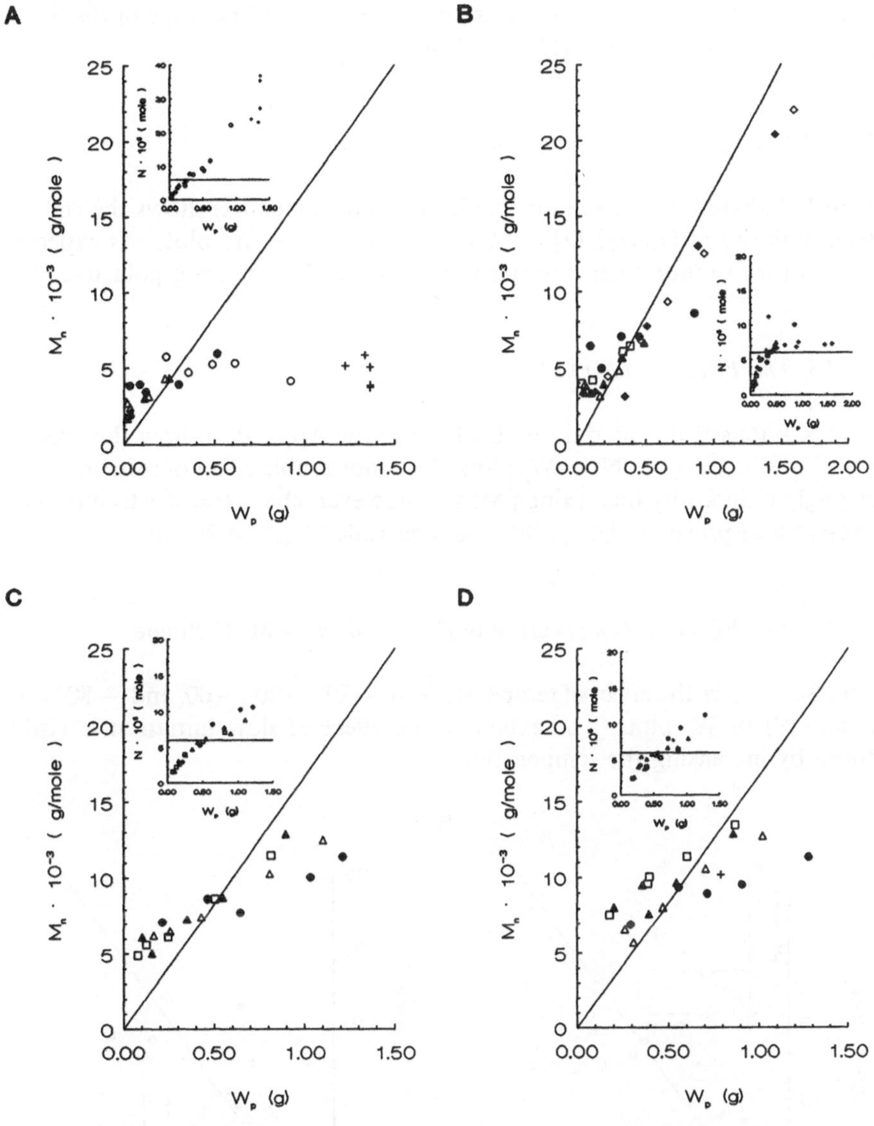

Fig. 32A–D. St polymerization. M_n (and N) vs W_p plots at various temperatures: **A** $-20\,°C$; **B** $-40\,°C$; **C** $-60\,°C$; **D** $-80\,°C$ and at various ED concentrations expressed by [TMPCl]/[TEA] ratios: + = 1:0; ○ = 1:0.083; ● = 1:0.17; △ = 1:0.5; ▲ = 1:1; □ = 1:2. ◇ = 1:3; ◆ = 1:4

Figure 32B (and Table 19 in the Appendix) show the effect of [TMPCl]/TEA] = 1:3 and 1:4 on $\bar{M}_n$ (and N) vs W_p at $-40\,°C$. At these relatively high [ED]s, the experimental points are not far from a straight line starting at the origin, indicating that chain transfer to monomer is virtually absent (cf Fig. 5A, B, plot C1; or Fig. 6A, B, plot C3b, which takes slow initiation into consideration which is present in this experiment.). The slope of the line in Fig. 32B is determined by $1/([I_0] + [H_2O])$.

4.2.1.2.2 The Effect of Medium Polarity at − 80 °C

Figure 33A (based on the data in Table 16 in the Appendix) shows the effect of solvent polarity at $[I_0]/[ED] = 1.0$ on $\bar{M}_n$ (and N) vs W_p plots. As expected, chain transfer to monomer appears to increase with increasing polarity.

4.2.1.2.3 The Effect of $[MtX_n]$

Figure 33B (based on the data in Table 17 in the Appendix) show the effect of $[MtX_n]$ on the $\bar{M}_n$ (and N) vs W_p plots. The undesirable effect of slow initiation is strongly reduced by increasing $[MtX_n]$, however, chain transfer to monomer is present and prevents the synthesis of controlled high MW PSt.

4.2.1.2.4 The Effect of Temperature in the − 20 to − 80°C Range

Figure 34A shows the effect of temperature at −20, −40, −60, and −80 °C on $\bar{M}_n$ (and N) vs W_p plots. The extent of the effect of slow initiation is visibly reduced by increasing the temperature.

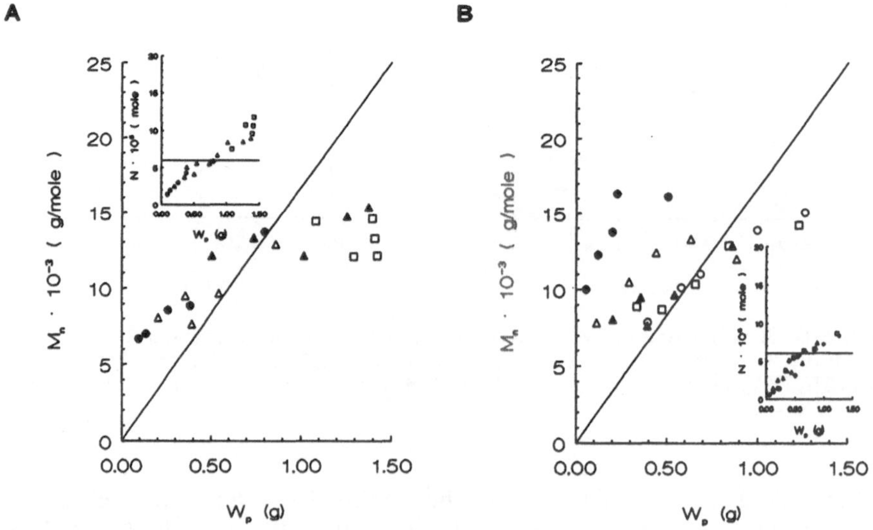

Fig. 33A, B. St polymerization. M_n (and N) vs W_p plots: **A** at various polarities at − 80 °C, CH_2Cl_2/Hex. (v/v) ● = 40/60; △ = 60/40; ▲ = 80/20; □ = 100/0; **B** various Friedel-Crafts acid concentrations at − 80°C, [TMPCl]/[$TiCl_4$] ● = 1:4.55; △ = 1:9.1; ▲ = 1:13.7; □ = 1:18.2; ○ = 1:22.7

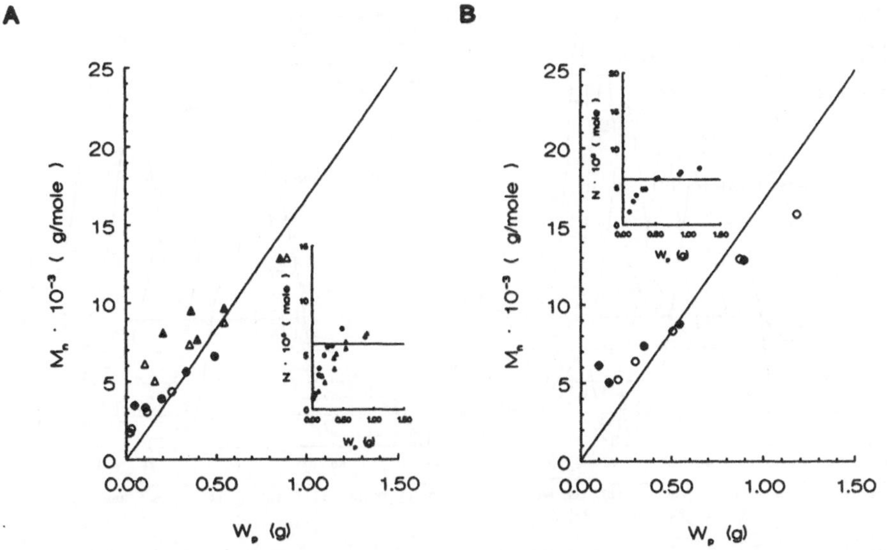

Fig. 34A, B. St polymerization. M_n (and N) vs W_p plots: **A** at various temperatures, ○ = − 20 °C; ● = − 40 °C; △ = − 60 °C; ▲ = − 80 °C; **B** ● = conventional and ○ = "aging" experiments at − 60 °C

4.2.1.2.5 The Effect of Aging at − 60 °C

Figure 34B (based on the data in Table 18 in the Appendix) shows the results of an aging experiment. In this instance, 4 min of aging discernibly reduces the extent of slow initiation at low conversions, similarly to IB polymerization (cf. Sect. 4.1.1.2.5), but shows little if any effect at higher conversions (insufficient data for meaningful comparison).

4.2.1.3 Apparent Rate Order in Friedel-Crafts Acid

To determine the reaction order of propagation in respect to the Friedel-Crafts acid, it is necessary to know k_p at various $[TiCl_4]$s. Since k_p could not be calculated due to the unavailability of control experiments at high conversion (see Sects. 2.3.3, 4.2.1.1, and 4.2.2), we could not derive the reaction order in regard to $[TiCl_4]$.

4.2.1.4 The Effects of Experimental Conditions on Molecular Weights ($\bar{M}_n$) and Molecular Weight Distributions (MWD)

4.2.1.4.1 The Effect of Time and ED Concentration

Figure 35A (based on the data in Table 12 in the Appendix) shows the effect of polymerization time on the GPC traces of PSts obtained at

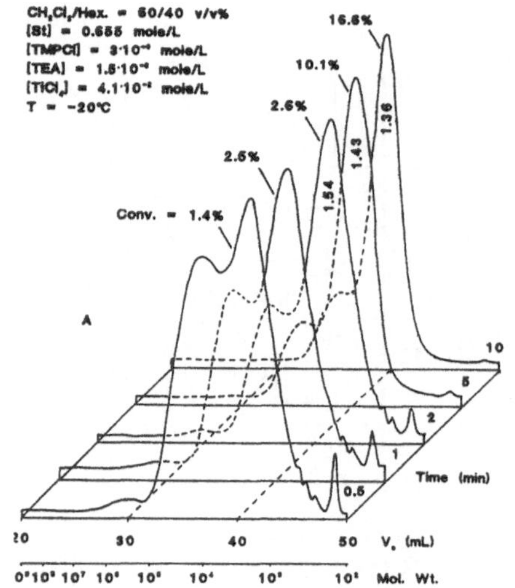

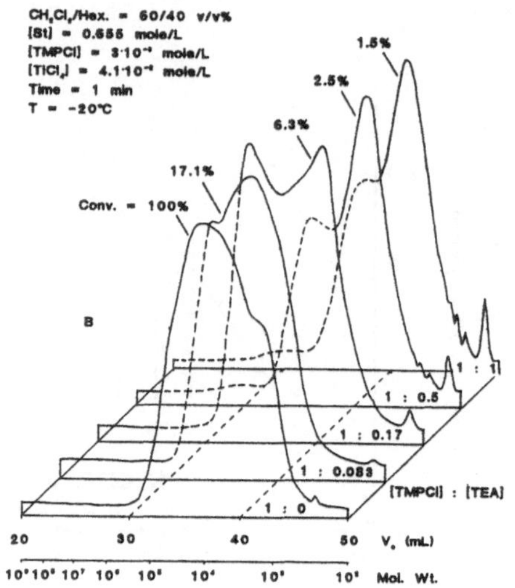

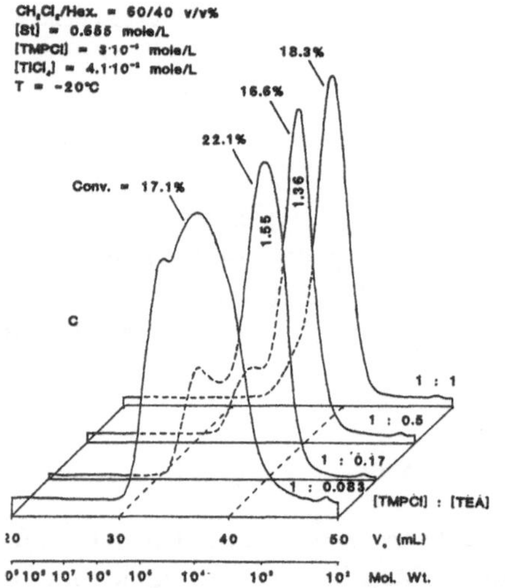

Fig. 35A–C. St polymerization at $-20\,°C$. GPC traces: **A** as a function of time; [TMPCl]/[TEA] ratios; **B** at the same time; **c** at a similar conversion

[TMPCl]/[ED] = 2 at $-20\,°C$. The traces are multimodal, however, the peaks can be assigned: The small peak appearing at the highest molecular weight ($\bar{M}_n > 10^6$) is most likely due to PSt formed by initiation with "H_2O" which leads to rapid nonliving propagation by free ionic species. With increasing conversions this trace disappears into the base line because the relative amount of this nonliving PSt decreases while the relatively slowly growing living PSt becomes more prominent. The second larger peak, $\bar{M}_n \approx 40\,000$, is most likely due to nonliving PSt formed by the TMPCl initiator and indicates rapidly propagating ionic species. This peak is also stationary and its contribution to the overall MWD gradually diminishes because the living species becomes more important with increasing conversions.

The third and tallest peak is due to the living PSt initiated by the TMP$^{\delta\oplus}$ ---[TiCl$_4^{\delta\oplus}$·TEA] ··· TiCl$_5^{\ominus}$ species, propagating relatively slowly toward higher molecular weights. During growth the MWD of this polymer becomes gradually narrower (i.e., $\bar{M}_w/\bar{M}_n$ decreases from 1.54 to 1.36 with conversions increasing from 2.6 to 16.6%, respectively). The three vertical arrows on the left

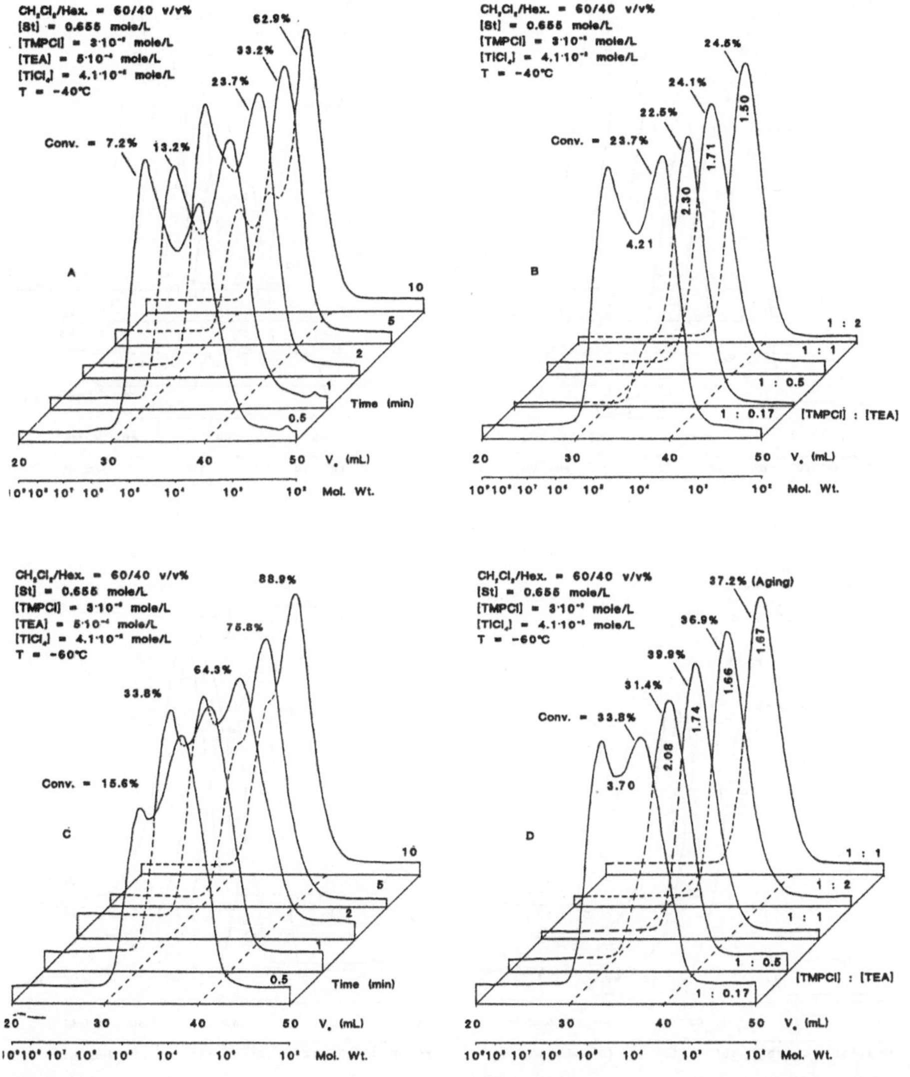

Fig. 36A–D. St polymerization: **A, C** GPC traces as a function of time; **B, D** [TMPCl]/[TEA] at a similar conversion level at −40 and −60 °C respectively. The last trace in Fig. 36D was obtained with an "aged" charge

side of Scheme 1 shows the routes of these three species toward the final polymer product.

The small peaks at lowest molecular weights are conceivably caused by chain transfer during the very early phases of the polymerization and become negligible with conversions beyond $\sim 10\%$.

These peak assignments were of general validity in this work and were of great value in the interpretation of molecular weight and MWD information.

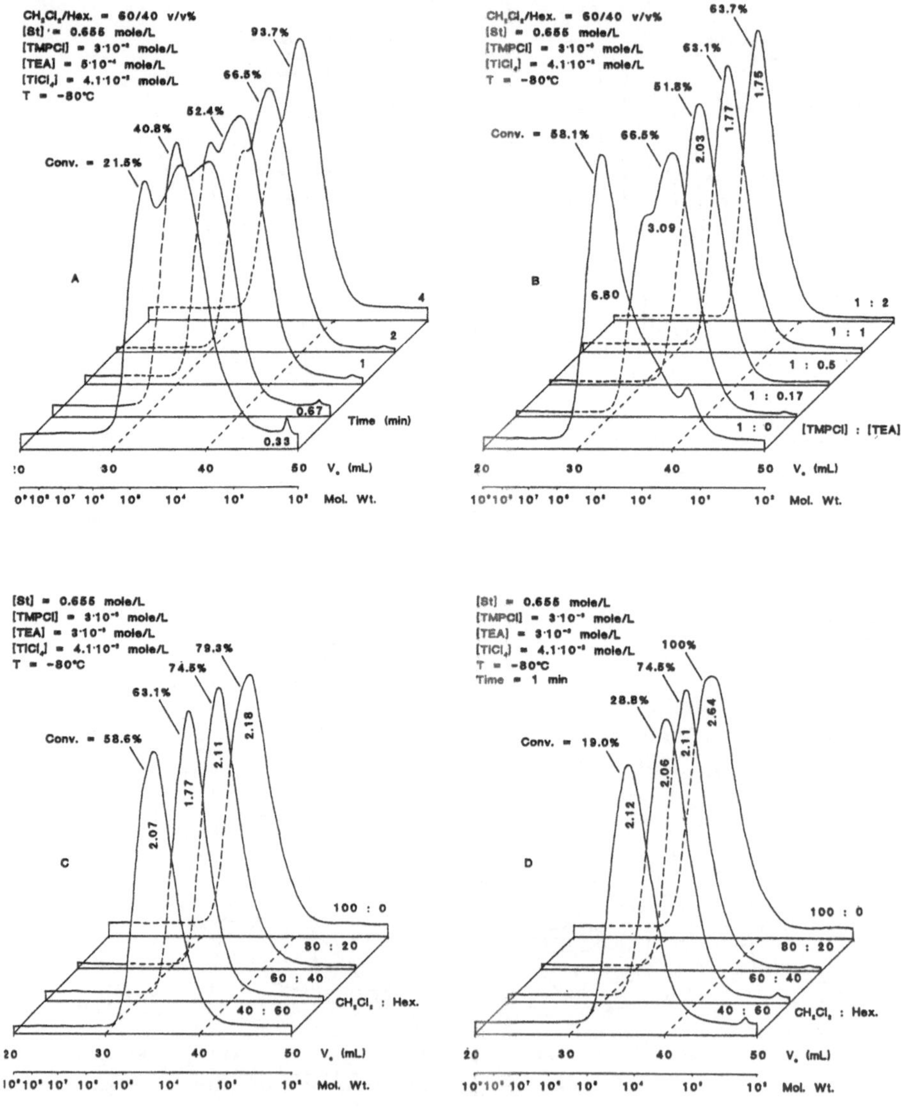

Fig. 37A–D. St polymerization at $-80\,°C$. GPC traces: **A** as a function of time; **B** [TMPCl]/[TEA] ratios at a similar conversion level; **C** solvent polarity at a similar relatively high conversion level; **D** at the same time

Figure 35B shows the GPC traces of PSt as a function of [ED] (that is, by changing the [TMPCl]/[TEA]) at the same polymerization time (1 min). In the absence of TEA, the rate is very high and conversion is complete in less than 1 min. With increasing [TEA], the rate is drastically reduced and with [TMPCl]/[TEA] = 1 the conversion drops below 2%. The GPC peaks are multimodal; nonetheless, the peak assigned to the living nonionic species is

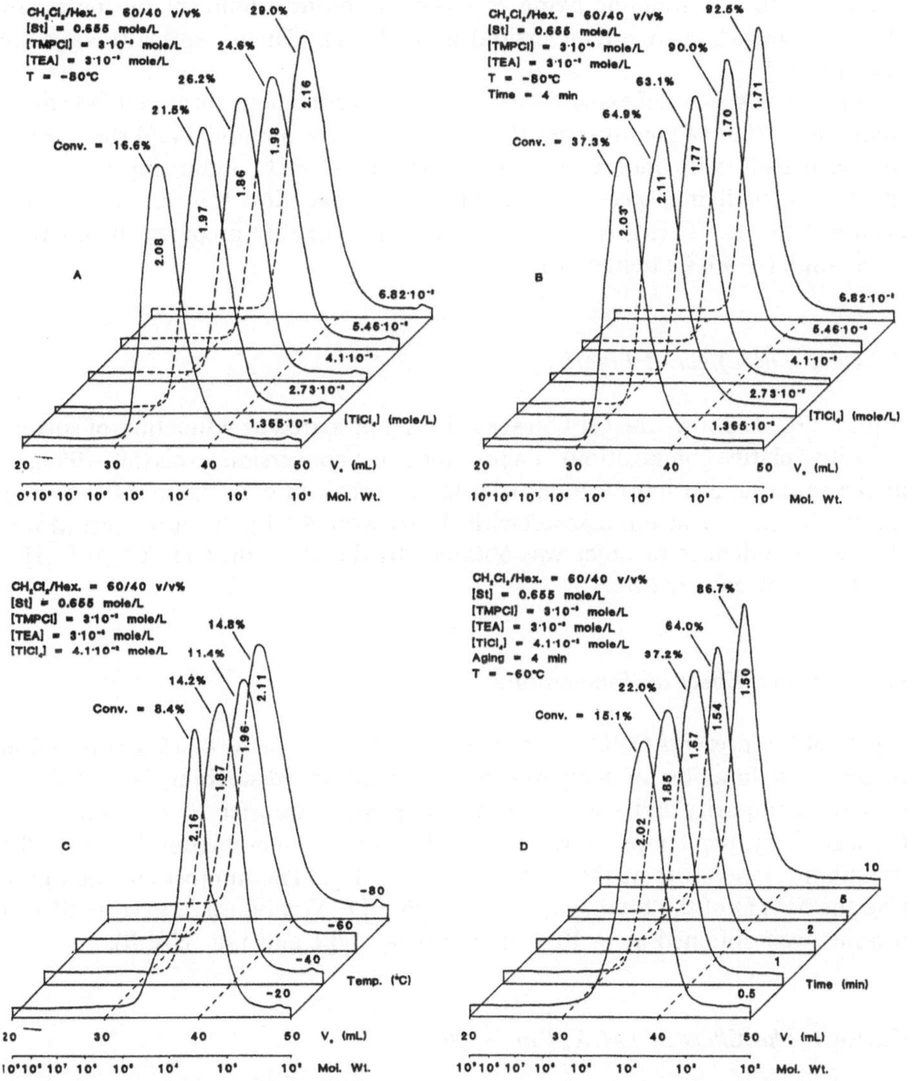

Fig. 38A–D. St polymerization: **A** GPC traces as a function of MtX_n concentration at similar "low" conversion; **B** at the same time at − 80°C; **C** as a function of temperature; **D** as a function of polymerization time in aging experiments at − 60°C

growing visibly more prominent with increasing [TEA] even though the conversions decline to ~2%. These findings are corroborated by the data in Fig. 35C, which shows GPC traces of PSts obtained at different [TEA]s but at about the same conversion levels. Evidently, the contribution of the relatively narrow MWD living species to the overall distribution increases with increasing [TEA].

Figures 36A–D, 37A, B, and 39A, B concern the effect of time and [ED] at −40, −60, and −80 °C and convey essentially the same message: the peaks attributed to the nonionic living PSt become more prominent by increasing the polymerization time (conversion) and [TEA]. This is especially valid for high [TEA].

Figure 39A, B is self explanatory. At ED concentrations three and four times higher than that of the initiator, the GPC traces are unimodal and the MWDs are decreasing with increasing conversions at −40 °C, indicating more and more uniform (living) species participating in propagation. It is noteworthy that such narrow MWD ($\bar{M}_w/\bar{M}_n \sim 1.35$) PSt can be readily prepared at this relatively high (−40 °C) temperature.

4.2.1.4.2 The Effect of Polarity

Figure 37C, D shows the GPC traces of PSts prepared as a function of solvent polarity (relative composition) at about the same conversion levels (58–79%), or at the same time (1 min) leading to a large spread of conversions (19–100%). Similar to the conclusion reached with IB (see Sect. 4.1.1.), the narrowest MWD PSt, i.e., best defined product, was obtained by the use of the $CH_2/Cl_2/n\text{-}C_6H_{14}$ 60/40 v/v solvent mixture.

4.2.1.4.3 The Effect of Temperature

Figure 38C represents GPC traces at $[I_0]/[ED] = 1.0$ in the 8–15% conversion range, as a function of temperature. Since all the disturbing factors (slow initiation, impurity induced initiation, and chain transfer to monomer) are influenced by the temperature, the results cannot unambiguously show the optimum temperature at this relatively low [ED]. (The numbers in the figure give the MWD of the total sample.) At higher [ED] uniform, a narrow MWD product was obtained at −40 °C (see Sect. 4.2.1.4.1 and Fig. 39A, B).

4.2.1.4.4 The Effect of $[MtX_n]$ at −80 °C

Figure 38A, B (based on the data in Table 17 in the Appendix) shows the effect of $[TiCl_4]$ on the GPC traces of PSts prepared under comparable conditions. Specifically, Fig. 38A focuses on PSts obtained at conversions below 30%, and

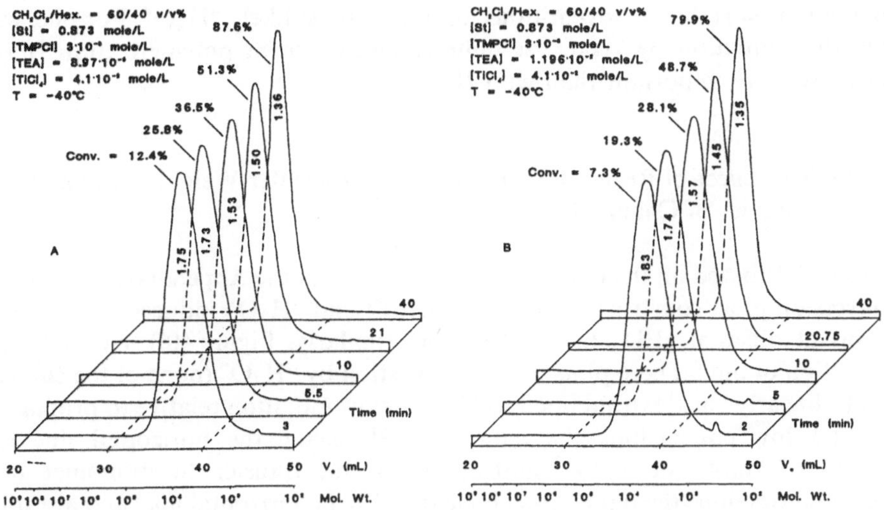

Fig. 39A, B. St polymerization. GPC traces as a function of time at two-different ED concentrations at $-40°C$

Fig. 38B on a series of conversions from 37 to 92% obtained at the same time (4 min).

Evidently, the effect of [$TiCl_4$] on the breadth of the GPC traces is insignificant under the conditions examined. Evidently the increase in [$TiCl_4$] reduces but does not eliminate the undersirable MWD broadening effect of slow initiation (see Fig. 33B). The relatively broad MWDs may also be due to the low temperature of the experiment where very rapid propagation by the nonliving ionic species is dominating.

4.2.1.4.5 The Effect of Aging

Whereas the effect of aging at $-60\,°C$ cannot be seen by GPC analysis (see Fig. 36D), it is noticeable in the $\bar{M}_n$ vs W_p plot (Fig. 33D). Figure 38D (and Table 18 in the Appendix) show the result of aging experiments at $[I_0]/[ED] = 1:1$. All the GPC traces are monomodal with relatively narrow MWDs ($\bar{M}_n = 15\,800$; $\bar{M}_w/\bar{M}_n = 1.50$), mainly because the extent of slow initiation is reduced. Even narrower MWDs were obtained at $-40\,°C$ at $[I_0]/[ED] = 1:3$ and $= 1:4$ (cf Fig. 39 and Table 19).

Further experiments should be carried out to elucidate the effect of aging.

4.2.2 Impurity-Induced Initiation by the "H_2O"/$TiCl_4$/St/TEA System

This section concerns an analysis of St polymerization experiments carried out in the absence of purposely added initiator (control experiments). We assume that initiation is due to adventitious impurities most likely "H_2O". For one, we find that initiation by "H_2O" is quite significant in St polymerization [65], certainly more important than with IB.

4.2.2.1 The Effect of Reaction Conditions on Molecular Weights ($\bar{M}_n$) and Number of Chains (N)

Figure 40A (based on the data in Tables 14 and 20 in the Appendix) shows N vs conversion plots constructed for both TMPCl and "H_2O" induced polymerizations at different [ED] at $-60\,°C$ and, similarly, Figure 40B shows N as a function of time. These plots indicate initiation by "H_2O" (none of the curves can be linearly back-extrapolated to $N = 0$), slow initiation relative to propagation (at low conversions the data are well below the horizontal line at $I_0 = 6 \cdot 10^{-5}$ mole (or $3 \cdot 10^{-3}$ mol/l)) and chain transfer to monomer (at higher conversions the data cross above the $N = I_0$ horizontal line, and are still increasing).

The theory for the determination of k_c and k_p in the presence of these complicating factors was discussed in Sect. 2.3.3.

N increases linearly and to reasonably high conversions in both TMPCl induced and "H_2O" induced polymerizations (indicated by ● and ○ in Fig. 40A), and shows clearly the difference between TMPCl and "H_2O" induced polymerizations. The data obtained in the latter experiment were back-extrapolated to zero conversion to obtain ["H_2O"]. The value, 10^{-3} mole "H_2O"/l, was used to calculate k_c and k_p (see Sect. 4.2.2.2).

Figure 40C shows that $\bar{M}_n$ and (N) vs W_p data at different [ED]s. Since the majority of the points are at low conversions, the differences are within experimental error (see also Table 20 in the Appendix).

Figure 40D (and Table 21 in the Appendix) show the effect of medium polarity on the $\bar{M}_n$ and (N) vs W_p curves. With increasing medium polarity, the conversion and the number of polymer chains are also increasing, indicating the increasing proportion of ionic species and the increased probability of chain transfer. Figure 40C, D indicates significant chain transfer in "H_2O" induced polymerizations.

4.2.2.2 The Determination of k_c and k_p

As developed in Sect. 2.3.3 k_c and k_p can be obtained for St polymerizations proceeding by relatively slow initiation by "H_2O" in the presence of chain transfer to monomer, by the plots shown in Fig. 41A. According to the control experiment shown in Fig. 40A, ["H_2O"] $= 10^{-3}$ mol/l. (This relatively high concentration may be due to less efficient drying of St than of IB.) The control experiment also indicates significant chain transfer (N has doubled at $< 50\%$, see Fig. 40A).

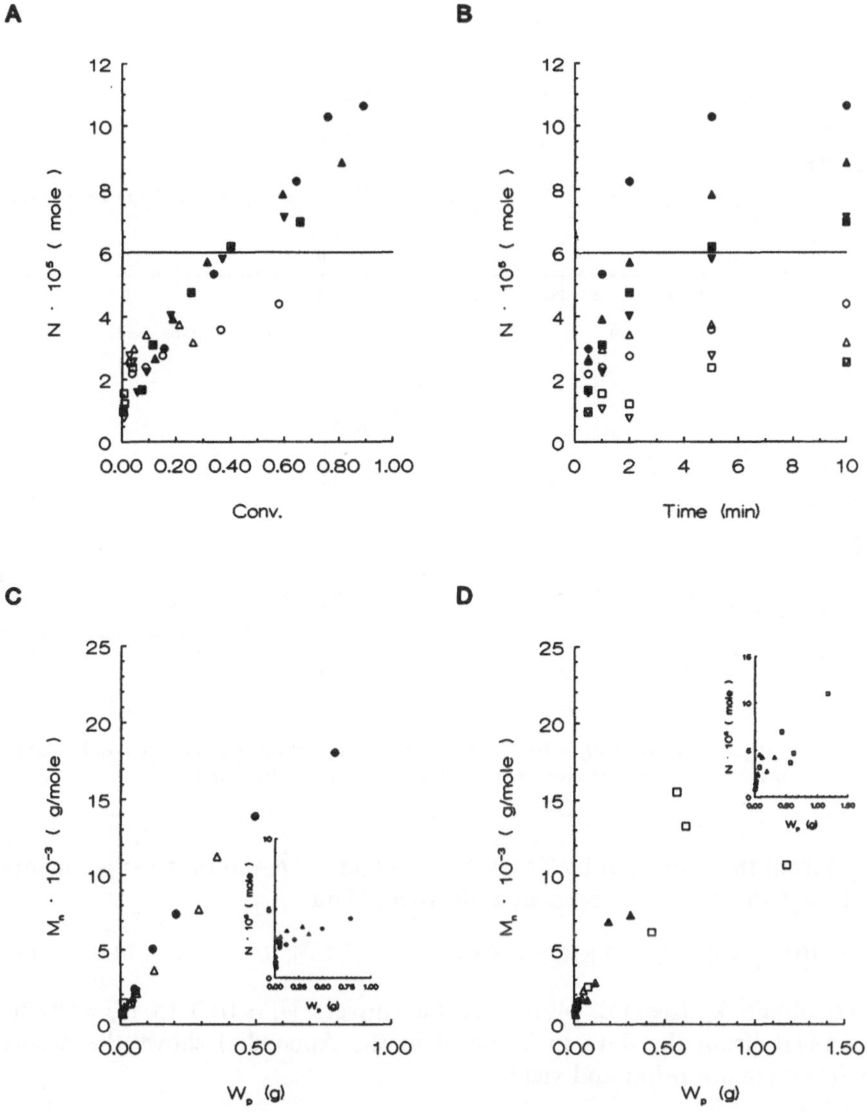

Fig. 40A–D. St polymerization at − 60 °C: **A** N vs conversion; **B** N vs time plots in the presence (filled signs) and in the absence (hollow signs) of initiator at various ED concentrations: [TEA] (mol/l) ●, ○ = $5 \cdot 10^{-4}$; ▲, △ = $1.5 \cdot 10^{-3}$; ■, □ = $3 \cdot 10^{-3}$; ▼, ▽ = $6 \cdot 10^{-3}$ at − 60 °C, and M_n (and N) vs W_p plots; **C** at various ED concentrations at − 60 °C ● = $5 \cdot 10^{-4}$; △ = $1.5 \cdot 10^{-3}$; ▲ = $3 \cdot 10^{-3}$; □ = $6 \cdot 10^{-3}$; **D** at various polarities at − 60 °C CH_2Cl_2/Hex (v/v) ● = 40/60; △ = 60/40; ▲ 80/20; □ = 100/0 in the absence of initiator

We can obtain the number of chains formed by purposeful initiation by TMPCl as a function of conversion, $N_i = f(C)$, (see Sect. 2.3.3) by computer fitting the experimental points generated in both TMPCl and "H_2O" induced polymerizations and subtracting the number of chains obtained by "H_2O"

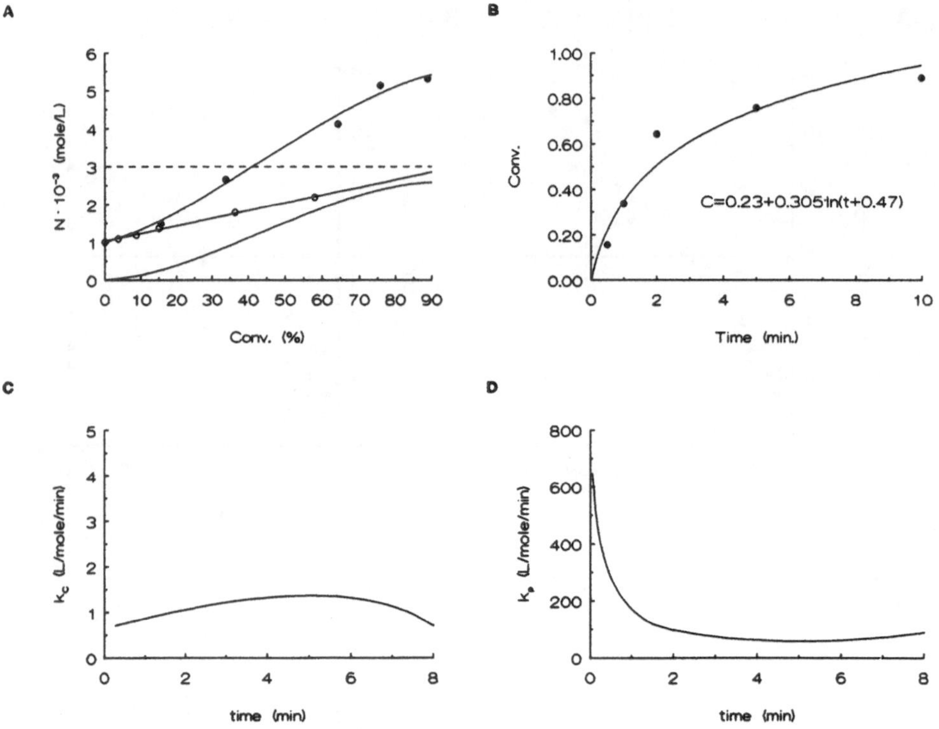

Fig. 41A–D. St polymerization at $-60\,°C$. The demonstration of the determination of k_c and k_p. (● = in the presence; ○ = in the absence of initiator) $[TEA] = 5 \cdot 10^{-4}$ mol

(N_{H_2O}) from that obtained by TMPCl (N_i). Figure 41A shows the experimental points and the computer-generated difference. Thus,

$$10^3 N_i = 0.61C + 7.81C^2 - 5.88C^3 \,. \tag{51}$$

To obtain k_c (see Eq. (49)), first we convert $N_i = f(C)$ to $N_i = f(\text{time})$. Figure 41B (from the data in Table 14 in the Appendix) shows the needed $C = f(\text{time})$ relationship and yields

$$C = 0.23 + 0.305 \ln(t + 0.47) \,. \tag{52}$$

Since we know the change in the monomer concentration as a function of conversion,

$$M = M_0\,(1 - C) \tag{53}$$

and also know M_0 (= 0.655 mol St/l) and I_0 ($= 3 \cdot 10^{-3}$ mol TMPCl/l), and since $d[N_i]/dt$ can be calculated by computer, all the terms in Eq. (49) are known. Thus, Fig. 41C shows the k_c vs time relationship between 0.5 and 8 min. (The very first and last parts of the curve are not shown because of the uncertainty of the data in these regions.) The information conveyed by this k_c vs time plot and that derived for the IB polymerization system (see Fig. 15C, [TEA] = $5 \cdot 10^{-4}$ mol/l) are comparable.

Similarly, the $k_p = f(\text{time})$ relationship was obtained by solving Eq. (50) (Sect. 2.3.3). The needed $d[M]/dt = f(\text{time})$ function was computer generated from Eqs. (52) and (53). Figure 41D shows the k_p vs time plot. Evidently, k_p is very high at very low conversions, which suggests a large contribution by rapidly propagating ionic species; after ~2 min (i.e., ~50% conversion) k_p levels off and becomes constant within experimental error.

It is of interest to compare the k_p values of St and IB polymerizations, that is Figs. 41D and 15D at [TEA] = $5 \cdot 10^{-4}$ mol/l. According to these results, the overall trends in both polymerizations are quite similar: after a burst of rapid ionic propagation, slow steady growth by the living species takes over. Significantly, the k_p of St polymerization is much larger (at least by a factor of two) than that of IB: $k_{p,IB} \sim 50$, $k_{p,St} \sim 100$ l/mol·min).

4.2.2.3 The Effect of Reaction Conditions on Polymerization Rate

4.2.2.3.1 The Effect of ED Concentration

Figure 42A, B shows conversion vs time and the corresponding $-\ln(1-C)$ vs time plots at various [ED]s. Assuming that initiation by "H_2O" is rapid and that chain transfer to monomer does not substantially influence the rate of polymerization, these data suggest polymerizations first order in St. As anticipated, the rate decreases with increasing [TEA].

The linear plots in Fig. 42B allow an independent estimation of k_p. Since the slopes of these $-\ln(1-C)$ vs time plots are equal to $k_p[N_{H_2O}]$ (see numbers next to the straight lines), and we known that $[N_{H_2O}] = 10^{-3}$ mol/l (see Fig. 40A in Sect. 4.2.2.1), we can readily obtain k_p at [TEA] = $5 \cdot 10^{-4}$ mol/l from:

$$k_p = \frac{0.0874}{10^{-3}} = 87.4 \text{ l/(mol} \cdot \text{min)} .$$

This value is in good agreement with that determined in St polymerization proceeding by slow initiation by "H_2O" in the presence of chain transfer (see Fig. 41D in Sect. 4.2.2.2).

4.2.2.3.2 The Effect of Solvent Polarity

Figure 42C, D (based on the raw data in Table 21, Appendix) shows the effect of polarity (i.e., relative $CH_2/Cl_2/n\text{-}C_6H_{14}$ composition) on the rates. Expectedly, the rates increase with increasing solvent polarity.

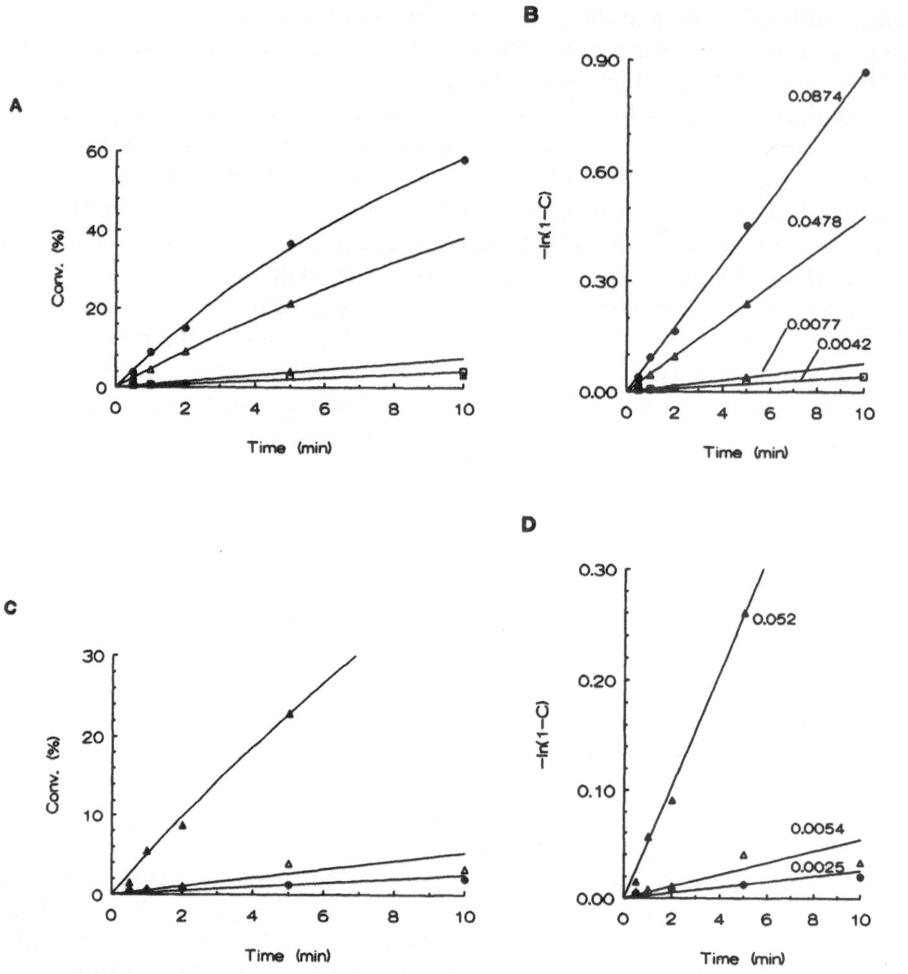

Fig. 42A–D St polymerization in the absence of TMPCl at $-60\,°C$. Conversion vs time and rate vs time plots: **A, B** at various ED concentrations [TEA] (mol/l) ● = $5 \cdot 10^{-4}$; △ = $1.5 \cdot 10^{-3}$; ▲ = $3 \cdot 10^{-3}$; □ = $6 \cdot 10^{-3}$; **C, D** at various medium polarity CH_2Cl_2/Hex. (v/v) ● = 40/60; △ = 60/40; ▲ = 80/20

It is worth emphasizing that the linear $-\ln(1-C)$ vs time plots starting at the origin prove the absence of irreversible termination in these polymerizations (see also Fig. 42B) [1, 21].

4.2.2.4 The Effect of Experimental Conditions on Molecular Weights ($\bar{M}_n$) and Molecular Weight Distributions (MWD)

The effect of time (conversion), [ED], medium polarity, and temperature have been studied in St polymerizations induced by "H_2O" in the absence of purposely added initiator. Specifically, we have investigated the effect of these experimental variables on the shape and position of RI GPC traces of PSt samples.

4.2.2.4.1 The Effect of the Time and [ED]

Figure 43A–D (based on the data shown in Table 20 in the Appendix) shows the effect of time (conversion) and [TEA] on $\bar{M}_n$ and MWD in $CH_2Cl_2/n\text{-}C_6H_{14}$

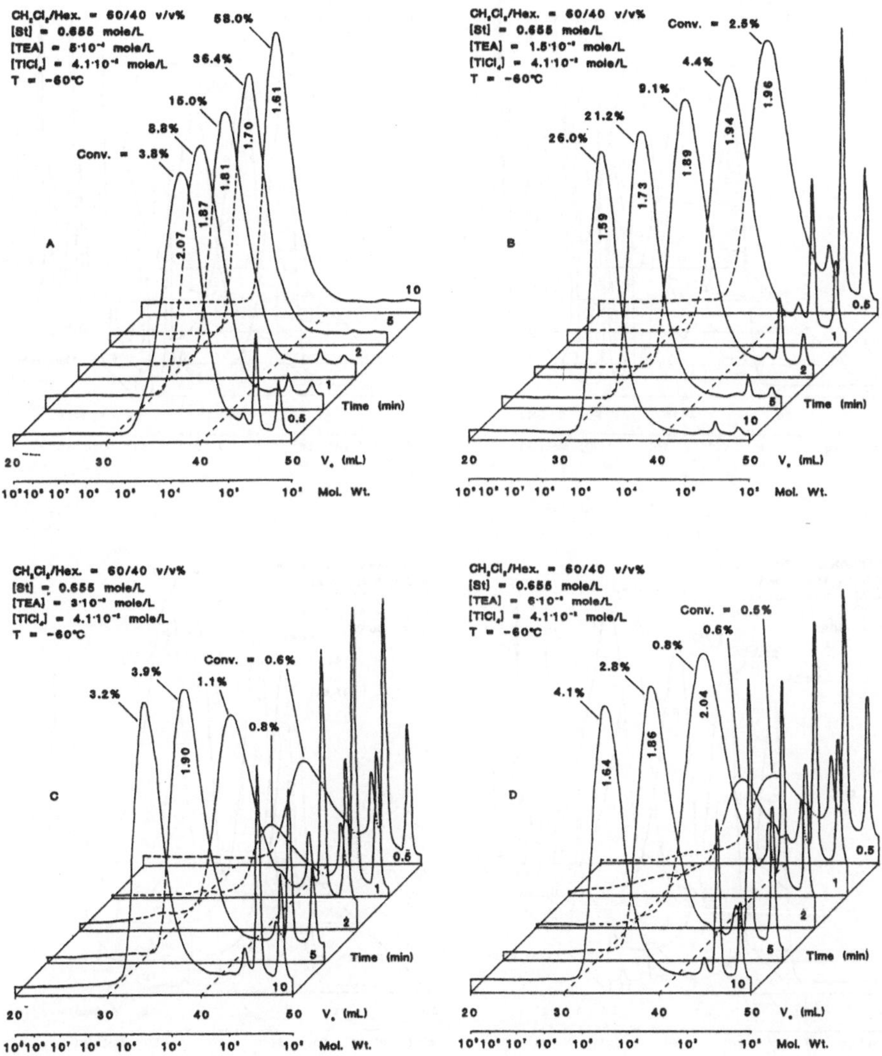

Fig. 43A–D. St polymerization in the absence of added initiator at $-60\,°C$. GPC traces as a function of time at various ED concentrations

60/40 v/v at $-60\,°C$. All the conclusions reached in Sects. 4.1.1.1.1, 4.1.2.3, and 4.2.1.4.1, in regard to the effect of [ED] on $\bar{M}_n$ and MWD are valid in these instances also (seen there). The very low molecular weight products showing prominent sharp peaks tend to diminish with increasing conversions.

In all cases the MWDs become narrower with time and can reach $\bar{M}_w/\bar{M}_n \sim 1.6$ at low [ED].

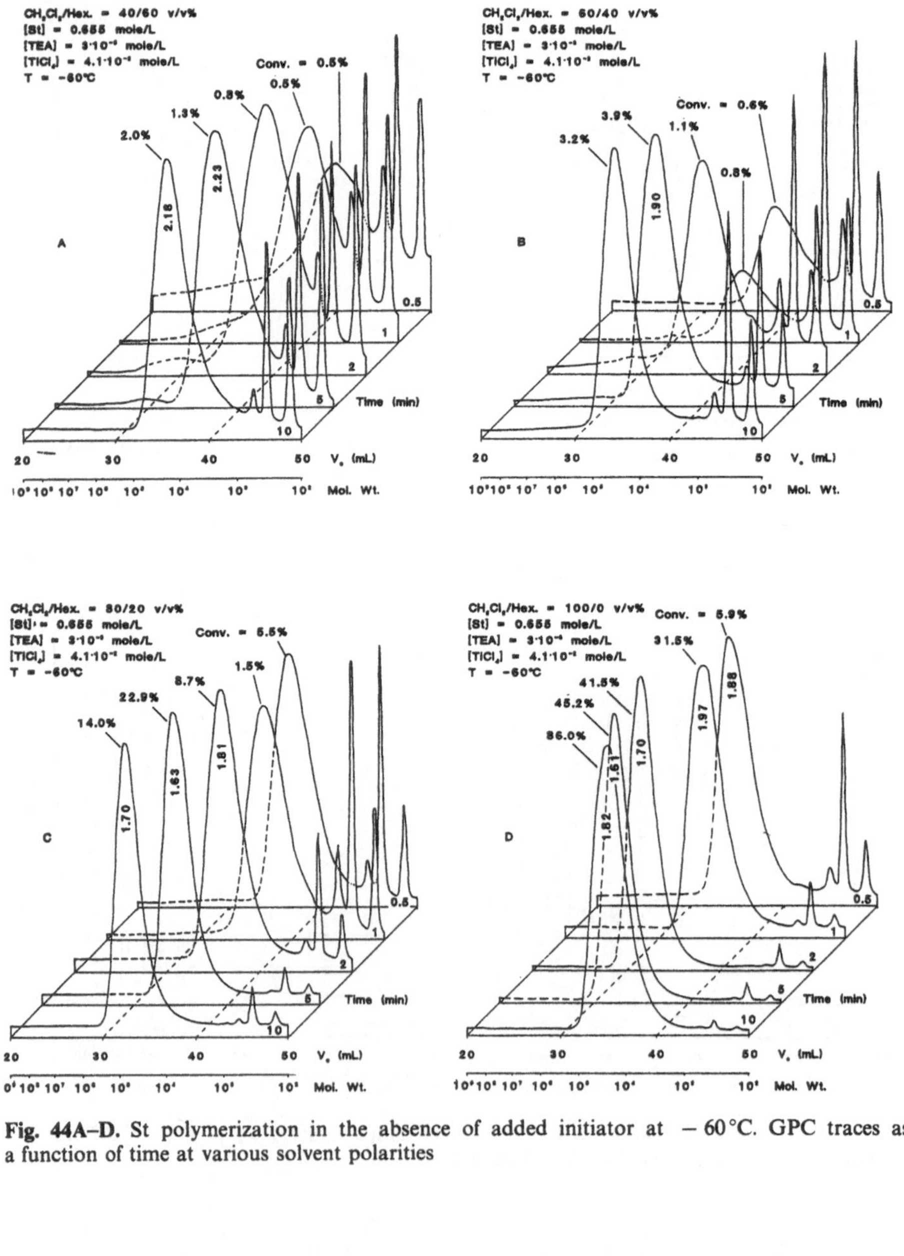

Fig. 44A–D. St polymerization in the absence of added initiator at $-60\,°C$. GPC traces as a function of time at various solvent polarities

4.2.2.4.2 The Effect of Solvent Polarity at − 60°C

Figure 44A–D (based on the data in Table 21 in the Appendix) shows the effect of changing the relative solvent composition, i.e., $CH_2Cl_2/n\text{-}C_6H_{14} = 40/60$, 60/40, 80/20, and 100/0 on the RI GPC traces at −60 °C. Figure 45A, B helps to compare the traces obtained at various [ED]s and polarities, respectively.

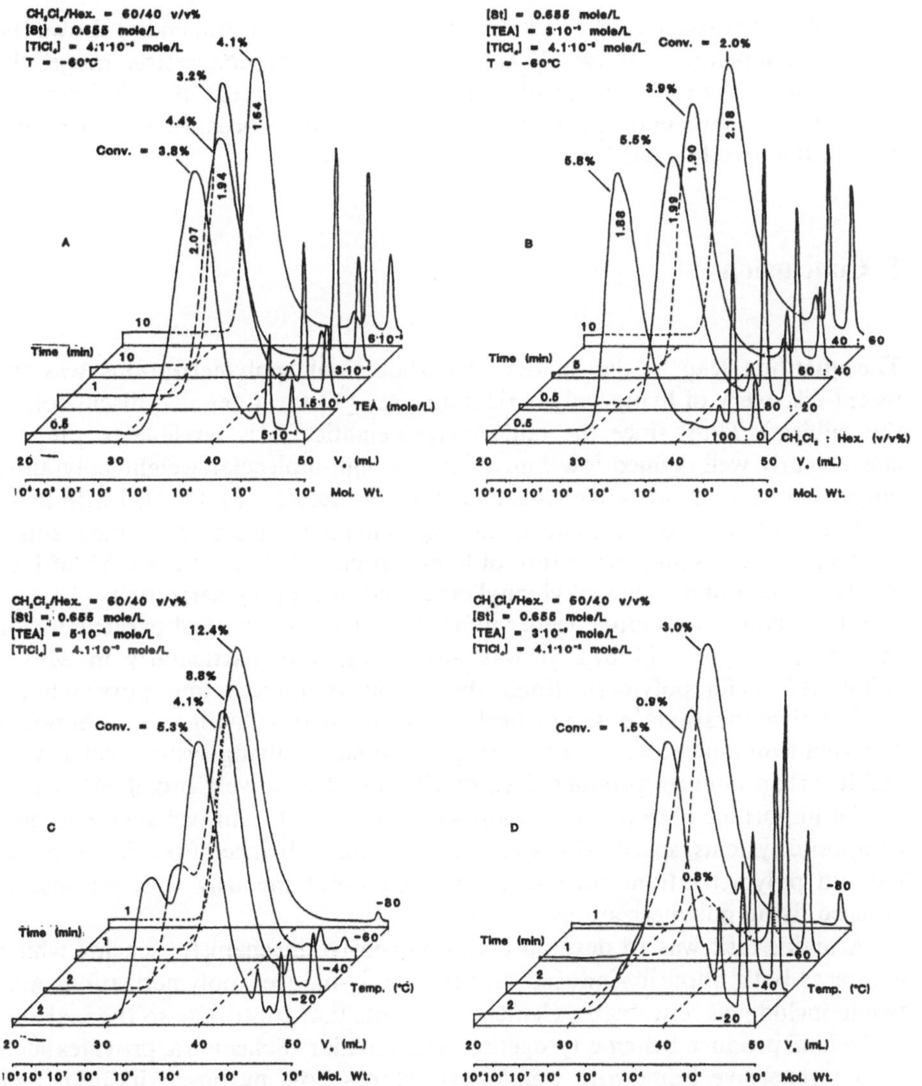

Fig. 45A–D. St polymerization in the absence of added initiator. GPC traces: **A** as a function of ED concentration at − 60 °C; **B** solvent polarity at − 60 °C; **C, D** as a function of temperature at two ED concentrations

The conclusions reached upon examining these data are very similar to those reached in similar experiments with IB and with St polymerizations induced by purposely added TMPCl initiator (see Sects. 4.1.1.2.2 and 4.2.1.4.2). Tthe MWD tend to get narrower with increasing [TEA] and the best-defined product is harvested in 60/40 $CH_2Cl_2/n\text{-}C_6H_{14}$.

4.2.2.4.3 The Effect of Temperature

Figure 45C, D (based on the raw data in Table 22, in the Appendix) shows the effect of temperature at two [TEA]s. In the 5–13% conversion range, the narrowest MWD product was obtained at $-60\,°C$, whereas at $> 3\%$ conversions, the comparison of the data is rendered unreliable because of the prominence of the very low $\bar{M}_n$ "trash" peaks.

5 Conclusions

The quantum leap in the history of carbocationic polymerizations was the recent discovery of living polymerizations (see [1] for a detailed discussion of this subject). Thus, since the mid-nineteen-eighties, it is possible to prepare conveniently well-defined low-, medium-, or high-molecular weight poly(alkyl vinyl ethers) and many polyolefins by cationic processes [1]. Parallel with these developments, considerable efforts have been made to elucidate the mechanism of living processes and the nature of living species [1, 6, 19–21, 45, 51, 66], to clarify and combine living alkyl vinyl ether and olefin polymerizations [13, 14], and in general to develop a deeper understanding of the critical parameters for achieving livingness [1, 67]. It was also recognized, particularly in MtX_n-coinitiated olefin polymerizations, that impurity-induced and purposely-induced polymerizations, either or both of which could be living or conventional occurred simultaneously so that even predominantly living systems often yielded less than uniform products (i.e., products with relatively broad MWDs).

An important objective of the present work was to unravel such complex composite systems, specifically, to understand the influence of protic impurity-induced polymerizations on and proceeding simultaneously with purposely-induced living polymerizations.

A major task was to develop a comprehensive mechanistic scheme which embraces both impurity-induced and purposely-induced polymerizations and would include *all* conceivable elementary events that contribute to the makeup of the final product. Scheme 1, together with auxiliary Scheme 1a, provides such a comprehensive framework and clearly shows how purposely-induced and impurity-induced polymerization systems are interconnected (see Sect. 1).

By use of this model we were able to account for a very large body of experimental observations. For example, the comprehensive model helped to recognize that the pathway connecting the two systems was proton expulsion/reprotonation, i.e., chain transfer (see Sect. 2). It was also recognized that the final products reflect the contributions of several growing species, and that in the absence of specific measures which would drive the system toward a well-defined or uniform growth center, the MWDs will reflect the reactivities/concentrations of many growing species. The contribution of the nonliving polymer arising from even minute concentrations of highly reactive ionic growing sites to the final product may be relatively more important than that forming from much higher concentrations of less reactive nonionized living species, and thus lead to MWD broadening or even to bimodal MWDs.

Regarding the comprehensive closed-loop model, a new methodology was developed to simulate kinetic responses of real-life carbocationic polymerizations to various events or combinations of events. Basic scenarios were set up to analyze numerically the above events and to demonstrate their effects on N and $\bar{M}_n$ vs W_p, and $-\ln(1-C)$ and C vs time diagnostic plots, respectively.

We were particularly interested in analyzing the effects of a) slow initiation relative to propagation, b) initiation by protic impurity (commonly "H_2O"), and c) mono- and bimolecular chain transfer to monomer on the apparent rate constants of cationation k_c and propagation k_p. Based on our method developed for the determination of k_c and k_p for events a, a + c, and a + b + c (see Sect. 2.3), numerical analyses of k_c and k_p obtained for IB and St polymerizations have shown that these quantities are in fact not constants but complex composites of rate constants of various species having different reactivities. With this insight, the enormous sensitivity of k_c and k_p, and beyond these the diagnostic plots ($\bar{M}_n$ or N vs W_p; rate or conversion vs time) and product characteristics (molecular weight, MWD) on reaction conditions (i.e., reagent concentrations, polarity, temperature, reagent addition sequence) became readily understandable.

Specifically, and in contrast to Faust et al. [4, 10, 37], who concluded that the role of EDs in IB polymerization is to trap protons, we found that EDs may exert several effects on the rate and molecular weights of PIBs. For one, we observed that with increasing [ED], k_c and k_p were decreasing, indicating the EDs are able to convert highly reactive species to less reactive species (see Sect. 4.1.1.1.1). Also, according to diagnostic $\bar{M}_n$ (or N) vs W_p plots the extent of chain transfer was reduced in the presence of ED at -45 and -60 °C, i.e., in the temperature range where the effect of slow initiation on the overall rate was minimum (see Section. 4.1.1.2.1).

RI (GPC) traces of PIB and PSt showed the existence of three major products at low [ED] and low conversion at -25 °C. The peak at highest $\bar{M}_n$s was attributed to products formed by initiation by "H_2O" and subsequent rapid ionic propagation; the peak at medium $\bar{M}_n$s was assigned to polymer formed by initiation by purposely added RX, while that at the relatively lowest $\bar{M}_n$s exhibiting the narrowest MWD was postulated to be due to product formed by living polymerization. The proportion of the two relatively higher $\bar{M}_n$ products decreases with increasing conversions and/or [ED], whereas the relatively low molecular weight product formed by living polymerization increases (see Sect. 4.1.1.4).

In control experiments (impurity-induced polymerizations) chain transfer is suppressed but not totally eliminated by increasing [ED] (see Sect. 4.1.2.1). Also with increasing [ED]s, the rate is reduced (Sect. 4.1.2.2) and the MWDs become narrower because of the increasing contribution of nondissociated living species to propagation (Sect. 4.1.2.3).

In initiator-induced St polymerization, the effect of ED on the diagnostic $\bar{M}_n$ (or N) W_p plots is visible only at [initiator]/[ED] = 1:3 and 1:4. At these higher [ED]s, the $\bar{M}_n$ data fall on a straight line which goes through the origin and whose slope is determined by $1/(I_0 + N_{H_2O})$, and thus indicates the absence of chain transfer but presence of significant initiation by "H_2O" (see Sect. 4.2.1.2.1). The contribution of the relatively narrow MWD living species to the overall distribution increases with increasing [ED] (especially at very high [TEA]) while the rate of polymerization decreases (see Sect. 4.2.1.4.1).

According to control experiments, initiation by "H_2O" is quite significant in St polymerization and chain transfer cannot be eliminated by increasing [ED]s (see Sect. 4.2.2.1).

The combined effects of slow initiation + initiation by "H_2O" + chain transfer to monomer on k_c and k_p of St polymerization were determined numerically (see Sect. 4.2.2.2). The overall trends in both IB and St polymerization are similar: The k_cs are comparable under similar conditions, but the k_ps are about twice as large for St as for IB.

In St control experiments (impurity-induced polymerizations), the rate decreases with increasing [ED]. Linear $-\ln(1-C)$ vs time plots indicate polymerizations first order in [St] (see Sect. 4.2.2.3.1), and allow an independent estimation of k_p; the k_p obtained by this route is in good agreement with that determined in Sect. 4.2.2.2. An anticipated, the GPC peaks become sharper by increasing [ED] and/or conversions (see Sect. 4.2.2.4.1).

The effect of solvent polarity on both initiator- and impurity-induced IB and St polymerizations is similar; thus with increasing polarity, k_c and k_p increase due to increasing solvation of the ionic species (see Sects. 4.1.1.1.2 and 4.1.2.2), the overall rate of polymerization increases (see Sects. 4.1.1.2.2, 4.1.2.1, 4.1.2.2, and 4.2.2.3.2) and the extent of chain transfer increases (see Sects. 4.1.2.1, 4.2.1.2.2, and 4.2.2.1). With increasing solvent polarity and MWDs broaden due to the increasing concentration of nonliving ionic species (see Sect. 4.1.1.4.2). In most cases, a solvent composition of 60/40 v/v $CH_2Cl_2/n\text{-}C_6H_{14}$ provided the narrowest MWDs (see Sects. 4.1.2.3, 4.2.1.4.2, and 4.2.2.4.1).

The linear $-\ln(1-C)$ vs time plots prove the absence of irreversible termination (see Sect. 4.2.2.3.2). The concentration of MtX_n affects both IB and St polymerization. At higher [$TiCl_4$], k_c and k_p are apparently higher, presumably because of the higher concentration of growing sites and lower concentration of dormant species (see Scheme 1, equilibria on the left in both sides, and also Sect. 4.1.1.1.3). Increasing [$TiCl_4$] increases the relative rate of initiation (see Sect. 4.1.1.2.3) and the rate of polymerization (see Sect. 4.1.1.4.4). Contrary to [37], we find first order behavior in IB polymerization in respect to $[TiCl_4]$ (see Sect. 4.1.1.3). Since we do not have control experiments at high conversions, we were unable to derive the reaction order in regard to $[TiCl_4]$ in St polymerization (see Section 4.2.1.3). Higher product uniformity was found at higher $[TiCl_4]$ (see Sect. 4.1.1.4.4); however, the undesirable MWD broadening due to slow initiation could not be totally eliminated (see Sect. (4.2.1.4.4), only strongly reduced (see Sect. 4.2.1.2.3).

Again (i.e., adding monomer to premixed initiator/coinitiator solutions) exerted various advantageous effects. Thus, aging decreases k_c at low conversion since the $R^{\delta\oplus}\text{---}\left[\overset{\delta\oplus}{MtX_n}\cdot ED\right]\cdots MtX_{n+1}^{\ominus}$ and similar nonionic species are more rapidly established. As anticipated, k_p is not affected by aging (see Sect. 4.1.1.1.4). Aging reduces the extent of slow initiation at low conversions. $\bar{M}_n$s are proportional to W_p at higher conversions, but lower N that I_0 may indicate irreversible initiator-consuming side reactions during aging (see Sect. 4.1.1.2.5 in IB polymerization). In St polymerization, aging visibly reduces the extent of slow initiation at low conversions; however, the data are insufficient for comparison at higher conversions (see Sect. 4.2.1.2.5). Aging improves product uniformity (MWD < 1.1) in IB polymerization (see Sect. 4.1.1.4.5) but has no significant effect in St polymerization (see Sect. 4.2.1.4.5). The optimum experimental conditions for aging have not yet been established.

The effect of temperature is complex because it may affect different reactions to a different degree. For example, in IB polymerization, both k_c and k_p increase with decreasing temperatures which suggest a higher contribution of ionic species because of increasing polarity of the medium (see Sect. 4.1.1.1.5). Chain transfer increases with increasing temperatures, while the extent of slow initiation becomes more significant at lower temperatures (see Sects. 4.1.1.2.4 and 4.2.1.2.4). MWDs broaden with decreasing temperatures (see Sect. 4.1.1.4.3, IB polymerization, where the optimum temperature range was from $-45°$ to $-60\,°C$). Because of these opposite effects, a generally preferred temperature for these polymerizations cannot be specified and has to be established empirically for each individual system.

6 Appendix

Table 2. Conversion vs time and N (number of PIB moles formed) for the TMPCl/$TiCl_4$/IB/TEA/CH_2Cl_2-Hex(60-40)/− 25°C system. ([TMPCl] = 3 10^{-3} mol/l; [$TiCl_4$] = 4.1 10^{-2} mol/l; [IB] = 1.16 mol/l.Total volume 20 ml)

Conditions			Results						
Number	[TEA], mol/l	time, min	Conv. %	$\bar{M}_n$, g/mol	$\bar{M}_w/\bar{M}_n$	$\bar{M}_n$, g/mol	$\bar{M}_w/\bar{M}_n$	[N] 10^3, mol/l	I_{eff}, %
				Total sample		Peak at highest V_e			
1	1.25 10^{-4}	2	99.6	11,650	4.66	–	–	5.57	185.8
2	1.25 10^{-4}	5	97.8	11,300	4.78	–	–	5.65	188.5
3	1.25 10^{-4}	10	98.5	9,500	5.11	–	–	6.75	225.5
4	1.25 10^{-4}	20	98.7	12,500	4.76	–	–	5.14	171.7
5	1.25 10^{-4}	40	98.6	8,900	4.79	–	–	7.19	240.1
6	3.75 10^{-4}	2	13.9	4,600	6.70	–	–	1.98	66.2
7	3.75 10^{-4}	5	28.1	5,200	5.06	1,600	1.30	2.95	98.5
8	3.75 10^{-4}	10	44.4	8,300	4.09	3,900	1.21	3.48	116.0
9	3.75 10^{-4}	20	68.2	9,700	3.56	4,600	1.22	4.58	152.8
10	3.75 10^{-4}	40	92.2	10,400	2.86	5,600	1.29	5.78	192.8
11	7.5 10^{-4}	2	3.6	1,700	12.13	1,000	1.32	1.38	46.1
12	7.5 10^{-4}	5	7.9	2,300	5.90	1,600	1.20	2.23	74.5
13	7.5 10^{-4}	10	13.8	3,500	3.04	2,900	1.14	2.55	85.0
14	7.5 10^{-4}	20	21.7	5,050	2.04	4,300	1.18	2.81	93.6
15	7.5 10^{-4}	40	41.0	6,900	1.91	7,000	1.13	3.88	129.4
16	1.5 10^{-3}	2	2.2	1,400	3.70	900	1.24	1.04	34.7
17	1.5 10^{-3}	5	5.2	1,900	8.13	1,400	1.19	1.81	60.3
18	1.5 10^{-3}	10	10.5	2,900	5.10	2,300	1.09	2.39	79.6
19	1.5 10^{-3}	20	18.8	4,100	1.07	4,100	1.07	2.91	96.9
20	1.5 10^{-3}	40	30.8	7,200	1.06	7,200	1.06	3.30	110.1
21	0.0	2	97.5	18,150	3.98	–	–	3.50	116.7
22	0.0	5	97.9	14,300	4.55	–	–	4.45	148.7
23	0.0	10	98.0	14,300	4.12	–	–	4.46	148.7
24	0.0	20	99.2	14,500	4.18	–	–	4.45	148.4
25	0.0	40	98.5	16,600	4.56	–	–	3.87	129.1

Table 3. Conversion vs time and N (number of PIB moles formed) for the TMPCl/$TiCl_4$/IB/TEA/CH_2Cl_2-Hex(60-40)/$-45\,°C$ system. ([TMPCl] = 3 10^{-3} mol/l; [$TiCl_4$] = 4.1 10^{-2} mol/l; [IB] = 1.205 mol/l. Total volume 20 ml)

Conditions			Results						
Number	[TEA], mol/l	time, min	Conv. %	$\bar{M}_n$, g/mol	$\bar{M}_w/\bar{M}_n$	$\bar{M}_n$, g/mol	$\bar{M}_w/\bar{M}_n$	[N] 10^3, mol/l	I_{eff}, %
				Total sample		Peak at highest V_e			
1	5.0 10^{-4}	1	5.2	2,500	2.96	1,800	1.37	1.41	46.9
2	5.0 10^{-4}	2	12.6	3,800	3.94	3,150	1.23	2.22	74.0
3	5.0 10^{-4}	4	16.9	4,400	2.49	3,900	1.17	2.59	86.5
4	5.0 10^{-4}	10	39.6	9,100	1.74	7,890	1.09	2.92	97.5
5	5.0 10^{-4}	20	61.0	13,300	1.39	12,300	1.06	3.08	102.9
6	1.5 10^{-3}	1	3.9	1,700	4.90	1,500	1.28	1.54	51.5
7	1.5 10^{-3}	2	11.0	3,000	3.13	3,050	1.21	2.43	81.8
8	1.5 10^{-3}	4	14.8	4,100	2.16	3,900	1.17	2.44	81.4
9	1.5 10^{-3}	10	28.7	7,150	1.44	6,750	1.13	2.71	90.3
10	1.5 10^{-3}	20	50.5	11,700	1.19	12,100	1.08	2.91	97.1
11	3.0 10^{-3}	1	2.9	1,800	10.99	1,700	1.38	1.07	35.6
12	3.0 10^{-3}	2	8.7	2,700	4.08	2,700	1.29	2.20	73.4
13	3.0 10^{-3}	4	13.3	3,800	5.30	3,700	1.23	2.36	78.6
14	3.0 10^{-3}	10	27.0	7,100	1.25	7,100	1.13	2.57	85.7
15	3.0 10^{-3}	20	42.8	10,600	1.12	10,700	1.10	2.72	90.7
16	6.0 10^{-3}	1	2.5	1,600	10.42	1,500	1.31	1.05	35.1
17	6.0 10^{-3}	2	5.9	2,200	6.15	2,200	1.35	1.83	60.9
18	6.0 10^{-3}	4	11.8	3,350	3.31	3,400	1.24	2.38	79.4
19	6.0 10^{-3}	10	19.2	5,100	1.20	5,200	1.14	2.54	84.6
20	6.0 10^{-3}	20	31.3	8,200	1.19	8,200	1.12	2.58	86.1

Table 4. Conversion vs time and N (number of PIB moles formed) for the $TMPCl/TiCl_4/IB/TEA/CH_2Cl_2$-Hex(60-40)/$-60\,°C$ system. ([TMPCl] = $3\,10^{-3}$ mol/l; $[TiCl_4] = 4.1\,10^{-2}$ mol/l; [IB] = 1.246 mol/l. Total volume 20 ml)

Conditions			Results						
Number	[TEA], mol/l	time, min	Conv. %	$\bar{M}_n$, g/mol	$\bar{M}_w/\bar{M}_n$,	$\bar{M}_n$, g/mol	$\bar{M}_w/\bar{M}_n$	[N] 10^3, mol/l	I_{eff}, %
				Total sample		Peak at highest V_e			
1	$5.0\,10^{-4}$	0.5	7.5	3,400	2.60	3,300	1.33	1.57	52.2
2	$5.0\,10^{-4}$	1	9.7	3,600	2.70	3,600	1.30	1.90	63.4
3	$5.0\,10^{-4}$	2	18.6	5,400	2.34	5,100	1.25	2.42	80.6
4	$5.0\,10^{-4}$	5	55.1	13,600	1.41	13,900	1.30	2.84	94.6
5	$5.0\,10^{-4}$	10	73.7	17,200	1.27	17,350	1.22	3.00	100.2
6	$1.5\,10^{-3}$	0.5	4.5	2,500	2.09	2,700	1.52	1.27	42.3
7	$1.5\,10^{-3}$	1	7.4	2,900	2.29	3,100	1.46	1.79	59.6
8	$1.5\,10^{-3}$	2	13.4	4,400	1.60	4,600	1.43	2.14	71.5
9	$1.5\,10^{-3}$	5	35.9	9,300	1.29	9,450	1.27	2.71	90.5
10	$1.5\,10^{-3}$	10	52.8	13,200	1.23	13,400	1.21	2.79	93.2
11	$3.0\,10^{-3}$	0.5	4.8	3,100	2.72	2,800	1.29	1.08	36.0
12	$3.0\,10^{-3}$	1	8.8	3,700	2.55	3,700	1.38	1.68	55.9
13	$3.0\,10^{-3}$	2	14.2	4,600	1.44	4,800	1.37	2.16	72.0
14	$3.0\,10^{-3}$	5	42.1	10,900	1.23	11,100	1.21	2.71	90.4
15	$3.0\,10^{-3}$	10	50.9	13,000	1.21	13,000	1.21	2.74	91.6
16	$6.0\,10^{-3}$	0.5	4.4	2,500	1.51	2,700	1.39	1.21	40.3
17	$6.0\,10^{-3}$	1	6.2	2,600	1.43	2,750	1.32	1.66	55.4
18	$6.0\,10^{-3}$	2	12.4	4,100	1.43	4,300	1.36	2.09	69.8
19	$6.0\,10^{-3}$	5	29.9	8,100	1.26	8,200	1.25	2.58	85.9
20	$6.0\,10^{-3}$	10	48.3	12,400	1.20	12,500	1.19	2.72	90.7

Table 5. Conversion vs time and N (number of PIB moles formed) for the TMPCl/$TiCl_4$/IB/TEA/CH_2Cl_2-Hex(60-40)/− 82 °C system. ([TMPCl] = 3 10^{-3} mol/l; [$TiCl_4$] = 4.1 10^{-2} mol/l; [IB] = 1.291 mol/l. Total volume 20 ml)

Conditions			Results						
Number	[TEA], mol/l	time, min	Conv. %	$\bar{M}_n$, g/mol	$\bar{M}_w/\bar{M}_n$	$\bar{M}_n$, g/mol	$\bar{M}_w/\bar{M}_n$	[N] 10^3, mol/l	I_{eff}, %
				Total sample		Peak at highest V_e			
1	5.0 10^{-4}	0.33	12.3	7,700	2.94	8,700	2.03	1.56	38.6
2	5.0 10^{-4}	0.67	19.6	7,200	2.08	7,600	2.08	1.96	65.5
3	5.0 10^{-4}	1	30.3	10,500	2.22	10,500	1.77	2.09	69.8
4	5.0 10^{-4}	2	54.6	13,400	1.93	14,050	1.61	2.95	98.3
5	5.0 10^{-4}	4	64.0	15,900	1.72	15,800	1.48	2.91	97.1
6	1.5 10^{-3}	0.33	11.2	8,100	1.97	8,500	1.79	1.00	33.2
7	1.5 10^{-3}	0.67	10.5	6,200	2.47	6,500	1.79	1.23	40.8
8	1.5 10^{-3}	1	22.2	8,800	1.83	9,000	1.66	1.82	60.7
9	1.5 10^{-3}	2	47.7	12,800	1.52	13,100	1.44	2.70	90.0
10	1.5 10^{-3}	4	60.3	15,100	1.34	15,100	1.33	2.90	96.6
11	3.0 10^{-3}	0.33	9.6	6,600	1.84	6,800	1.79	1.06	35.4
12	3.0 10^{-3}	0.67	15.0	7,400	1.72	7,600	1.67	1.47	48.8
13	3.0 10^{-3}	1	22.7	9,300	1.71	9,250	1.69	1.78	59.4
14	3.0 10^{-3}	2	43.4	12,400	1.44	12,500	1.39	2.55	85.0
15	3.0 10^{-3}	4	54.5	14,200	1.40	14,500	1.37	2.79	93.1
16	6.0 10^{-3}	0.33	6.3	5,250	1.81	5,500	1.71	0.87	29.0
17	6.0 10^{-3}	0.67	14.2	7,100	1.71	7,400	1.63	1.45	48.4
18	6.0 10^{-3}	1	21.4	8,700	1.66	8,900	1.63	1.79	59.6
19	6.0 10^{-3}	2	31.9	9,750	1.59	10,100	1.50	2.37	79.1
20	6.0 10^{-3}	4	60.2	14,700	1.25	14,800	1.25	2.96	98.7
21	0.0	0.33	96.5	16,000	7.67	–	–	4.36	146.0
22	0.0	0.67	97.5	20,300	5.74	–	–	3.49	116.0
23	0.0	1	99.1	23,700	4.96	–	–	3.04	101.0
24	0.0	2	99.7	21,200	5.73	–	–	3.41	114.0
25	0.0	4	99.4	24,850	4.44	–	–	2.90	96.8

Table 6. Conversion vs time and N (number of PIB moles formed) for the TMPCl/$TiCl_4$/IB/TEA/−80 °C system. ([TMPCl] = 3 10^{-3} mol/l; [TEA] = 3 10^{-3} mol/l; [$TiCl_4$] = 4.1 10^{-2} mol/l; [IB] = 1.284 mol/l. Total volume 20 ml)

Conditions			Results				
Number	CH_2Cl_2/Hex (v/v)	time, min	Conv. %	$\bar{M}_n$, g/mol	$\bar{M}_w/\bar{M}_n$	[N] 10^3 mol/l	I_{eff}, %
					Total sample		
1	40/60	0.5	0.3	1,600	7.34	0.13	4.3
2	40/60	1	1.2	1,800	4.27	0.49	16.3
3	40/60	2	4.0	2,700	1.89	1.09	36.2
4	40/60	4	9.7	3,650	1.80	1.86	62.1
5	40/60	10	24.6	7,050	1.34	2.52	84.0
6	60/40	0.33	4.7	5,700	1.82	0.60	20.1
7	60/40	0.67	10.9	7,700	1.75	1.02	34.1
8	60/40	1	15.7	8,000	1.65	1.41	47.1
9	60/40	2	30.2	10,900	1.52	2.01	66.9
10	60/40	4	50.3	13,100	1.34	2.76	92.1
11	80/20	0.5	24.3	10,200	1.59	1.72	57.2
12	80/20	1	41.7	14,350	1.62	2.10	70.0
13	80/20	2	56.8	14,100	1.60	2.90	96.6
14	80/20	4	68.8	16,000	1.66	3.10	103.3
15	80/20	10	73.4	15,000	1.53	3.52	117.4
16	100/0	0.5	73.4	17,400	2.35	3.04	101.4
17	100/0	1	83.6	16,100	2.45	3.75	125.0
18	100/0	2	85.4	21,600	1.84	2.85	95.0
19	100/0	4	88.3	21,500	1.90	2.96	98.9
20	100/0	10	95.8	21,300	2.09	3.25	108.4

Table 7. Conversion vs time and N (number of PIB moles formed) for the TMPCl/$TiCl_4$/IB/TEA/CH_2Cl_2-Hex(60-40)/−80 °C system. ([TMPCl] = $3\,10^{-3}$ mol/l; [TEA] = $3\,10^{-3}$ mol/l; [IB] = 1.291 mol/l. Total volume 20 ml)

Conditions			Results				
Number	$[TiCl_4]$, mol/l	time, min	Conv. %	$\bar{M}_n$, g/mol	$\bar{M}_w/\bar{M}_n$	[N] 10^3 mol/l	I_{eff}, %
					Total sample		
1	$5.47\,10^{-2}$	0.5	10.7	6,950	1.74	1.11	37.1
2	$5.47\,10^{-2}$	1	19.3	7,550	1.57	1.84	61.3
3	$5.47\,10^{-2}$	2	37.2	10,500	1.46	2.54	84.9
4	$5.47\,10^{-2}$	4	57.5	14,300	1.25	2.89	96.5
5	$5.47\,10^{-2}$	10	68.7	17,100	1.17	2.90	96.7
6	$4.10\,10^{-2}$	0.33	4.7	5,700	1.82	0.60	20.1
7	$4.10\,10^{-2}$	0.67	10.9	7,700	1.75	1.02	34.1
8	$4.10\,10^{-2}$	1	15.7	8,000	1.65	1.41	47.1
9	$4.10\,10^{-2}$	2	30.2	10,900	1.52	2.01	66.9
10	$4.10\,10^{-2}$	4	50.3	13,100	1.34	2.76	92.1
11	$2.73\,10^{-2}$	0.5	2.5	4,900	1.73	0.37	12.3
12	$2.73\,10^{-2}$	1	9.6	10,050	1.77	0.69	23.0
13	$2.73\,10^{-2}$	2	18.9	10,500	1.79	1.30	43.3
14	$2.73\,10^{-2}$	4	34.8	14,200	1.60	1.76	58.7
15	$2.73\,10^{-2}$	10	46.7	13,600	1.43	2.48	82.6
16	$1.37\,10^{-2}$	0.5	0.8	6,100	1.67	0.10	3.3
17	$1.37\,10^{-2}$	1	1.3	9,200	1.86	0.10	3.3
18	$1.37\,10^{-2}$	2	6.1	13,650	1.80	0.33	10.0
19	$1.37\,10^{-2}$	4	19.9	15,400	1.78	0.93	31.0
20	$1.37\,10^{-2}$	10	36.2	13,000	1.56	2.01	67.1

Table 8. Conversion vs time and N (number of PIB moles formed) for the TMPCl/$TiCl_4$/IB/TEA/CH_2Cl_2-Hex(60-40)/−60 °C system. ([TMPCl] = $3\,10^{-3}$ mol/l; [TEA] = $3\,10^{-3}$ mol/l; $[TiCl_4]$ = $5.4\,10^{-2}$ mol/l; [IB] = 1.24 mol/l. Aging time = 4 min. Total volume 20 ml)

Conditions		Results						
Number	time, min	Conv. %	$\bar{M}_n$, g/mol	$\bar{M}_w/\bar{M}_n$	$\bar{M}_n$, g/mol	$\bar{M}_w/\bar{M}_n$	[N] 10^3 mol/l	I_{eff}, %
			Total sample		Peak at highest V_e			
1	1	5.8	2,350	1.43	2,350	1.32	1.715	56.9
2	2	10.3	3,500	1.29	3,500	1.28	2.06	68.2
3	5	32.0	9,500	1.15	9,500	1.15	2.335	77.3
4	10	56.8	16,800	1.10	16,800	1.10	2.345	77.6
5	20	84.4	21,950	1.07	21,900	1.07	2.67	88.5

Table 9. Conversion vs time and N (number of PIB moles formed) for the "H_2O"/$TiCl_4$/IB/TEA/CH_2Cl_2-Hex(60-40)/− 60 °C system. ([$TiCl_4$] = 4.1 10^{-2} mol/l; [IB] = 1.15 mol/l. Total volume 20 ml)

Conditions			Results				
Number	[TEA], mol/l	time, min	Conv. %	$\bar{M}_n$, g/mol	$\bar{M}_w/\bar{M}_n$	[N] 10^4 mol/l	W_p g
					Total sample		
1	5.0 10^{-4}	0.5	0.5	4,600	15.1	0.75	0.0069
2	5.0 10^{-4}	1	1.0	4,100	17.7	1.55	0.0124
3	5.0 10^{-4}	2	2.6	5,100	11.5	3.30	0.0337
4	5.0 10^{-4}	5	6.2	8,300	4.3	4.85	0.0801
5	5.0 10^{-4}	10	13.3	14,700	2.4	5.85	0.1712
6	1.5 10^{-3}	0.5	0.6	3,700	14.7	1.10	0.0081
7	1.5 10^{-3}	1	0.8	4,200	26.7	1.20	0.0103
8	1.5 10^{-3}	2	1.3	5,000	23.6	1.65	0.0162
9	1.5 10^{-3}	5	3.5	6,600	6.4	3.45	0.0455
10	1.5 10^{-3}	10	5.5	13,200	3.3	2.70	0.0711
11	3.0 10^{-3}	0.5	0.5	5,300	21.4	0.65	0.0068
12	3.0 10^{-3}	1	0.6	6,200	26.3	0.60	0.0076
13	3.0 10^{-3}	2	0.8	7,200	23.9	0.70	0.0103
14	3.0 10^{-3}	5	1.2	9,100	11.2	0.80	0.0150
15	3.0 10^{-3}	10	3.1	11,100	4.2	1.80	0.0397
16	6.0 10^{-3}	0.5	0.4	7,600	30.9	0.35	0.0053
17	6.0 10^{-3}	1	0.6	6,500	30.0	0.55	0.0074
18	6.0 10^{-3}	2	0.7	7,400	24.2	0.55	0.0084
19	6.0 10^{-3}	5	1.2	8,900	17.1	0.90	0.0158
20	6.0 10^{-3}	10	2.2	9,600	6.3	1.50	0.0286
21	0.0	2	4.4	9,800	15.0	2.90	0.0568

Table 10. Conversion vs time and N (number of PIB moles formed) for the "H_2O"/$TiCl_4$/IB/TEA/ − 60 °C system. ([TEA] = 3 10^{-3} mol/l; [$TiCl_4$] = 4.1 10^{-2} mol/l; [IB] = 1.15 mol/l. Total volume 20 ml)

Conditions			Results				
Number	CH_2Cl_2/Hex (v/v)	time, min	Conv. %	$\bar{M}_n$, g/mol	$\bar{M}_w/\bar{M}_n$	[N] 10^4 mol/l	W_p g
					Total sample		
1	40/60	0.5	0.5	3,900	19.3	0.9	0.0069
2	40/60	1	0.6	6,100	24.4	0.6	0.0071
3	40/60	2	0.6	6,500	24.1	0.6	0.0078
4	40/60	5	0.9	7,000	20.5	0.8	0.0113
5	40/60	10	1.3	6,900	12.2	1.15	0.0162
6	60/40	0.5	0.5	5,300	21.4	0.65	0.0068
7	60/40	1	0.6	6,200	26.3	0.6	0.0076
8	60/40	2	0.8	7,200	23.9	0.7	0.0103
9	60/40	5	1.2	9,100	11.2	0.8	0.0150
10	60/40	10	3.1	11,100	4.2	1.8	0.0397
11	80/20	0.5	1.3	4,100	13.0	2.0	0.0165
12	80/20	1	2.2	3,500	15.9	4.15	0.0290
13	80/20	2	6.7	10,600	4.7	4.1	0.0865
14	80/20	5	6.7	11,800	4.5	3.65	0.0858
15	80/20	10	23.7	13,800	2.0	11.1	0.3056
16	100/0	0.5	3.4	5,900	5.3	3.65	0.0434
17	100/0	1	7.7	7,800	4.9	6.35	0.0993
18	100/0	2	20.9	15,500	2.8	8.7	0.2696
19	100/0	5	58.9	17,800	2.2	21.3	0.7588
20	100/0	10	20.6	15,600	3.8	8.5	0.2652

Table 11. Conversion vs time and N (number of PIB moles formed) data for "H_2O"/$TiCl_4$/IB/TEA/CH_2Cl_2-Hex(60-40) at different temperatures. ([$TiCl_4$] = 4.1 10^{-2} mol/l; [IB] = 1.15 mol/l. Total volume 20 ml)

Conditions				Results				
Number	T °C	[TEA] mol/l	time, min	Conv. %	$\bar{M}_n$ g/mol	$\bar{M}_w/\bar{M}_n$	[N] 10^4 mol/l	W_p g
						Total sample		
1	− 20	5 10^{-4}	10	0.8	3,300	6.0	1.55	0.0105
2	− 40	5 10^{-4}	4	1.6	3,500	12.4	2.95	0.0209
3	− 60	5 10^{-4}	2	2.6	5,100	11.5	3.30	0.0337
4	− 80	5 10^{-4}	1	4.5	14,400	3.7	2.00	0.0577
5	− 20	3 10^{-3}	10	0.4	7,700	4.4	0.35	0.0052
6	− 40	3 10^{-3}	4	0.8	9,900	12.1	0.55	0.0107
7	− 60	3 10^{-3}	2	0.8	7,200	23.9	0.70	0.0103
8	− 80	3 10^{-3}	1	2.0	10,800	3.7	1.20	0.0259

Table 12. Conversion vs time and N (number of PSt moles formed) for the TMPCl/$TiCl_4$/St/TEA/CH_2Cl_2-Hex(60-40)/ − 20 °C system. ([TMPCl] = 3 10^{-3} mol/l; [$TiCl_4$] = 4.1 10^{-2} mol/l; [St] = 0.655 mol/l. Total volume 20 ml)

Conditions			Results				
Number	[TEA], mol/l	Time, min	Conv. %	$\bar{M}_n$ g/mol	$\bar{M}_w/\bar{M}_n$	[N] 10^3 mol/l	I_{eff}, %
					Total sample		
1	2.50 10^{-4}	0.5	17.1	4,700	4.88	3.77	125.7
2	2.50 10^{-4}	1	26.1	5,700	4.13	2.03	67.7
3	2.50 10^{-4}	2	35.9	5,200	3.96	4.68	155.8
4	2.50 10^{-4}	5	44.9	5,300	3.83	5.78	192.7
5	2.50 10^{-4}	10	67.6	4,100	3.76	11.16	372.0
6	5.00 10^{-4}	0.5	2.4	3,900	5.86	0.42	13.8
7	5.00 10^{-4}	1	6.3	3,900	5.63	1.10	36.5
8	5.00 10^{-4}	2	8.9	3,400	5.20	1.76	58.7
9	5.00 10^{-4}	5	22.1	3,900	3.80	3.83	127.5
10	5.00 10^{-4}	10	37.8	5,900	2.55	4.35	145.0
11	1.50 10^{-3}	0.5	1.4	2,700	5.40	0.335	11.2
12	1.50 10^{-3}	1	2.4	2,400	3.88	0.695	23.2
13	1.50 10^{-3}	2	2.6	2,100	3.82	0.84	28.0
14	1.50 10^{-3}	5	10.1	3,150	2.71	2.19	72.8
15	1.50 10^{-3}	10	16.6	4,300	2.22	2.62	87.2
16	3.00 10^{-3}	0.5	1.3	1,700	9.09	0.49	16.3
17	3.00 10^{-3}	1	1.5	1,700	6.32	0.605	20.2
18	3.00 10^{-3}	2	2.1	2,000	5.40	0.715	23.8
19	3.00 10^{-3}	5	8.4	3,000	2.16	1.88	62.7
20	3.00 10^{-3}	10	18.3	4,300	1.56	2.875	95.8
21	0.00	0.33	89.9	5,100	3.96	11.98	399.0
22	0.00	0.67	98.8	5,800	4.02	11.52	384.0
23	0.00	1	100	3,700	4.51	18.33	611.0
24	0.00	1.5	100	5,000	3.81	13.55	452.0
25	0.00	2	100	3,900	5.07	17.60	587.0

Table 13. Conversion vs time and N (number of PSt moles formed) for the TMPCl/$TiCl_4$/St/TEA/CH_2Cl_2-Hex(60-40)/ − 40 °C system. ([TMPCl] = 3 10^{-3} mol/l; [$TiCl_4$] = 4.1 10^{-2} mol/l; [St] = 0.655 mol/l. Total volume 20 ml)

Conditions			Results				
Number	[TEA], mol/l	Time, min	Conv. %	$\bar{M}_n$, g/mol	$\bar{M}_w/\bar{M}_n$	[N] 10^3 g/mol	I_{eff}, %
					Total sample		
1	5.00 10^{-4}	0.5	7.2	6,400	4.51	0.765	25.5
2	5.00 10^{-4}	1	13.2	4,900	5.25	1.83	61.0
3	5.00 10^{-4}	2	23.7	7,000	4.21	2.31	77.0
4	5.00 10^{-4}	5	33.2	7,000	3.23	3.24	108.2
5	5.00 10^{-4}	10	62.9	8,500	2.66	5.03	167.5
6	1.50 10^{-3}	0.5	4.0	3,900	4.08	0.695	23.2
7	1.50 10^{-3}	1	5.7	3,500	3.85	1.11	37.1
8	1.50 10^{-3}	2	12.3	3,100	2.93	2.68	89.4
9	1.50 10^{-3}	5	22.5	4,800	2.30	3.18	106.2
10	3.00 10^{-3}	0.5	3.4	3,500	2.09	0.67	22.2
11	3.00 10^{-3}	2	7.6	3,300	2.04	1.57	52.3
12	3.00 10^{-3}	2.5	14.2	3,900	1.87	2.48	82.7
13	3.00 10^{-3}	5	24.1	5,600	1.71	2.93	97.6
14	3.00 10^{-3}	10	35.9	6,600	1.57	3.71	123.7
15	6.00 10^{-3}	0.5	2.4	4,000	1.86	0.42	14.0
16	6.00 10^{-3}	1	4.4	3,350	2.05	0.90	30.0
17	6.00 10^{-3}	2	8.3	4,200	1.66	1.35	45.1
18	6.00 10^{-3}	5	24.5	6,000	1.50	2.78	92.6
19	6.00 10^{-3}	10	28.4	6,400	1.46	3.04	101.4

Table 14. Conversion vs time and N (number of PSt moles formed) for the TMPCl/$TiCl_4$/St/TEA/CH_2Cl_2-Hex(60-40)/ − 60 °C system. ([TMPCl] = 3 10^{-3} mol/l; [$TiCl_4$] = 4.1 10^{-2} mol/l; [St] = 0.655 mol/l. Total volume 20 ml)

Conditions			Results				
Number	[TEA], mol/l	Time, min	Conv. %	M_n, g/mol	M_w/M_n	[N] 10^3 g/mol	I_{eff}, %
					Total sample		
1	5.00 10^{-4}	0.5	15.6	7,100	3.45	1.49	49.6
2	5.00 10^{-4}	1	33.8	8,700	3.70	2.66	88.6
3	5.00 10^{-4}	2	64.3	7,800	3.86	4.12	137.3
4	5.00 10^{-4}	5	75.8	10,100	2.64	5.14	171.2
5	5.00 10^{-4}	10	88.9	11,400	2.51	5.31	177.0
6	1.50 10^{-3}	0.5	12.2	6,200	2.16	1.33	44.3
7	1.50 10^{-3}	1	18.8	6,500	2.25	1.96	65.3
8	1.50 10^{-3}	2	31.4	7,500	2.08	2.85	95.1
9	1.50 10^{-3}	5	59.2	10,300	1.81	3.92	130.6
10	1.50 10^{-3}	10	80.9	12,500	1.68	4.41	147.1
11	3.00 10^{-3}	0.5	7.4	6,100	1.76	0.83	27.6
12	3.00 10^{-3}	1	11.4	5,000	1.96	1.55	51.5
13	3.00 10^{-3}	2	25.6	7,350	1.81	2.37	79.0
14	3.00 10^{-3}	5	39.9	8,800	1.74	3.10	103.2
15	3.00 10^{-3}	10	65.7	12,900	1.58	3.48	115.9
16	6.00 10^{-3}	0.5	5.7	4,900	1.73	0.79	26.3
17	6.00 10^{-3}	1	9.2	5,600	1.66	1.12	37.2
18	6.00 10^{-3}	2	18.1	6,100	1.69	2.01	67.0
19	6.00 10^{-3}	5	36.9	8,700	1.66	2.90	96.7
20	6.00 10^{-3}	10	59.8	11,500	1.68	3.55	118.2

Table 15. Conversion vs time and N (number of PSt moles formed) for the $TMPCl/TiCl_4/St/TEA/CH_2Cl_2$-Hex(60-40)/ − 80 °C system. ([TMPCl] = $3\ 10^{-3}$ mol/l; $[TiCl_4] = 4.1\ 10^{-2}$ mol/l; [St] = 0.655 mol/l. Total volume 20 ml)

Conditions		Results					
Number	[TEA], mol/l	Time, min	Conv. %	$\bar{M}_n$, g/mol	$\bar{M}_w/\bar{M}_n$	[N] 10^3 g/mol	I_{eff}, %
					Total sample		
1	$5.0\ 10^{-4}$	0.33	21.5	6,900	4.63	2.12	70.7
2	$5.0\ 10^{-4}$	0.67	40.8	9,400	3.74	2.97	99.0
3	$5.0\ 10^{-4}$	1	52.4	9,000	3.47	3.98	132.8
4	$5.0\ 10^{-4}$	2	66.5	9,500	3.09	4.76	158.5
5	$5.0\ 10^{-4}$	4	93.7	11,300	2.58	5.63	187.7
6	$1.5\ 10^{-3}$	0.33	19.2	6,600	2.39	1.99	66.2
7	$1.5\ 10^{-3}$	0.67	22.7	5,700	2.68	2.72	90.6
8	$1.5\ 10^{-3}$	1	34.4	8,100	2.11	2.90	96.7
9	$1.5\ 10^{-3}$	2	51.8	10,600	2.03	3.34	111.3
10	$1.5\ 10^{-3}$	4	75.0	12,700	1.84	4.02	133.9
11	$3.0\ 10^{-3}$	0.33	14.8	8,100	2.11	1.25	41.7
12	$3.0\ 10^{-3}$	0.67	26.2	9,500	1.86	1.88	62.6
13	$3.0\ 10^{-3}$	1	28.8	7,700	2.06	2.56	85.3
14	$3.0\ 10^{-3}$	2	39.9	9,700	1.91	2.81	93.7
15	$3.0\ 10^{-3}$	4	63.1	12,850	1.77	3.35	111.5
16	$6.0\ 10^{-3}$	0.33	13.0	7,600	1.89	1.17	39.0
17	$6.0\ 10^{-3}$	0.67	28.4	9,600	1.96	2.02	67.2
18	$6.0\ 10^{-3}$	1	29.1	10,100	1.88	1.97	65.7
19	$6.0\ 10^{-3}$	2	44.2	11,400	1.86	2.65	88.2
20	$6.0\ 10^{-3}$	4	63.7	13,400	1.75	3.23	107.7
21	0.0	0.33	58.1	10,200	6.80	3.89	129.5

Table 16. Conversion vs time and N (number of PSt moles formed) for the TMPCl/$TiCl_4$/St/TEA/ − 80 °C system. ([TMPCl] = 3 10^{-3} mol/l; [$TiCl_4$] = 4.1 10^{-2} mol/l; [St] = 0.655 mol/l; [TEA] = 3 10^{-3} mol/l. Total volume 20 ml)

Conditions			Results				
Number	CH_2Cl_2/Hex (v/v)	Time, min	Conv. %	$\bar{M}_n$, g/mol	$\bar{M}_w/\bar{M}_n$	[N] 10^3 mol/l	I_{eff}, %
					Total sample		
1	40/60	0.33	6.8	6,700	2.05	0.70	23.3
2	40/60	0.67	9.8	7,000	2.18	0.96	32.1
3	40/60	1	19.0	8,600	2.12	1.51	50.3
4	40/60	2	28.2	8,900	2.19	2.17	72.2
5	40/60	4	58.6	13,700	2.07	2.92	97.4
6	60/40	0.33	14.8	8,100	2.11	1.25	41.8
7	60/40	0.67	26.2	9,500	1.86	1.88	62.6
8	60/40	1	28.8	7,700	2.06	2.56	85.3
9	60/40	2	39.9	9,700	1.91	2.81	93.7
10	60/40	4	63.1	12,850	1.77	3.35	111.6
11	80/20	0.33	37.0	12,100	1.97	2.08	69.4
12	80/20	0.67	54.1	13,300	1.92	2.78	92.6
13	80/20	1	74.5	12,100	2.11	4.19	139.6
14	80/20	2	92.0	14,700	2.00	4.27	142.2
15	80/20	4	100.0	15,300	1.81	4.48	149.2
16	100/0	0.33	79.3	14,400	2.18	3.76	125.4
17	100/0	0.67	94.8	12,100	2.76	5.36	178.7
18	100/0	1	100.0	14,600	2.64	4.78	159.4
19	100/0	2	100.0	12,100	2.36	5.88	195.8
20	100/0	4	100.0	13,200	2.54	5.32	177.3

Table 17. Conversion vs time and N (number of PSt moles formed) for the TMPCl/$TiCl_4$/TEA/TEA/CH_2Cl_2-Hex(60-40)/ − 80 °C system. ([TMPCl] = 3 10^{-3} mol/l; [St] = 0.655 mol/l; [TEA] = 3 10^{-3} mol/l. Total volume 20 ml)

Conditions		Results					
Number	[$TiCl_4$], mol/l	Time, min	Conv. %	$\bar{M}_n$, g/mol	$\bar{M}_w/\bar{M}_n$	[N] 10^3 mol/l	I_{eff}, %
					Total sample		
1	1.37 10^{-2}	0.33	4.1	10,000	2.18	0.28	9.3
2	1.37 10^{-2}	0.67	8.9	12,300	2.15	0.50	16.5
3	1.37 10^{-2}	1	14.7	13,750	2.07	0.73	24.2
4	1.37 10^{-2}	2	16.6	16,300	2.08	0.70	23.2
5	1.37 10^{-2}	4	37.3	16,100	2.03	1.58	52.5
6	2.73 10^{-2}	0.33	8.3	7,900	1.98	0.72	24.1
7	2.73 10^{-2}	0.67	21.5	10,500	1.97	1.40	46.5
8	2.73 10^{-2}	1	32.4	12,400	2.02	1.78	59.3
9	2.73 10^{-2}	2	46.4	13,300	2.00	2.38	79.4
10	2.73 10^{-2}	4	64.9	12,000	2.11	3.69	123.0
11	4.10 10^{-2}	0.33	14.8	8,100	2.11	1.25	41.8
12	4.10 10^{-2}	0.67	26.2	9,500	1.86	1.88	62.6
13	4.10 10^{-2}	1	28.8	7,700	2.06	2.56	85.3
14	4.10 10^{-2}	2	39.9	9,700	1.91	2.81	93.7
15	4.10 10^{-2}	4	63.1	12,850	1.77	3.35	111.6
16	5.46 10^{-2}	0.33	24.6	8,900	1.98	1.88	62.6
17	5.46 10^{-2}	0.67	34.8	8,700	1.94	2.72	90.6
18	5.46 10^{-2}	1	48.3	10,400	1.88	3.18	105.9
19	5.46 10^{-2}	2	61.6	12,900	1.77	3.27	109.0
20	5.46 10^{-2}	4	90.0	14,200	1.70	4.32	144.1
21	6.82 10^{-2}	0.33	29.0	7,900	2.16	2.50	83.4
22	6.82 10^{-2}	0.67	42.7	10,100	1.97	2.87	95.6
23	6.82 10^{-2}	1	50.4	11,000	1.80	3.12	103.9
24	6.82 10^{-2}	2	73.1	13,900	1.68	3.60	119.8
25	6.82 10^{-2}	4	92.5	15,050	1.71	4.19	139.7

Table 18. Conversion vs time and N (number of PSt moles formed) for the TMPCl/$TiCl_4$/St/TEA/CH_2Cl_2-Hex(60-40)/ − 60 °C system. Time of aging; 4 min. ([TMPCl] = 3 10^{-3} mol/l; [$TiCl_4$] = 5.46 10^{-2} mol/l; [St] = 0.655 mol/l; Total volume 20 ml)

Conditions		Results					
Number	[TEA], mol/l	Time, min	Conv. %	$\bar{M}_n$, g/mol	$\bar{M}_w/\bar{M}_n$	[N] 10^3 mol/l	I_{eff}, %
					Total sample		
1	3.00 10^{-4}	0.5	15.1	5,200	2.02	1.96	65.5
2	3.00 10^{-4}	1	22.0	6,400	1.85	2.36	78.6
3	3.00 10^{-4}	2	37.2	8,300	1.67	3.04	101.5
4	3.00 10^{-4}	5	64.0	13,000	1.54	3.36	112.2
5	3.00 10^{-4}	10	86.7	15,800	1.50	3.74	124.5

Table 19. Conversion vs time and N (number of PSt moles formed) for the TMPCl/$TiCl_4$/St/TEA/CH_2Cl_2-Hex(60-40)/ − 40 °C system. ([TMPCl] = 3 10^{-3} mol/l; [$TiCl_4$] = 4.1 10^{-2} mol/l; [St] = 0.873 mol/l; Total volume 20 ml)

Conditions			Results				
Number	[TEA], mol/l	Time, min	Conv. %	$\bar{M}_n$, g/mol	$\bar{M}_w/\bar{M}_n$	[N] 10^3 mol/l	I_{eff}, %
				Total sample			
1	9.0 10^{-3}	3.0	12.4	4,400	1.75	2.54	84.5
2	9.0 10^{-3}	5.5	25.8	6,700	1.73	3.52	117.3
3	9.0 10^{-3}	10.0	36.5	9,300	1.53	3.55	118.3
4	9.0 10^{-3}	21.0	51.3	12,500	1.50	3.74	124.6
5	9.0 10^{-3}	40.0	87.6	22,000	1.36	3.62	120.4
6	1.2 10^{-2}	2.0	7.3	3,400	1.83	1.96	65.4
7	1.2 10^{-2}	5.0	19.3	3,100	1.74	5.60	186.6
8	1.2 10^{-2}	10.0	28.1	7,700	1.57	3.33	110.9
9	1.2 10^{-2}	20.75	48.7	13,000	1.45	3.40	113.4
10	1.2 10^{-2}	40.0	79.9	20,400	1.35	3.56	118.7

Table 20. Conversion vs time and N (number of PSt moles formed) for the "H_2O"/$TiCl_4$/St/TEA/CH_2Cl_2-Hex(60-40)/ − 60 °C system. ([$TiCl_4$] = 4.1 10^{-2} mol/l; [St] = 0.655 mol/l. Total volume 20 ml)

Conditions			Results				
Number	[TEA], mol/l	Time, min	Conv. %	$\bar{M}_n$ g/mol	$\bar{M}_w/\bar{M}_n$	[N] 10^3 mol/l	W_p, g
					Total sample		
1	5.0 10^{-4}	0.5	3.8	2,400	3.97	1.09	0.0515
2	5.0 10^{-4}	1	8.8	5,000	2.78	1.19	0.1197
3	5.0 10^{-4}	2	15.0	7,450	2.37	1.375	0.2049
4	5.0 10^{-4}	5	36.4	13,900	1.86	1.79	0.4964
5	5.0 10^{-4}	10	58.0	18,000	1.63	2.19	0.7903
6	1.5 10^{-3}	0.5	2.5	1,300	5.03	1.30	0.0346
7	1.5 10^{-3}	1	4.4	2,000	4.95	1.48	0.0603
8	1.5 10^{-3}	2	9.1	3,600	4.45	1.71	0.1236
9	1.5 10^{-3}	5	21.2	7,750	3.63	1.87	0.2896
10	1.5 10^{-3}	10	26.0	11,200	3.14	1.585	0.3551
11	3.0 10^{-3}	0.5	0.6	800	7.23	0.48	0.0079
12	3.0 10^{-3}	1	0.8	700	8.52	0.78	0.0104
13	3.0 10^{-3}	2	1.1	1,200	9.75	0.615	0.0147
14	3.0 10^{-3}	5	3.9	2,300	10.44	1.185	0.0535
15	3.0 10^{-3}	10	3.2	1,700	17.50	1.265	0.0441
16	6.0 10^{-3}	0.5	0.5	700	5.99	0.49	0.0068
17	6.0 10^{-3}	1	0.6	800	6.33	0.53	0.0085
18	6.0 10^{-3}	2	0.8	1,500	6.21	0.385	0.0113
19	6.0 10^{-3}	5	2.8	1,400	10.54	1.375	0.0388
20	6.0 10^{-3}	10	4.1	2,200	13.57	1.275	0.0563
21	0.0	2.67	72.0	11,900	2.06	4.135	0.9816

Table 21. Conversion vs time and N (number of PSt moles formed) for the "H_2O"/$TiCl_4$/St/TEA/ − 60 °C system. ([TEA] = 3 10^{-3} mol/l; [$TiCl_4$] = 4.1 10^{-2} mol/l; [St] = 0.655 mol/l. Total volume 20 ml)

Conditions			Results				
Number	CH_2Cl_2/Hex (v/v)	Time, min	Conv. %	$\bar{M}_n$, g/mol/l	$\bar{M}_w/\bar{M}_n$	[N] 10^3 mol/l	W_p, g
					Total sample		
1	40/60	0.5	0.5	800	6.54	0.375	0.0062
2	40/60	1	0.5	900	4.53	0.375	0.0068
3	40/60	2	0.8	1,200	4.54	0.445	0.0107
4	40/60	5	1.3	1,200	6.54	0.690	0.0171
5	40/60	10	2.0	1,600	10.64	0.860	0.0269
6	60/40	0.5	0.6	.800	7.23	0.480	0.0079
7	60/40	1	0.8	700	8.52	0.780	0.0104
8	60/40	2	1.1	1,200	9.75	0.615	0.0147
9	60/40	5	3.9	2,300	10.44	1.185	0.0535
10	60/40	10	3.2	1,700	17.50	1.265	0.0441
11	80/20	0.5	5.5	1,600	7.32	2.265	0.0744
12	80/20	1	1.5	1,300	7.77	0.780	0.0203
13	80/20	2	8.7	2,800	6.03	2.115	0.1188
14	80/20	5	22.9	7,400	4.47	2.105	0.3126
15	80/20	10	14.0	7,000	9.97	1.370	0.1911
16	100/0	0.5	5.9	2,500	12.54	1.590	0.0796
17	100/0	1	31.5	6,200	4.92	3.470	0.4291
18	100/0	2	41.5	15,500	8.65	1.820	0.5654
19	100/0	5	45.2	13,300	6.59	2.325	0.6166
20	100/0	10	86.0	10,700	2.57	5.485	1.1731

Table 22. Conversion vs time and N (number of PSt moles formed) data for the "H_2O"/$TiCl_4$/St/TEA/CH_2Cl_2-Hex(60-40) at different temperatures. ([$TiCl_4$] = 4.1 10^{-2} mol/l; [St] = 0.655 mol/l. Total volume 20 ml)

Conditions				Results				
Number	T °C	[TEA] mol/l	Time, min	Conv. %	$\bar{M}_n$, g/mol	$\bar{M}_w/\bar{M}_n$	[N] 10^4 mol/l	W_p, g
						Total sample		
1	− 20	5 10^{-4}	2	5.3	2,100	5.95	1.73	0.0725
2	− 40	5 10^{-4}	2	4.1	2,600	5.37	1.07	0.0561
3	− 60	5 10^{-4}	1	8.8	5,000	2.78	1.19	0.1197
4	− 80	5 10^{-4}	1	12.4	14,300	3.46	0.595	0.1694
5	− 20	3 10^{-3}	2	1.5	1,200	3.67	0.88	0.0203
6	− 40	3 10^{-3}	2	0.9	1,200	5.74	0.495	0.0123
7	− 60	3 10^{-3}	1	0.8	700	8.52	0.78	0.0104
8	− 80	3 10^{-3}	1	3.0	7,050	4.61	0.29	0.0408

Acknowledgements. The authors wish to thank the National Science Foundation (Grant 89-20826) for financial support and Dr. H.-C. Wang (Exxon Chemicals Co.) for valuable discussions and helpful comments. A generous gift by the Exxon Corp. is also gratefully acknowledged.

7 References

1. Kennedy JP, Ivàn B (1991) Designed polymers by carbocationic macromolecular engineering: theory and practice. Hanser, Munich
2. Vairon JP, Rives A, Bunel C (1992) Makromol Chem, Macromol Symp 60: 97
3. Cho CG, Feit BA, Webster OW (1990) Macromolecules 23: 1918
4. Balogh L, Faust R (1992) Polym Bull 28: 367
5. Kennedy JP, Midha S, Keszler B (1993) Macromolecules 26: 424
6. Puskàs JE, Kaszàs G, Litt M (1991) Macromolecules 24: 5278
7. Pratap G, Heller JP (1992) J Polym Sci Part A: Polym Chem 30: 163
8. Kennedy JP (1991) Makromol Chem, Macromol Symp 47: 55
9. Kennedy JP (1992) Makromol Chem, Macromol Symp 60: 1
10. Györ M, Balogh L, Wang H-C, Faust R (1992) Polymer Preprints 33(1): 158
11. Plesch PH (1992) Makromol Chem, Macromol Symp 60: 11
12. Kennedy JP, Hayashi A (1991) J Macromol Sci-Chem A28(2): 197
13. Lubnin AV, Kennedy JP (1992) Polym Bull 29: 9
14. Lubnin AV, Kennedy JP (1992) Polym Bull 29: 247
15. Pernecker T, Kelen T, Kennedy JP: publication in preparation
16. Deàk Gy, Zsuga M, Kelen T (1992) Polym Bull 29: 239
17. Faust R, Ivan B, Kennedy JP (1991) J Macromol Sci-Chem A28(1): 1
18. Kennedy JP, Kelen T, Tüdös F (1982–83) J Macromol Sci-Chem A18: 1189
19. Szwarc M (1992) Makromol Chem, Rapid Common 13: 141
20. Sigwalt P (1991) Makromol Chem, Macromol Symp 47: 179
21. Penczek S, Kubisa P, Szymanski R (1991) Makromol Chem, Rapid Comm 12: 77
22. Kamigaito M, Yamacka K, Sawamoto M, Higashimura T (1992) Macromolecules 25: 6400
23. Ivàn B (1993) Makromol Chem, Macromol Symp 67: 311
24. Ivàn B, Kennedy JP (1990) Macromolecules 23: 2880
25. Plesch PH (1988) Polym Bull 19: 145
26. Sawamoto M (1991) Prog Polym Sci 16: 111
27. Higashimura T, Sawamoto M (1984) Advances in Polymer Sci 62: 49
28. Ohtori T, Hirokawa Y, Higashimura T (1979) Polymer J 11(6): 471
29. Zsuga M, Kennedy JP, Kelen T (1989) J Macromol Sci-Chem A26: 1305
30. Ivàn B, Zsuga M, Gruber F, Kennedy JP (1988) Polymer Preprints 29(2): 33
31. Matyjaszewski K, Chih-Kwa Lin (1991) Makromol Chem, Macromol Symp 47: 221
32. Penczek S (1992) Makromol Chem, Rapid Commun 13: 147
33. Szwarc M, Zimm BH (1983) Macromolecules 16: 1918
34. Kennedy JP (1992) Polym Preprints 33(1): 150
35. Gandini A, Martinez A (1988) Makromol Chem, Macromol Symp 13/14: 211
36. Kaszàs G, Puskàs J, Chen CC, Kennedy JP (1988) Polym Bull 20: 413
37. Györ M, Wang H-C, Faust R (1992) J Macromol Sci-Pure Appl Chem A29(8): 639
38. Pernecker T, Kennedy JP (1991) Polymer Bulletin 26: 305
39. Nuyken O, Kroner H (1990) Macromol Chem 191: 1
40. Pernecker T, Kennedy JP (1992) Polym Bull 29: 15
41. Sawamoto M, Higashimura T (1991) 10th International Symp. Balatonfured, Hungary
42. Kaszàs G, Puskàs J, Kennedy JP (1987) Polym Bull 18: 123
43. Kennedy JP, Marechal E (1982) Carbocationic polymerization. Wiley-Interscience, New York, 1982
44. Kennedy JP, Smith RA (1980) J Polym Sci, Polym Chem Ed 18: 1523
45. a. Szwarc M, Van Beylen M, Van Hoyweghen D (1987) Macromolecules 20: 445
    b. Szwarc M (1990) Macromolecules 23: 4616

46. Nuyken O, Pask SD, Vischer A (1983) Makromol Chem 184: 553
47. Nuyken O, Pask SD, Vischer A, Walter M (1985) Makromol Chem 186: 173
48. Faust R, Fehèrvàri A, Kennedy JP (1982–83) J Macromol Sci-Chem A18: 1209
49. Puskàs J, Kaszàs G, Kennedy JP, Kelen T, Tüdös F (1982–83) J Macromol Sci Chem A18: 2229
50. Kennedy JP, Thomas RM (1960) J Polym Sci 46: 233
51. Leleu FM, Tardi M, Polton A, Sigwalt P (1991) Makromol Chem, Macromol Symp 47: 253
52. Faust R, Kennedy JP (1988) Polym Bull 19: 21
53. Faust R, Kennedy JP (1988) Polym Bull 19: 29
54. Faust R, Kennedy JP (1988) Polym Bull 19: 35
55. Faust R, Kennedy JP (1987) J Polym Sci Part A, Polymer Chemistry, 25: 1847
56. Ivàn B, Kennedy JP (1990) J Polym Sci Part A: Chem Ed 28: 89
57. Heroguez V, Deffieux A, Fontanille M (1990) Makromol Chem, Macromol Symp 32: 199
58. Litt M (1962) J Polym Sci 58: 429
59. Szwarc M (1968) Carbanious, living polymers and electron transfer processes. Wiley-Interscience, New York, 1968
60. Kennedy JP, Chen FJ-Y (1986) Polymer Bulletin 15: 201
61. Tsunogae Y, Majoros I, Kennedy JP (1993) J Macromol Sci-Chem A30: 253
62. Kaszàs G, Puskàs J, Kennedy JP (1988) Makromol Chem, Macromol Symp 13/14: 473
63. Bui L, Nguyen HA, Marechal E (1987) Polym Bull 17: 157
64. Higashimura T, Hira M, Hasegawa H (1979) Macromolecules 12: 217
65. Zsuga M, Kelen T (1986) Polym Bull 16: 285
66. Thomas L, Polton A, Tardi M, Sigwalt P (1992) Macromolecules 25: 5886
67. Matyjaszewski K (1993) J Polym Sci, Polym Chem Ed 31: 995

Heterophase Polymerizations: A Physical and Kinetic Comparison and Categorization

D. Hunkeler[1], F. Candau[2], C. Pichot[3], A.E. Hemielec[4], T.Y. Xie[5], J. Barton[6], V. Vaskova[6], J. Guillot[7], M.V. Dimonie[8], K.H. Reichert[9]

A set of physical and chemical criteria have been identified which provide a basis for the categorization of heterophase free radical polymerization processes, and the development of a systematic nomenclature. These include three quantitative transitional thresholds: a surface tension driving force, the type of stability (kinetic, thermodynamic), and the level of surfactant with respect to the critical micelle concentration. The nucleation mechanism and the average number of macroradicals within the polymer particles provide a secondary criterion for process distinction. Four mutually exclusive domains have been identified: I. Macroemulsion Polymerization, consisting of subdomains where Suspension (Ia) and Emulsion (Ib) behavior dominate, II. Inverse-Macroemulsion Polymerization, a water-in-oil analogy to the preceding encompassing Inverse-Suspension (IIa) and Inverse-Emulsion (IIb), and the thermodynamically stable Microemulsion (III) and Inverse-Microemulsion (IV) polymerizations. Within each domain a unique set of physical and chemical phenomena dominate. A reaction mechanism or model of colloidal behavior must, therefore, be specific for each regime. This is particularly important for water-in-oil polymerizations due to the unique chemistry of various organic phases and interfacial recipes. It is recommended that further development proceed independently for Inverse-Suspension, Inverse-Emulsion and Inverse-Microemulsion polymerizations, as has been the precedent for the more extensively researched oil-in-water polymerizations. A set of criteria is also developed to define and distinguish physically or kinetically identical "emulsion" and "suspension" polymerization processes.

[1] Department of Chemical Engineering, Vanderbilt University, Nashville, TN, 37235 USA
[2] Institut Charies Sadron (CRM-EAHP), Strasbourg, 67083 Cedex, France
[3] Unité-Mixte, CNRS-Biomérieux, ENSL-Lyon, 69364 Cedex, France
[4] Institute for Polymer Production Technology, McMaster University, Hamilton, Ontario, Canada
[5] Department of Chemical Engineering, Queen's University, Kingston, Ontario, Canada
[6] Polymer Institute, Slovak Academy of Sciences, Bratislava, Slovakia
[7] Laboratoire de Chemie et Procedes de Polymerization (CNRS), Vernaison, 69390 France
[8] Department of Macromolecular Chemistry, Central Institute of Chemistry, Bucharest, Romania
[9] Institüt fur Technische Chemie, Technische Universität Berlin, Berlin, Germany

Advances in Polymer Science, Vol. 112

1 Introduction

Polymers synthesized through free radical techniques are inherently heterogeneous due to the stochastic nature of chain addition and termination reactions. At the molecular level heterogeneity is represented by the stereochemical configuration, chain composition, chain architecture, monomer sequence length distribution, molecular weight and branching distributions, and intra- and intermolecular interactions. As the length scale of interest expands these microscopic differences are obscured by physical inhomogeneities, such as changes in state and differences in solubility. It is these macroscopic or bulk properties which shall be considered in this paper. Specifically, the chemical and colloidal characteristics of various heterophase polymerizations will be compared with the objective of identifying phenomena which can be used as unambiguous criteria to distinguish processes and establish a nomenclature.

This paper focuses on heterophase free radical polymerizations. It is limited to processes where multiple phases, distinguished by the insolubility of reagents, exists at the onset of the reaction. It therefore does not consider "precipitation" polymerization [1], which occurs when the polymer is insoluble in the monomer and precipitates out from an initially homogeneous solution. It also does not address emulsifier-free polymerization or "dispersion" polymerization. This rather general nomenclature is now accepted as applying to specific systems where the heterophase nature is produced at the onset of the reaction by homogeneous nucleation of oligomers or polymer chains which have exceeded their solubility limit [2].

2 Classification of Heterophase Polymerizations

2.1 Heterophase Polymerization Domains and Subdomains

Two primary criteria exist for categorizing heterophase processes: a surface tension driving force, to delineate the emulsion structure, and a stability threshold. The former is the difference (Δ) in surface tension (γ) between the aqueous phase and the hydrophilic moiety of the emulsifier (γ_{A-H}) and the organic phase-lipophile (γ_{O-L}):

$$\Delta\gamma = \gamma_{A-H} - \gamma_{O-L} \tag{1}$$

If the driving force ($\Delta\gamma$) is less than zero an oil-in-water dispersion forms, while if $\Delta\gamma$ is of the opposite sign an inverse (water-in-oil) dispersion is produced. The stability threshold represents a critical emulsifier concentration below which a kinetically stable "macroemulsion" is produced. These can be transformed,

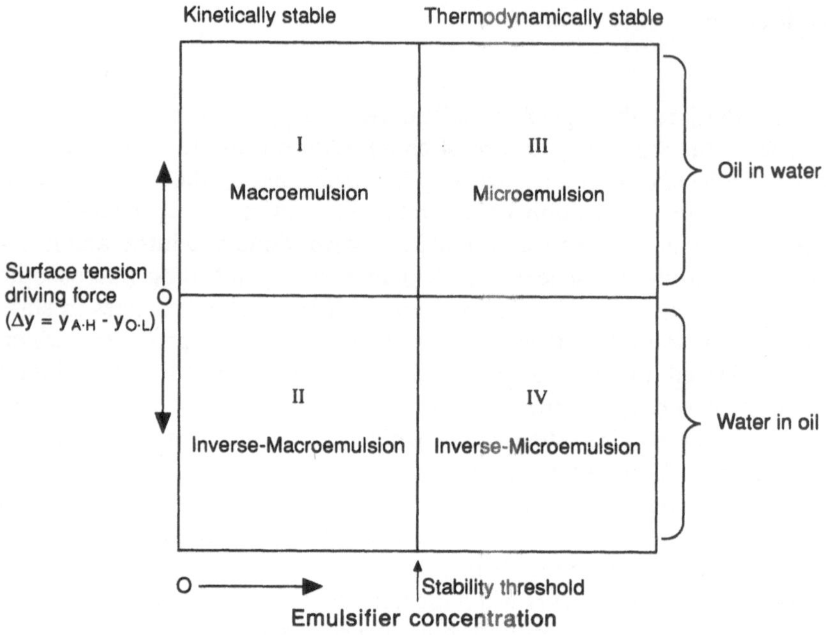

Fig. 1. The four heterophase polymerization domains are shown as a function of the emulsifier concentration and the surface tension driving force

under the appropriate conditions, to thermodynamically stable[1] "microemulsions" with the further addition of emulsifier above the critical concentration. This threshold is dependent on the composition of the aqueous and organic phases, with most microemulsion formulations requiring a cosurfactant to produce a monophasic system at reasonable emulsifier levels[2]. Four heterophase polymerization domains have therefore been identified: I. Macroemulsion, II. Inverse-Macroemulsion, III. Microemulsion, and IV. Inverse-Microemulsion. These are shown in Fig. 1. For domain I the prefix 'macro' has historically been omitted in reference to "Emulsion Polymerizations". The nomenclature inverse-emulsion was suggested by Vanderhoff in 1962 for domain II in his pioneering work on heterophase water-in-oil polymerizations [4]. The excellent example of process designation is particularly impressive in view of the uncertainty in the colloidal nature of such water-in-oil dispersions prevalent at that time. Specifically, the existence of inverse-micelles was postulated but not yet detected, although they have subsequently been identified by

[1] Thermodynamically stable systems form spontaneously, have negative free energies and will not coalesce over indefinite periods. Kinetic stability implies a colloidal system which can be stabilized by the application of external force, such as agitation. The free energy is greater than zero and while they can be stable for periods of years they will eventually coalesce. Therefore, for large time frames, a kinetically stable system can be considered metastable

[2] The emulsifier level can be further reduced by matching the properties of the hydrophobic moiety of the emulsifier (molar volume, solubility parameters) with those of the organic phase, and those of the hydrophilic head with the aqueous phase [3]

Bartelt [5]. Therefore, due to historic precedent the description of water-in-oil polymerizations proceeds through an analogy, either colloidal or kinetic, to a process with a continuous aqueous phase. The prefix 'inverse' is generally accepted for water-in-oil emulsions in contrast to 'direct' or 'conventional' oil-in-water emulsions/microemulsions for which the prefix is implied but not often explicitly stated.

Macroemulsions and inverse-macroemulsions can be further subdivided by a micellar transition at the critical micelle concentration. At surfactant levels below the CMC a suspension of a dispersed monomer phase in an insoluble continuous medium exists with nucleation proceeding predominantly in the monomer droplets [6]. In such cases each particle behaves as an isolated microbatch polymerization reactor containing all the reagents (initiator and monomer) plus the polymer product [7]. The kinetics resemble bulk polymerization [8] with $\bar{n}$ on the order of 10^{2-6} and the continuous phase serves primarily to reduce the viscosity and dissipate heat. The nomenclature "Suspension Polymerization"[3] was originally employed [6] and continues to be used today. By contrast, at emulsifier levels above the micellar threshold, micelles or inverse-micelles are produced which may have a significant role in nucleation [10]. These "emulsion" polymerizations are distinguished by nucleation proceeding outside the monomer droplet, and an average number of radicals per particle on the order of one [11]. The low $\bar{n}$ is a consequence of the small size of the polymer particles. Therefore, when as few as two radicals coexist in a particle, very large macroradical concentrations are generated. This leads to extremely high bimolecular termination rates which greatly exceed the radical entry rate and limit the concentration to 0 or 1 radicals per particle. A secondary factor for the low $\bar{n}$ can be the desorption of radicals from particles [12]. The macroemulsion domain (I) can therefore be divided into suspension (Ia) and emulsion (Ib) subdomains. By analogy, the inverse-macroemulsion domain (II) may be sectioned into inverse-suspension (IIa) and inverse-emulsion (IIb) subdomains. Figure 2 provides a complete categorization of the heterophase polymerization domains discussed herein.

Table 1 compares the Suspension, Emulsion and Microemulsion regimes. The level of surfactant with respect to the CMC and the stability threshold, the locus of nucleation, and the particle size clearly distinguish these three regimes. This table also illustrates the analogies between "direct" oil-in-water and "inverse" water-in-oil polymerization processes with respect to the aforementioned attributes. To summarize: As the concentration of emulsifier increases, at a fixed temperature, monomer level, and aqueous and organic phase ratio, a transition between the suspension, emulsion and microemulsion domains ensues. This occurs with a corresponding decrease in particle size. Within the

[3] In the initial stages of the reaction a suspension polymerization is a liquid-in-liquid system and from a colloidal point of view is a classical emulsion, i.e. macroemulsion. However, the system undergoes a transition during polymerization and in the final stages becomes a solid-in-liquid "suspension", in analogy with the sol of gold [9].

Table 1. Comparison of oil-in-water and water-in-oil polymerizations

Nomenclature	Thermo-dynamic stability	Surfactant concentration relative to cmc	Existence of micelles/ inverse-micelles	Primary Locus of Nucleation		
				Monomer droplet	Micelle/ inverse-micelle	Homo-geneous
Suspension/ microsuspension	No	<	No	Yes		Yes
Inverse-suspension	No	<	No	Yes		
Emulsion	No	>	Yes	Yes	Yes	Yes
Inverse-emulsion	No	> [a]	Yes[a]	Yes[a]	No[b]	No[b]
Microemulsion	Yes	> >	Yes		Yes	
Inverse-microemulsion	Yes	> >	Yes		Yes	

[a] Inverse-emulsions can also be formed with monomer droplets very similar in size to the monomer-swollen micelles, due to the large emulsifier levels needed to stabilize these systems. Distinguishing such species may be both arbitrary and rather semantic.

[b] Not detected based on experiments to date, however in some systems such as acrylamide/toluene it is thought to occur [13].

Average number of radicals per polymer particle ($\bar{n}$)	Particle size (μm)	Influence of Agitation on Particle Diameter (d_p)	Emulsion structure O: Oil W: Water	Description
10^{2-4}	$50–10^4$	d_p is inversely related to the rate of agitation (droplets formed via a breakup coalescence mechanism)	O-in-W	Organically soluble monomer dispersed in a continuous aqueous phase. Kinetics resemble bulk polymerization. Microsuspension is a subdomain with higher surfactant concentrations, smaller particle sizes and smaller $\bar{n}$.
10^{2-4}	10^{2-4}		W-in-O	Water soluble monomer in solution dispersed in a continuous organic phase. Kinetics resemble solution polymerization. Inverse-microsuspension is a subdomain with a higher surfactant concentration, smaller particle size (10^{0-1} μm) and smaller n (10^{1-2}).
$10^{-3} \rightarrow +1$[c]	$10^{-1 \rightarrow 0}$	d_p is independent of the rate of agitation	O-in-W	Dispersion of an organically soluble monomer in a continuous aqueous phase. Kinetics are dominated by radical isolation. Miniemulsion is a subdomain employing an additive, lowering particle size.
$10^{-3} \rightarrow +1$[d]	$10^{-1 \rightarrow 0}$		W-in-O	Dispersion of a water soluble monomer in a continuous organic phase. Kinetics are dominated by radical isolation.
< 1[e]	$10^{-2,-1}$	d_p is independent of the rate of agitation	O-in-W	Thermodynamically stable, monophasic dispersion of an oil soluble monomer in a continuous aqueous phase.
< 1[f]	$10^{-2,-1}$		W-in-O	Thermodynamically stable, monophasic dispersion of a water soluble monomer, in solution, in a continuous organic phase.

[c] This is a general categorization, including examples of Smith-Ewart Case I kinetics [14–16] Case II kinetics ($n = 1/2$) and Case III kinetics [17, 18].
[d] For inverse-emulsions low n's have been reported by Vanderhoff (0.006-0.2) [19] corresponding to a Case I type Smith-Ewart scheme. Baade has also observed Case III kinetics [13].
[e] Continuous particle nucleation with a few polymer chain per particle (2–3) and some coagulation in the latter stages of the reaction [20].
[f] Each nucleated particle contains one collapsed macromolecule either growing or in its final form. However, all particles are not active at any given time and the average n is less than one.
[g] Provided coagulation is negligible and the mass transfer rate of monomer is much faster than the polymerization rate.

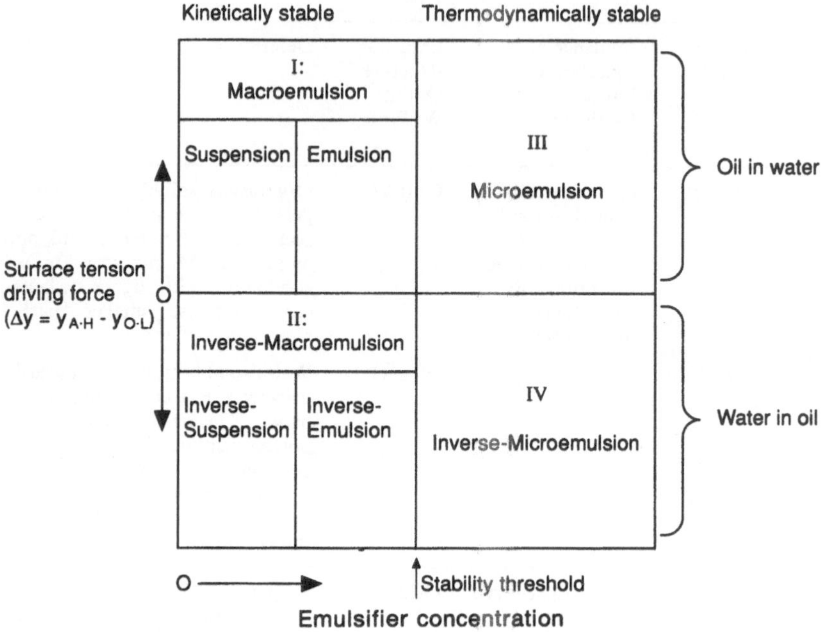

Fig. 2. A complete categorization of the heterophase polymerization domains and subdomains

macroemulsion domain the average number of radicals per particle ($\bar{n}$) decreases with a reduction in diameter. However, in the microemulsion domain each active particle contains one collapsed macromolecule and $\bar{n}$, averaged over the entire population, is less than one [21]. The transition from suspension to emulsion to microemulsion is also accompanied by a change in the locus of nucleation from the monomer droplets to either micelles or a homogeneous process, with emulsions occurring in an intermediate domain where all mechanisms can be operative. For the processes where nucleation in monomer droplets dominates, the particle diameter is invariant or very slightly dependent on the conversion level provided there is no agglomeration[4]. The size does however vary with the agitation conditions which affect the Weber number and the droplet breakup coalescence mechanism [23]. Conversely, in micellar systems the nucleation mechanism is thermodynamic in origin [10]. Therefore, particle diameters increase with the polymerization but are not significantly affected by agitation. The agitation does, however, affect the particle size distribution.

It is further evident from Table 1 that an exact analogy between direct oil-in-water emulsions and water-in-oil inverse-emulsions is not obtained. For

[4] In suspension polymerization, polymer particle size decreases with the conversion of monomer due to the conversion of a less dense monomer to a higher density polymer. Another effect on particle diameter is agglomeration which causes a dramatic increase in size after a certain conversion is obtained [22].

example, where direct emulsions are usually nucleated through micellar or homogeneous mechanisms, inverse-emulsions generally favor nucleation in monomer droplets particularly when commercial emulsifiers are used [19, 24–26]. Water-in-oil systems are also inherently less stable than emulsions which employ aqueous continuous phases. This is a consequence of the low dielectric constant of organic compounds, which generates thick electrical double layers (several microns) which are ineffective for electrostatic stabilization. This necessitates the use of steric repulsive forces. Such stabilizers often have broad molecular weight distributions, and are block copolymers, for example polyvinylalcohol based systems. Therefore, to provide enhanced kinetic stability in water-in-oil systems, emulsifier blends are often used [3] since they maximize the interfacial entropy of mixing by forming a condensed surfactant layer.

2.2 Distinguishing Emulsion and Suspension Polymerizations

It is generally agreed that for a polymerization to be considered an emulsion *two* criteria must be satisfied [27]:

1. The kinetics, as defined by the average number of macroradicals per particle ($\bar{n}$) must not be significantly larger than one. This implies that if $\bar{n}$ is low (Smith-Ewart case I or II kinetics) the system is considered an "emulsion". Similarly, if $\bar{n}$ is very large, on the order of 10^{2-6}, the polymerization is kinetically a suspension. The categorization of Smith-Ewart case III kinetics will be discussed in the following section of this paper.

2. An emulsion polymerization requires the mechanism of polymer particle nucleation to reside outside the monomer droplets. This physical-chemical process involves a series of radical reactions in the continuous phase followed by homogeneous or micellar particle formation. Either of these mechanisms require the initiator to be insoluble in the monomer phase, such as a water soluble initiator and an organically soluble monomer, or vice versa. If, in contrast, the initiator is soluble in the monomer phase, all the components of the reaction are contained in the dispersed phase and the continuous phase serves only to decrease the viscosity and dissipate heat. Such polymerizations are categorized as suspensions. The second definition, however, makes no statement as to the magnitude of $\bar{n}$ and therefore the two criteria are mutually exclusive.

The preceeding categorization is unambiguous provided we recognize that neither the initiator nor monomer are absolutely insoluble in either aqueous or organic media and partition to some extent. For example, styrene and methyl methacrylate have solubilities of 0.007 wt % and 1.6 wt % respectively in water [28], and acrylamide has a solubility of 1.6% in isoparaffinic solvents [29] and 2% in toluene [24]. Therefore, in referring to solubility we must define the phase of primary solubility.

A classical emulsion experiment involves the polymerization of styrene dispersed in water with a high HLB emulsifier such as sodium dodecyl sulfate. A predominantly water soluble species, potassium persulfate, is often employed, segregating the monomer and initiator and resulting in a nucleation mechanism which proceeds outside the monomer droplets and a low $\bar{n}$. That is, the system is an emulsion by the categorization presented above. Alternatively, if a predominantly oil soluble initiator is employed, such as azobisisobutyronitrile (AIBN), the monomer and initiator are localized in the same phase restricting the particle nucleation to the monomer droplets and rendering the polymerization a suspension.

By analogy to the proceeding two rules used to designate emulsions, we can assign the nomenclature for inverse, water-in-oil, processes. A common example is the polymerization of a nonionic monomer such as acrylamide in an aqueous solution suspended in an organic phase with a low HLB steric stabilizer. If the organic soluble initiator AIBN is employed the polymerization is an "inverse-emulsion" due to the heterophase nucleation mechanism caused by the separation of monomer and initiator. Inverse-emulsions have also been generated with homogeneous nucleation mechanisms. This occurs with water soluble monomers such as acrylamide, which have marginal solubilities in organic media [24], and can react with primary radicals in the oil phase to generate oligoradicals [30]. Contrarily, if the water soluble potassium persulfate is employed nucleation occurs in the monomer droplets, the kinetics resemble those of a solution process, and $\bar{n}$ is high, all of which characterize the polymerization as a inverse-suspension.

An ideal emulsion or suspension is difficult to physically, chemically or kinetically realize. Mixed mechanisms exist. Nonetheless categorizations can still be made such as the two styrene and two acrylamide examples discussed in the previous paragraph. These are distinguishable by the primary nucleation mechanism and the primary phase of initiator and monomer solubility. However, the possibility of having mixed emulsion and suspension mechanisms, which are largely indistinguishable, also exists. Consider two scenarios:

1. A polymerization kinetically resembles as an emulsion such that $\bar{n}$ is approximately one, with nucleation occurring in the small monomer droplets, a suspension like characteristic. An example of this phenomena is "miniemulsion" polymerization [31] which will be discussed in further detail in a later section of the paper.

2. A polymerization with $\bar{n}$ very large (10^{2-6}) indicating suspension-like (bulk polymerization) kinetic behavior, and a particle nucleation mechanism residing outside the monomer droplets, which delineates an emulsion process. Smith-Ewart case III systems are examples of this type of behavior provided they have evolved from case I and/or case II polymerization at low conversions, which is common.

Both of these polymerizations can be described as either emulsions or suspensions depending on a preference for a kinetic or nucleation based nomenclature scheme. This duality in the nomenclature is the present state of affairs in heterophase polymerizations, particularly water-in-oil systems, and it is sum-

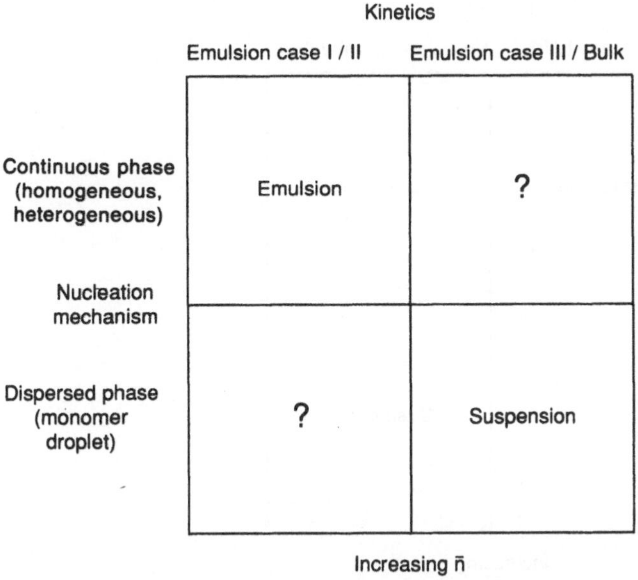

Fig. 3. Uncertainty in the nomenclature of various macroemulsion polymerization subdomains as a function of the nucleation mechanism and the average number of macroradicals per particle. The off-diagonal elements, designated with question marks, have properties of both emulsion and suspension polymerizations

marized in Fig. 3. Figure 3 is a transformation of the macroemulsion (I) and inverse-macroemulsion (II) domains of Figs. 1 and 2 with Kinetic ($\bar{n}$) and Nucleation coordinates replacing the Emulsifier Concentration and Surface Tension coordinates respectively. The two mixed emulsion and suspension systems are represented by the off diagonal elements and are designated with question marks. To attempt to reach a clear and systematic definition of emulsions and suspensions we refer to the founder of inverse heterophase polymerization, John Vanderhoff. Through his pioneering research and nomenclature Vanderhoff has always categorized systems as emulsions, or inverse-emulsions, if *either* the kinetic or nucleation mechanisms resembled that of a classical emulsion [4, 19]. That is to say an emulsion-like behavior takes precedence over a suspension-like behaviour. If we adopt this as a model then the off-diagonal elements in Fig. 3 can be considered emulsions. The same logic can be applied to inverse water-in-oil systems. If *either* the kinetics are emulsion-like ($\bar{n}$ not significantly larger than one) *or* the nucleation proceeds outside the monomer droplets then the polymerization can be referred to as an inverse-emulsion. Inverse-suspension is reserved as a nomenclature *only if both* nucleation occurs in the monomer droplets *and* $\bar{n}$ is very high. Figure 4 summarizes this discussion. Therefore miniemulsion polymerizations are categorized and named as emulsions due to the low $\bar{n}$, which takes precedence over the partial nucleation in monomer droplets. Furthermore, Smith-Ewart case III emulsions

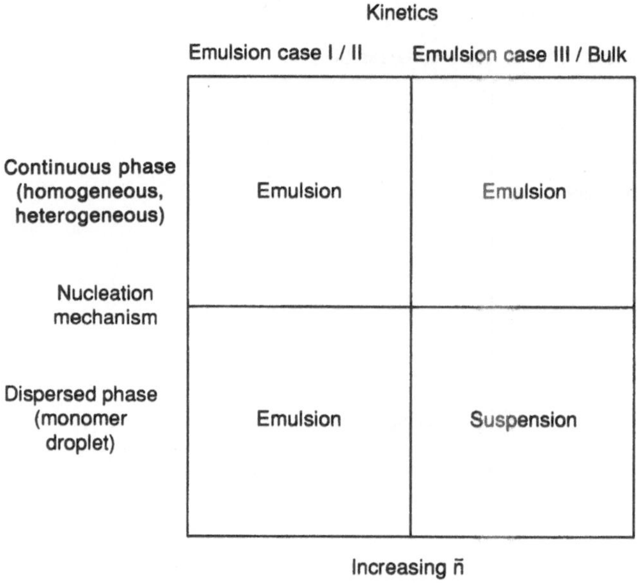

Fig. 4. Proposed nomenclature for macroemulsion polymerization subdomains

are indeed emulsion polymerizations provided the polymer particles were nucleated outside the monomer droplet, since this is an emulsion-like behavior and it again takes precedence over the suspension-like property of high values of $\bar{n}$.

Using this categorization we can also identify some existing nomenclature: Hunkeler and Hamielec's "inverse-microsuspension" refers to two system: 1) potassium persulfate/acrylamide-water/organic/low HLB stabilizer [32] which is an inverse-suspension by the proposed scheme, and 2) AIBN/acrylamide-water/organic/low HLB stabilizer [29] which is an inverse-emulsion.

2.3 Explanation of some Existing Nomenclature

Microemulsion Polymerization. Microemulsions are generally stabilized with a classical emulsifier (soap) and a weak amphiphile cosurfactant such as a short chain alcohol. The level of emulsifier is greatly augmented with respect to that in suspensions or emulsions, and the monomer concentration is usually lower. Therefore, small micelles of a uniform size are always in excess throughout the mixture. These act to distribute the monomer uniformly. Furthermore, their high total interfacial area relative to nucleated particles implies the micelles preferentially capture primary radicals generated in the continuous phase. This leads to a process of continuous particle nucleation with each particle formed in a single step. At any given conversion a particle contains either one or zero radicals [20, 33].

A growth in micellar size is always observed during the reaction due to the internal dynamics of microemulsions and inverse-microemulsions. This takes the form of either coagulation of active and inactive micelles or the diffusion of monomer from the unreacted micelles to the nucleated particles. Each final particle contains a number of macromolecules, on the order of one, in a collapsed state [33], with the particle size independent of the nature of the free radical initiator [34]. These features lead to a unique kinetic mechanism relative to the other heterophase polymerizations discussed herein [33, 35, 36]. A more detailed discussion of microemulsion and inverse-microemulsion polymerizations are given in two recent reviews [37, 38].

Miniemulsion Polymerization. The physical and chemical nature of emulsions and suspensions are sensitive to the interfacial recipe. The addition of an additive such as hexadecane or cetyl alcohol to an "emulsion" polymerization is such an example [31]. The long chain oils penetrate less of the interface than the small molecules (monomer). Consequently, the emulsifier molecules move closer together, decreasing the interfacial radius of curvature. This results in the formation of smaller monomer droplets, causing their total interfacial area to approach that of the monomer-swollen micelles and polymer particles. All three species can therefore compete for radical capture. This combination of nucleation in micelles and monomer droplets confers characteristics of both emulsion and suspension behavior to the polymerization. These have historically been referred to as "miniemulsion" polymerizations [31, 39], with the prefix designating the reduced particle size relative to emulsion polymerization. The transition between miniemulsion and emulsion is ambiguous to some extent. A useful criterion is the rate of nucleation by either homogeneous or micellar mechanisms relative to the rate of nucleation in the monomer droplets. Miniemulsion polymerization is preferred as a nomenclature if the latter is significant. Miniemulsions are categorized in the Macroemulsion subdomain (Ib) since the initiator is soluble in the continuous phase and the polymerization follows Case II and III Smith-Ewart kinetics.

Microsuspension and Inverse-microsuspension. In suspension polymerizations, particle formation occurs through a droplet breakup-coalescence mechanism, with the diameter controlled by the temperature, interfacial tension, agitation intensity and conversion. Suspension polymerizations have typically been characterized by an initiator soluble in the monomer phase and particle diameters in the 50–1000 μm range [40]. Smaller particles (0.2–20 μm) have been produced at higher agitation speeds (lower interfacial tensions) [41] and in such cases a prefix 'micro' has been added to the nomenclature (microsuspension) to reflect both the dominant synthesis conditions (suspension) and the nominal particle size (1 micron). Therefore, microsuspension polymerization has historically referred to a subdomain of suspension polymerization occurring at smaller particle sizes. Based on an analogy to this nomenclature, inverse-microsuspension polymerization has been proposed for similar water-in-oil processes [42]. In these systems particles with diameters between 0.1 and 20 μm are produced by utilizing higher emulsifier concentrations. During inverse-microsuspension polymerizations the small droplets and high interfacial areas favor radical reactions at the interface which can compete with propagation and termination occurring in the droplet. An example of such a phenomenon is unimolecular termination with emulsifier, which dominates over the bimolecular process for certain classes of emulsifiers [43, 44]. Clearly there is no distinct transition between microsuspension and suspension, and both microsuspension and inverse-microsuspension can be respectively grouped under macroemulsion (Ia) and inverse-macroemulsion (IIa) polymerization.

3 Modelling Heterophase Polymerizations

The theoretical development of direct oil-in-water polymerizations has proceeded independently for the suspension and emulsion processes. However, even within each of these broad subdomains a general set of phenomena has not been identified which can be applied to a range of monomers or emulsifier recipes[5]. Although such a general mechanism can perhaps be postulated its utility would be limited. The specific chemistry between individual reagents would lead to a huge set of elementary reactions which could not be solved in general and would require severe simplifications and extensive assumptions to be applied to a specific monomeric system. These computational difficulties are exacerbated by the superposition of physical effects such as the chain length dependence of rate constants and diffusion controlled termination. Such is the present state of affairs in emulsion polymerization that specific kinetic models are required for the polymerization of styrene [45], methyl methacrylate [17, 18], vinyl acetate [15], vinyl chloride [16] and vinylidene chloride [46]. Suspension polymerization of vinyl chloride [47], styrene [7] and methyl methacrylate [48] are further examples of similar processes which cannot be generalized.

Over the past decade there has been extensive interest in the kinetics and colloidal behavior of water-in-oil polymerizations. However, these efforts have focused on the elucidation of a general set of phenomena to describe water-in-oil processes, without distinguishing inverse-emulsion and inverse-suspension subdomains. A confounding factor is certainly the inconsistent nomenclature; inverse-suspensions (IIa) are within the inverse-macroemulsion polymerization domain (II), and are often described as inverse-emulsions, where the prefix 'macro' has been omitted for brevity. However, inverse-emulsion (IIb) is itself a

[5] The difficulty in generalizing heterophase free radical polymerizations is not surprising given the large variation in physical and chemical behaviour present in relatively uncomplicated homogeneous processes such as bulk or solution polymerizations

subdomain of inverse-macroemulsion polymerization, thereby revealing the significance of the prefix 'macro' to describe regime II. It is recommended that inverse-suspension (IIa), inverse-emulsion (IIb) and inverse-microemulsion (IV) be developed independently, as has been the useful precedence for 'direct' oil-in-water suspension (Ia), emulsion (Ib) and microemulsion (III) polymerizations.

It must further be recognized that the development of the fundamental understanding of water-in-oil polymerization phenomena will be more difficult than for the corresponding oil-in-water process. While conventional emulsion polymerizations employ aqueous phases virtually exclusively with sodium dodecyl sulfate as a common emulsifier, for water-in-oil polymerizations both aliphatic and aromatic continuous phases are common. Further, a broad range of steric stabilizers and initiators are employed, with some initiators, such as persulfates, themselves generating unique kinetic phenomena [32]. Both the interfacial composition and the nature of the dispersion media influence micellization. Therefore water-in-oil polymerizations are characterized by an inherent ability to shift between inverse-suspension (IIa) and inverse-emulsion (IIb) regimes, and vice versa, for apparently minor modifications to the chemical composition of the dispersion, such as a change in emulsifier concentration! This is a significant deviation in behavior from conventional macroemulsions and complicates generality considerably.

An example of this sensitivity is evident in the heterophase water-in-oil polymerization of acrylamide with fatty acid esters of sorbitan as emulsifiers. Table 2 summarizes the effect of the level of emulsifier and type of organic phase. In the latter case the volume ratio of the organic and aqueous phases is identical. However, substituting an aromatic for a paraffinic dispersion medium causes the polymerization to change from inverse-suspension (Isopar-M) to inverse-emulsion (toluene). This is due to the different aggregation behavior of emulsifiers in aliphatic and aromatic solvents [5]. The transition from inverse-suspension to inverse-emulsion is accompanied by a change in the kinetics (Table 3), with the rate order with respect to emulsifier concentration increasing from -0.2 to $+0.45$. This kinetic order also varies within a polymerization domain for different emulsifiers, due to differences in the concentration and functionality of radically active groups. Furthermore, increasing the concentration of a sorbitan emulsifier from 6% to 10%, and appropriately selecting the HLB, can produce a shift in the polymerization domain to inverse-microemulsion (Table 2). Therefore, the development of a mechanism or kinetic model for water-in-oil polymerizations must be restricted to a single subdomain (inverse-suspension, inverse-emulsion or inverse-microemulsion). Within each domain and subdomain, the elementary reaction scheme will be specific to the type of organic phase and emulsifier, as well as a range of emulsifier concentrations. Such an observation explains much of the apparent disparity in the heterophase polymerization of acrylamide [13, 19, 42, 49, 50]. It is also clear, from Tables 2 and 3, that the assignment of a polymerization domain cannot be based on the formulation. Correct nomenclature requires post experimental latex and kinetic characterization.

Table 2. A comparison of heterophase polymerizations of acrylamide

Organic phase	Emulsifier	Initiator	Phase ratio (Water : Oil)	Emulsifier concentration (wt % of organic phase)	Monomer concentration (wt % of aqueous phase)	Temperature (°C)	Reference	Characteristics of the polymerization	Polymerization domain
Isopar-M	SMO	AIBN	1 : 1	6	30–50	47	Baade and Reichert [49]	Inverse micelles not detected. Polymerization in monomer droplets. Kinetic latex stability. Solution-like kinetics, with interfacial reactions	Inverse-Suspension
Toluene	SMO	AIBN	1 : 1	6	30–50	47	Baade and Reichert [13]	Inverse micelles may be present. Kinetic latex stability. Kinetics can follow the Smith-Ewart scheme	Inverse-Emulsion
Isopar-M	SSO-POEHO	AIBN	1 : 1	> 10	50	uv	Holtzscherer and Candau [3]	Bicontinuous microemulsion. Thermodynamic stability	Inverse-Microemulsion

Isopar-M: An isoparaffinic mixture; SMO: Sorbitanmonooleate; SSO-POEHO: Sorbitansesquioleate-polyoxyethylenehexaoleate; AIBN: Azobisisobutryonitrile

Table 3. Dependence of the polymerization rate on emulsifier concentration in heretophase polymerizations of acrylamide

Emulsifier	Polymerization domain	"a" in $R_p \propto E_a$	Reference
OP-10	Inverse-Emulsion	1.0	Kurenkov [50]
Tetronic-1120	Inverse-Emulsion	1.0	Vanderhoff [19]
SMO	Inverse-Emulsion	0.45	Baade [13]
SMO	Inverse-Suspension	– 0.2	Baade [49]

OP-10: A mixture of mono and dialkylphenols (10% oxyethylation); SMO: Sorbitanmonooleate

Other factors distinguishing heterophase polymerizations include the solubility of the initiator, aqueous or organic, and the partitioning and mass transfer of species between the dispersed phase, continuous phase and emulsifier layer [51]. In polymerizations where either the initiator or monomer has a significant solubility in the continuous phase, a complete set of elementary reactions must be considered in each phase [43]. While such a reaction scheme is simple to construct, verification is difficult due to a confounding of rate effects. For example, for water soluble monomers with marginal solubility in organic media, initiated with species soluble in the organic phase, propagation can occur in both the continuous and dispersed phases. Furthermore, the propagation rate constant may differ in the aqueous and organic phases due to differences in the ionic strength or pH. This causes difficulties in experimentally isolating rate phenomena and obtaining uncoupled parameter estimates.

4 Recommendations

4.1 Nomenclature

Heterophase processes should be primarily distinguished based on their emulsion structure (oil-in-water or water-in-oil) and type of stability (kinetic or thermodynamic). This identifies four mutually independent polymerization regimes, each with unique colloidal and chemical behavior: I. Macroemulsion, II. Inverse-Macroemulsion, III. Microemulsion, IV. Inverse-Microemulsion. The macroemulsion and inverse-macroemulsion domains can be further subdivided into Suspension (Ia), Emulsion (Ib), Inverse-suspension (IIa) and Inverse-emulsion (IIb) subdomains based on a transition at the critical micelle concentration.

4.2 The Transition Between Emulsion and Suspension Polymerizations

Emulsions and suspensions can be distinguished by the particle nucleation mechanism and the kinetics. A polymerization is considered an "emulsion" if *either* of the following criteria are satisfied: 1) the kinetics, as defined by the average number of macroradicals per polymer particle ($\bar{n}$), are not be significantly larger than one, *or* 2) the mechanism of particle nucleation occurs outside the monomer droplet. The nomenclature "suspension" is reserved for systems where *both* nucleation occurs in the monomer droplets *and* $\bar{n}$ is very high (10^{2-6}). This logical distinction, first used by John Vanderhoff, can also be applied to water-in-oil systems.

4.3 Theoretical Development of Water-in-Oil Polymerizations

Inverse-suspension, inverse-emulsion and inverse-microemulsion polymerizations should be developed independently as has been the precedent for oil-in-water polymerizations. This includes explicitly considering the unique chemistry of various emulsifiers, organic phases, monomers and initiators. Furthermore, the chemical and colloidal models for each of the three water-in-oil polymerizations will be specific to a given type of organic phase and a restricted family of emulsifiers.

4.4 New Heterophase Polymerizations

Investigations of water-in-oil polymerizations employing new monomers or emulsifiers for which kinetic or colloidal characterization is incomplete, require careful nomenclature designation. Under such circumstances a general description such as "Water-in-Oil Polymerization" or "Heterophase Polymerization" is recommended until the physical and chemical nature of the polymerizations can be identified. The designations inverse-suspension, inverse-emulsion and inverse-microemulsion should be reserved for processes for which a relatively advanced level of understanding exists.

5 References

1. Abere J, Goldfinger G, Naidus H, Mark HF (1945) J Phys Chem 49: 211
2. Barrett KEJ (ed) (1975) Dispersion polymerization in organic media. John Wiley, New York
3. Holtzscherer C, Candau F (1988) Colloids and Surfaces 29: 411
4. Vanderhoff JW, Bradford EB, Tarkowski HL, Shaffer JB, Wiley RM (1962) Adv Chem Ser 34: 32
5. Bartelt G (1990) PhD Thesis, Technical University of Berlin, FRG
6. Hofman F, Delbruck K (1909) Ger Pat 250: 690
7. Hohenstein WP, Mark HF (1946) J Polym Sci 1: 127
8. Farber E, Koral M (1968) Poly Eng Sci 8: 11
9. Barton J, Capek I (1991) Radical Polymerization in Disperse Systems, VEDA, Publishing House of the Slovak Academy of Sciences, Bratislava, p 25
10. Fikentscher L, Angew Z (1938) Chem 51: 433
11. Harkins WD (1947) J Am Chem Soc 69: 1428

12. Stockmayer WH (1957) J Polym Sci 24: 314
13. Baade W, Reichert KH (1986) Makromol Chem Rapid Commun 7: 235
14. French DM (1958) J Polym Sci 32: 395
15. El-Aasser MS, Vanderhoff JW (eds) (1981) Emulsion Polymerization of Vinyl Acetate, Applied Science Publishers, New York
16. Ugelstad J, Mork PC, Hansen EF (1981) Pure Appl Chem 53: 323
17. Zimmt WS (1959) J Appl Polym Sci 1: 323
18. Stickler M, Panke D, Hamielec AE (1984) J Polym Sci, Polym Chem Ed 22: 2243
19. Vanderhoff JW, DiStefano FV, El-Aasser MS, O'Leary R, Shaffer OM, Visioli DL (1984) J Dispersion Sci Tech 5: 323
20. Guo JS, Sudol ED, Vanderhoff JW El Aasser MS (1992) J Polym Sci Polym Chem Ed 30: 691
21. Candau F, Leong YS, Fitch RM. (1985) J Polym Sci Polym Chem Ed 23: 193
22. Konno M, Arai K, Salto S (1982) J Chem Eng Japan 15: 131
23. Lee JC, Tasakorn P (1979) Proc Third European Conference on Mixing, BHRA Fluid Publishers, Cranfield, U.K
24. Glukhikh V, Graillat C, Pichot C (1987) J Polym Chem Ed 25: 1127
25. Pichot C, Graillat C, Glukhikh V, Lauro MF (1985) Polymer Latex II, 11/1-10, Plastics and Rubber Institute, London
26. Graillat C, Pichot C, Guyot A, El-Aasser MS (1986) J Polym Sci, Polym Chem Ed 24: 427
27. Gerrens H (1980) Polymerisationstechnik, Ullmann's Enclylopedia of Industrial Chemistry, 4th edition, vol 19, pp 109, Weinheim, FRG
28. Gardon JL (1977) In: Schildknecht CE, Skeisteds I (eds) Polymerization processes. Wiley-Interscience, New York
29. Hunkeler D, Hamielec AE (1991) Polymer 32: 2626
30. Barton J (1991) Makromol Chem, Rapid Commun 12: 675
31. Delgado J, El-Aasser MS, Silebi CA, Vanderhoff JW (1989) J Polym Sci, Polym Chem Ed 27: 193
32. Hunkeler D (1991) Macromolecules 24: 2160
33. Candau F, Leong YS, Pouyet G, Candau S (1984) J Colloid Interf Sci 101: 167
34. Vaskova V, Juranicova V, Barton J (1990) Makromol Chem 191: 717
35. Carver MT, Dreyer U, Knoesel R, Candau F, Fitch RM (1989) J Polym Sci, Polym Chem Ed 27: 2161
36. Carver MT, Candau F, Fitch RM (1989) J Polym Sci Polym Chem Ed 27: 2179
37. Candau F (1987) In: Mark HF, Bikales NM, Overberger CG, Menges G (eds) Encyclopedia of Polymer Science and Technology, V9, p 718. John Wiley, New York
38. Candau F (1992) In: C. Paleos (ed) Polymerization in Organized Media. Gordon and Breach, New York, NY
39. Ugelstad J, El-Aasser MS, Vanderhoff JW (1973) J Polym Sci, Polym Lett Ed 11: 503
40. Grulke EA (1986) In: Mark HF, Bikales NM, Overberger CG, Merges G (eds) Suspension Polymerization. Encyclopedia of Polymer Science and Technology, 2nd edn. John Wiley, New York
41. Langson M (1986) In: Nass LI, Heiberger CA, Marcel Dekker (eds) PVC Processes and Manufacture. Encyclopedia of PVC, 2nd edn. John Wiley, New York
42. Hunkeler D, Baade W, Hamielec AE (1987) Polym Mat Sci Eng 57: 854
43. Hunkeler D, Hamielec AE, Baade W (1989) Polymer 30: 127
44. Dimonie MV, Boghina GM, Marinescu NN, Cincu CJ, Opescu OG (1982) Eur Polym J 18: 639
45. Smith WV, Ewart RH (1948) J Chem Phys 16: 592
46. Hay PM, Light JC, Marker L, Murray RW, Santonicola AT, Sweeting OJ, Wepsic JW (1961) J Appl Polym Sci 5: 23
47. Xie TY, Hamielec AE, Wood PE, Woods DR (1991) Polymer 32: 537
48. Balke ST, Hamielec AE (1973) J Appl Polym Sci 17: 905
49. Baade W, Reichert KH (1984) Eur Polym J 20: 505
50. Kurenkov VF, Verzhnikova AS, Myagchenkov VA (1984) Doklady Academii Nauk SSR, 278: 1173
51. Hunkeler D (1992) Polymer International, 27: 23

Editor: Prof. H.H. Kausch
Received July 1993

Bases of the Axiomatic Theory of Addition Polymerization

V.V. Shamanin
The Institute of Macromolecular Compounds of the Russian Academy of Sciences, St. Petersburg, Russia

A new qualitative quantum-chemical concept of the elementary act of addition polymerization has been proposed as the development of the polymerization theory. An extensive set of various data on the kinetics and the mechanism of polymer structure controlling has been found to have a new explanation from an uniform viewpoint. This concept is developed in the framework of the axiomatic approach to the general polymerization theory and is based on five postulates, namely: the principle of the intermediate, the principle of intermediate cyclicity, the principle of correspondence, the principle of local symmetry and the spin exclusion principle.

Advances in Polymer Science, Vol. 112

1 Introduction

The principal problem of polymer science is the mastering of methods for obtaining materials with predetermined properties. Consequently, the main task of polymer chemistry is the development of the methods of synthesis of polymers which exhibit, in principle, any theoretically possible primary structure of macromolecules. At present, many "empirical" data have been accumulated in the field of experimental chemistry, which in many cases, make it possible to synthesize polymers with a predetermined primary microstructure. In particular, the most interesting in this sense is the available experience of the synthesis of stereoregular polymers. On the whole, the successful fulfilment of this program requires a reasonable and harmonious combination of empirical and theoretical approaches.

However, at present the disproportion in the development of experimental and theoretical chemistry of polymers is so great, that the absence of the appropriate theoretical concepts of the mechanism of stereoregular polymers formation has led to a certain stagnation in the experimental field as well. Although the above disproportion is inherent in the chemistry of polymers from time immemorial, so to speak, nevertheless, as the extensive paths of development of polymer synthesis are being exhausted, its effect becomes increasingly negative. The overcoming of this unfavourable tendency requires not only greater attention to the problems related to the polymerization mechanism but also, predominantly, a qualitatively higher level of investigations.

What may be considered to be a qualitatively higher level of investigations of the polymerization mechanism from the present-day standpoint? The answer to this question may be only relative. Therefore it is necessary to generalize the evaluation of the existing results and the tendency of development in the field of chemistry.

The essence of the polymerization process may be represented as follows:

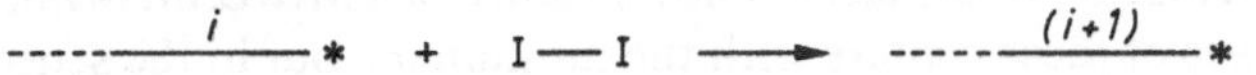

Scheme 1

where i is the degree of polymerization of the growing polymer chain.

Up till now, the predominant and, it should be mentioned, successfully solved problems have been related to the determination of the nature (cationic, free-radical or anionic) and the structure of the active center of the growing polymer chain represented by an asterisk * in Scheme 1. However, the investigation of the process of the direct insertion of the monomer in the polymer chain, i.e. everything represented in Scheme 1 by an arrow → was considered to be of secondary importance, with the exception of anionic coordination polymerization. It is usually a priori assumed that this is an elementary single-stage activation transition in the literal sense without any peculiar features, and if these features even exist, they are completely predetermined by * (Fig. 1).

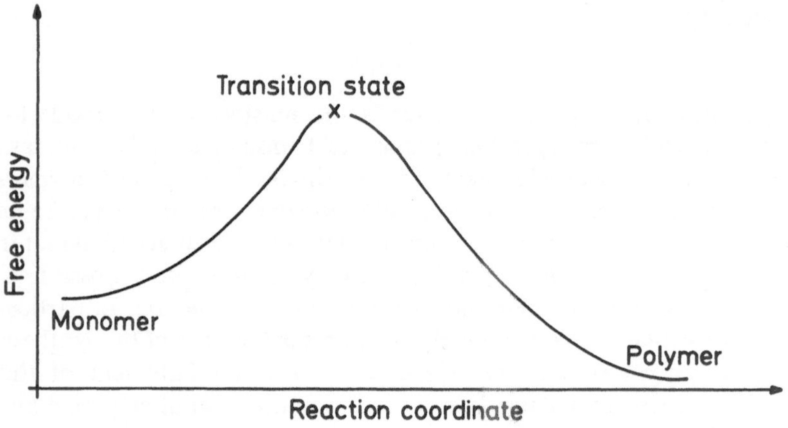

Fig. 1. Generally used free energy profile for monomer insertion into the polymer chain

Attempts at a detailed determination of → made in anionic coordination polymerization may be characterized as "steric-geometric" approaches. Their distinctive peculiarity is the introduction of an intermediate stage of monomer complexing with the active center. It is assumed to be the geometric features of the complex that determine the stereospecificity of the entire process. The fact of complex formation is considered to be an indispensable and even almost sufficient reason for the existence of stereospecificity. This is assumed in spite of the fact that the process of monomer insertion in the polymer chain from the postulated complex of a certain structure is usually not considered at all.

Summing up the above consideration, it is possible to assert that an important defect of this tradition in polymer chemistry is the complete neglect of the electronic aspect of polymerization. This aspect should not be understood in the sense of computer calculation of the electronic structure of active centers and the hypothetical complexes of these centers with the monomers, but in the sense of the development of fundamental qualitative concepts of the motive forces, the character and paths of rearrangement of electron shells of the reagents, i.e. the active center and the monomer, during polymerization. It seems evident a priori that without taking into account certain electronic factors it is impossible to give an exhaustive answer to the question: why in a given situation just this primary structure of the macromolecule is formed and not any other structure?

In the opinion of the author, the development of fundamental qualitative concepts should be a primary problem because without an even simplified but general understanding of the physical basis of the "elementary act" of polymerization, it is difficult to develop a general theory of polymerization. The process of generalised comprehension of experimental facts cannot be replaced by any most universal and detailed quantum-chemistry calculations. Their results are in many respects predetermined by the initial formulation of the problem and the choice of the method of its numerical analysis. All these considerations have induced the author to dare to develop his own approach to the problem of the polymerization mechanism [1, 2].

2 Ideology

2.1 Formulation of the Problem

The general theory of polymerization should include a) the theory of chemical reactions, b) the reaction rates theory and c) the kinetic statistical theory of polymerization.

The problem of the theory of chemical reactions being applied to any specific system under consideration is the presentation of an extensive scheme for chemical reactions, its main features reflecting the essence of chemical transformations actually occurring or being possible in this system. Moreover, it is implied that not only the reactions are described but also the detailed mechanism of the stages, i.e. all possible intermediates are indicated.

Hence, the problem of the theory of reaction rates consists of the determination of the numerical values of rate constants for all chemical reactions considered and thus in the transformation of the proposed scheme for chemical reactions into a kinetic scheme.

The problem of the kinetic statistical theory of polymerization, in turn, is the establishment of the quantitative relationship between the primary microstructure of macromolecules and the initial state of the system and the chosen conditions of the process on the basis of the given kinetic scheme.

Naturally, it is not actually possible to carry out the above program at present. This is the ideal to which one should aspire and which, luckily (in the opposite case, polymer chemistry would no longer exist as a science), will probably be attained in a very remote future or never. Consequently, in this paper only those problems will be considered that are related to the mechanism of monomer insertion in the growing polymer chain. In other words, the above scheme for chemical reactions will be actually replaced by the consideration of only one reaction: propagation. Doubtless, this reaction is the "principal" polymerization reaction. The formation of the macromolecule with all its properties is the result of this reaction.

From the chemical viewpoint, all polymerization processes are only a subclass of a wider class of addition reactions. There is nothing specifically "macromolecular" in the polymerization "elementary act" itself. However, from the viewpoint of investigations of the propagation mechanism, the chain being formed in the course of the process is actually a material store of information, a kind of natural "memory device" fixing the final results of the periodically breaking out "electron storm" accompanying each act of monomer insertion in the growing polymer chain. Hence, the problem of investigations of the polymerization mechanism may be regarded as that of investigation of the mechanism of an active center attachment to the monomer. The information about this reaction has been partially provided by nature in the form of the primary structure of the corresponding macromolecule.

The limitations of modern experimental physico-chemical methods in the direct and detailed study of the monomer insertion in the polymer chain on the one hand, and the large amount of information about the dependence of the macromolecules primary structure on the reaction conditions on the other hand should be taken into account. Consequently, it seems that the only real path of the search for relationships between the electronic aspects of the reaction and the observed microstructural statistics of polymers is the "heuristic" method of formulation of definite rules. It is most suitable for carrying out an adequate formulation of these "imaginary discoveries" in the framework of an axiomatic approach.

2.2 The Axiomatic Approach

If the axiomatic approach is used, any new theoretical schemes (including the concept of the "elementary act" of polymerization) are developed according to the model of the axiomatic theory, i.e. in such a way that all its results should be rigorous consequences of a single system of "fundamentals", in this case physico-chemical, assumptions or axioms. The essence of the axiomatic method has been figuratively expressed by its consistent ideologist, the famous German mathematician Hilbert who said: "one should attempt to make it possible to replace the words: a point, a straight line, a plane in all geometrical statements by the words: a table, a chair, a cup" [3].

In our case the quotation marks framing the word "fundamental" imply that the hypotheses considered in the initial stage of development of the theory as postulates can later themselves become the subject of independent investigations and appear as consequences of a more "fundamental" system of axioms.

It is possible to find in the history of science many vivid examples illustrating the relativity of the concept "fundamental". For instance, the Planck postulate of energy quantization and the Bohr postulate on the quantization on angular momentum made a revolution in physics and were actually axioms at that time. At present from the formal viewpoint, they are only "ordinary" consequences of Schroedinger's equation [4]. Another vivid example is provided by the four famous Maxwell electrodynamic equations which, as was found later, can be derived from Coulomb's law and Einstein's relativity principle [5].

A considerable advantage of the axiomatic approach is its universality (for instance, the author previously succeeded in developing the kinetic statistic theory of molecular weight distributions of polymers formulated according to the model for the axiomatic theory [6]). Moreover, in the initial stage of the formation of a concept using the axiomatic approach, i.e. in the search and formulation of the necessary basic postulates, if desired, it is possible (and even probably advisable) to "forget" for a certain time all theoretical concepts existing in the literature about the polymerization mechanism. In other words, the axiomatic method makes it possible to abandon easily the framework of stable stereotypes. In the initial stage of a work the only leading criteria should be intuition as well as the simplicity, elegance, internal coordination and the logical completion of the construction being developed.

The adequacy and the predicting ability of new concepts are the objective criteria for the fact that the energy of the investigator is not wasted. In the present work the ideas of symmetry form the basis of the new concept, act as a kind of "lode star" in the search and formulation of the postulates.

2.3 The Role of Symmetry

The fundamental and promising character of any concept is determined not only by its form (although it is doubtless very important in itself) but rather, by its content. In our case the form is caused by the axiomatic approach, whereas the content is based on the ideology of symmetry. Hence, it seems desirable to characterize briefly the sources, development and historical prospects of the ideas determining the objective regularity of applying the proposed approach to the investigation of the polymerization mechanism.

What is symmetry? In physics and mathematics, symmetry is understood as the invariance of some properties of the object being investigated with respect to all the transformations considered. In chemistry, symmetry is usually identified with the invariance of the Hamiltonian of the system with respect to spatial transformations of the object (molecule). The knowledge of symmetry makes it possible to draw certain conclusions on the behavior of the system without its complete description in the formal terms of the quantum theory [7]. The group theory is the mathematical theory of symmetry.

The general features of the mathematical concept of the group were first developed by Galois in 1832 [8]. In the 1840s, the works of Cayley and Sylvester led to the development of the theory of algebraic invariants [8]. In 1872, Klein showed that any geometry may be considered as a theory of invariants of a special group of transformations. The classifications of groups of transformations give the classifications of geometries, whereas the theory of algebraic and differential invariants of each group characterizes the analytical structure of the corresponding geometry [8]. In the 1880s, Lie discovered contact transformations, and from that time Hamiltonian dynamics was considered to be a component of the theory of continuous groups [8]. At the beginning of the 20th century, on the basis of theoretic group considerations, Hamel, Klein and Noether proposed a very elegant and fine method of derivation of conservation laws existing in classical physics [8]. Hence, in the course of the development of mathematics it has been shown that the great laws of the conservation of energy, momentum and angular momentum are based on the symmetry of the main principles of the structural organization of nature.

Poincaré was probably the first to understand the euristic value of the invariance principle for the development of physics as a whole [9]. However, until the special relativity theory appeared in 1905, it had been usual to derive the invariance principles from the laws of motion. After Einstein's work it became a general tendency to derive and sometimes even to discover the laws of nature with the aid of a priori postulated invariance principles. The general relativity theory is the first example of a consistent development of a physical theory on the basis of invariance ideas [10]. However, a drastic change in the attitude of physicists to the ideas of symmetry took place after the publication of Weyl and Wigner's papers. At the end of the 1920s these authors were the first to show the efficiency of applying the "abstract" apparatus of the group theory to quantum mechanics that had just been discovered [11]. From the 1960s, after Gell-Mann and Neeman's papers had appeared in which the quark theory of elementary particles was developed on the basis of postulated and highly abstract unitarian symmetry [12], invariance became figurately speaking the "absolute sovereign" of theoretical physics.

From 1965, the ideas of symmetry begin also to penetrate gradually into the theory of chemical reactions. The principle of orbital symmetry conservation formulated by Woodward and Hoffmann [13] became a powerful incentive to the development of the theoretical group concepts of the quantum theory of chemical reactions. Hence, it should be expected, proceeding from the general tendencies of science development, that the theoretical chemistry of the future will follow mathematics and physics and will be to a considerable extent a science on symmetry or, in other words, a science about the "harmony of nature".

What conclusions can be drawn from the above considerations? First, as we have seen, a constant "drift" exists of the fundamental, conceptual ideas from mathematics into physics and from physics into chemistry. Secondly, in the development of the ideas of symmetry, theoretical chemistry lags behind theoretical physics by about forty years and, correspondingly, polymerization science lags behind theoretical chemistry by at least another twenty years. Thirdly, it seems useful from the methodological viewpoint to draw an analogy with a certain physical process which from the viewpoint of its inherent relationships including those of symmetry, would completely represent the polymerization process.

In the opinion of the author, the process of light absorption by an atom or a molecule is a good model for stereospecific polymerization in the above sense. Only a few transitions among all possible transitions in an atom or a molecule are spectrally active. Similarly, only some possible microstructures are formed in stereospecific polymerizations.

In other words, polymer chain propagation should not be represented as a single reaction channel (Scheme 1) but, rather, as a combination of several concurrent reaction channels. Each of these channels has its own transition state and provides its partial contributions to the formation of various structures of the terminal units of the chain:

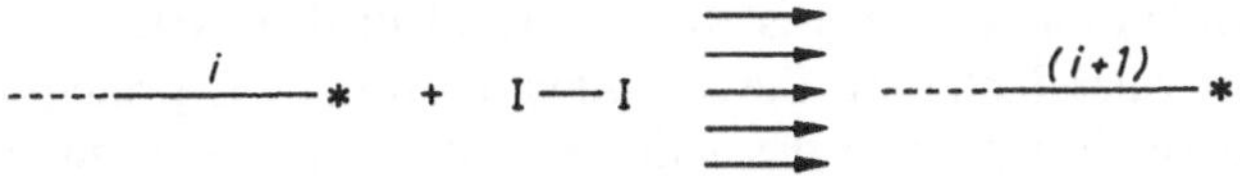

Scheme 2

Hence, it should be expected that, just as in photonics, in stereospecific polymerization and, consequently, in polymerization in general there act certain "selection rules", i.e., some principles controlling the course of the process.

2.4 The Nature of "Selection Rules"

According to the general concepts of quantum mechanics [14], the state of any system including that consisting of reagents and products can be represented at any time moment t by the ket-vector $|\psi(t)\rangle$ belonging to a certain abstract vector space E. In the absence of any external effects, the dynamic state of the system evolves with time in a rigorously causative manner obeying Schroedinger's equation

$$i\hbar\frac{\partial}{\partial t}|\psi(t)\rangle = H|\psi(t)\rangle \tag{1}$$

Moreover, if any operator A representing in space E a certain observable commutes with a Hamiltonian H, i.e.

$$A\cdot H = H\cdot A \tag{2}$$

the observable A retains a certain constant average value a all the time. If the ket-vector $|\psi(t)\rangle$ is the eigenvector of the operator A then the observable A retains a definite constant eigenvalue a. In this case the value a is said to be a good quantum number. In other words, in this case the ket-vector $|\psi(t)\rangle$ remains constantly in the subspace of the eigenvalue a, i.e., at any time moment

$$A|\psi(t)\rangle = a|\psi(t)\rangle \tag{3}$$

In the language of chemical reactions this equation implies that under the given conditions formally characterized by Eqs. 1 and 2, these reagents may yield only products with the ket-vector satisfying Eq. 3 in which the eigenvalue a is determined by the initial conditions, in particular, by the structure of the initial reagents. The greater the number of the observables of type (2), the greater the selectivity and stereospecificity of the chemical reaction considered.

Hence, from the standpoint of invariance, any chemical reaction including polymerization is characterized by a set of observables (e.g., the elements of symmetry with operators commuting with the Hamiltonian of the reaction) $\{A\}$. The processes during which all the combination of properties of the reacting system (formally characterized by a set of eigenvalues $\{a\}$) remains invariant are "allowed" in the sense of symmetry. This is the nature of both physical and chemical kinetic selection rules including the Woodward-Hoffmann principle. Hence, the specific feature of all selection rules is the fact that they "allow" much less than "forbid". In other words, each of them exhibits a kind of "veto" right on the occurrence of chemical reactions. At the same time, the processes "allowed" with respect to symmetry may be "forbidden" by thermodynamic or steric factors.

Furthermore, it is also necessary to take into account those distortions of symmetry which take place as a result of the action of external conditions, in particular, when the chemical reaction is carried out in the condensed phase. Consequently, strictly speaking all the "veto" rules based on symmetry should be distorted to a certain extent in chemical reactions. However, taking into account that the time of the elementary chemical act is approximately 10^{-13} s [15], it should be expected that in the "decisive" stages of the process most chemical reactions are quasiclosed. In this case the effect of the medium may be taken into account either by the introduction of additional initial conditions or by further detailing of the process. Hence, it may be expected that at least the general tendencies in chemical reactions based on symmetry considerations will be predicted correctly in most cases.

2.5 Presentation of the Problem

The necessity of considering the propagation stage of the chain as a complex of simultaneously occurring reactions is a consequence not only of general considerations (Scheme 2) but also of the Murrel-Laidler theorem stating that only one reaction coordinate corresponds to each transition state regardless of its structure [16]. It follows from this theorem that the variety of monomer unit configurations in the chain implies the variety of reaction coordinates and the transition states for the stage of monomer insertion in a given polymer chain. This variety is equivalent to the existence of a complex of parallel channels of propagation.

The question naturally arises about the role and the contribution of a certain channel to the entire process. It is probably advisable to divide this problem into three stages. The first qualitative stage is the formulation of the above-mentioned selection rules, i.e. the development of the qualitative concepts of the mechanism, the physical basis and the motive factor of polymerization. The second semiquantitative stage should evaluate the probability of the monomer insertion into the polymer chain via each of the possible reaction channels. The third quantitative stage should lead to the calculation of absolute values of kinetic rate constants of these reactions.

The study, in its present state, certainly does not extend further than the first stage. What properties of homogeneous addition polymerization may help in the search for and the formulation of selection rules? The main feature of the homogeneous addition polymerization of non-cyclic monomers is the fact that the addition of the following molecule of the monomer to the growing polymer chain proceeds through the opening of one π-bond with the transfer of the active center of the monomer molecule. Since the σ-backbone is invariable, the configuration of the terminal chain unit is similar in many respects to that of the initial monomer. In particular, the configurations of individual fragments virtually coincide.

Furthermore, if we do not consider the increase in chain length, then, as already mentioned, from the topological standpoint, the elementary propagation act will appear to be the addition of the active center to the monomer, i.e. as the opening of some bonds and the formation of some other bonds in the course of the monomer approach to the active center of the chain considered as an independent low molecular weight compound.

The formal presentation of the problem was an attempt to develop a qualitative quantum-chemical model of the elementary polymerization act – adequately describing the complex of available experimental facts taking into account the above properties of propagation, the theoretical achievements of organic chemistry and the known principles of invariance in the framework of the axiomatic approach on the ideological basis of symmetry concepts. It should be emphasized that we do not mean to achieve in the present work a global interpretation of the entire experimental material accumulated in the literature. The aim of the author is mainly to demonstrate the promising character of this approach as a whole.

2.6 The LCFMOF Approximation

Generally speaking, it is probably impossible to give any general recommendations concerning the search for the observables, the operators of which commute with the Hamiltonian. The exceptions are the cases when the observable to be found characterizes the properties of the spatial symmetry of the ket-vector $|\psi(t)\rangle$ (in Schroedinger's spatial presentation this vector is called the wave function). It should be noted that a more or less precise pattern of the wave function $\psi(t)$ is known for very few molecular system. At present the most widespread is the $\psi(t)$ presentation in the Hartree-Fock approximation as a symmetrized linear combination of atomic orbitals (LCAO) [16].

However, classifying the wave function of polymerization by the spatial symmetry features and taking into account the above specificity of addition polymerization, it is advisable for simplicity to introduce as a supplement to LCAO the approximation of the "polymerization wave function" in the form of a linear combination of "molecular orbitals of fragments" (LCMOF). The validity of introduction of this approximation is based on the general quantum-mechanical principle of superposition.

From this standpoint, for instance, any vinyl monomer may be regarded as a certain "subatomic" particle consisting of two fragments (F): $F_1 = F_2$, and the polymerization may be analyzed as the interaction between the corresponding frontier "molecular orbitals" of these fragments:

$$\cdots F^* + F_1 = F_2 \rightarrow \cdots\cdot F - F_1 - F_2^*$$

Consequently, the symmetry of the wave function which is intermolecular in nature may be conveniently evaluated in the approximation of a linear combination of frontier "molecular orbitals" of fragments (LCFMOF). It should be noted that the method of perturbations of molecular orbitals developed by Dewar [17] is actually based on the LCFMOF approximation.

3 Postulates of the Theory

A qualitative quantum-chemical concept of the elementary polymerization act is proposed as a development of the concepts of the polymerization mechanism on the basis of the available experimental data generalizes in the framework of quantum-mechanical concepts. Up to the present, this concept has been based on five postulates.

Before passing to the formulation of these postulates, we will try to answer the question why the postulates are meant and not the selection rules. When one speaks about the postulates, something "primary" is meant, whereas the selection rules imply something "secondary", i.e. that the rules themselves are the consequences of some more fundamental principles. For instance, the relationship between the heuristic principle of orbital symmetry conservation formulated by Woodward and Hoffmann and a number of selection rules derived on its basis is of this type. On the qualitative level, these rules make it possible to predict the stereochemical direction of concerted reactions of various types [18].

The principles formulated below and coinciding in their meaning with selection rules are essentially postulates because they do not follow from anything. They should not be understood as something quite fixed and invariable. This is only the first and possibly not the best attempt to formulate tentatively the physical bases of catalytic polymerization on the basis of the axiomatic method.

3.1 Principle of the Intermediate

Principle of the intermediate. The insertion of the monomer into the polymer chain usually passes through the stage of the formation of at least one intermediate.

As will be shown below, there are probably virtually no exceptions to these postulates in dark processes. Hence, the words "usually" in the formulation of the principle of the intermediate should be understood in the sense "always". For the thermal reaction to occur, at the moment of the physical contact between the reagents at least the following three conditions should be obeyed [19]: a) the molecule should have sufficient energy for the reaction, b) the energy should be redistributed in the proper manner between the degrees of freedom contributing to the reaction coordinate, c) the changes in the corresponding degrees of freedom are correlated to the necessary extent.

Let us consider this situation from the viewpoint of the quantum field theory [20]. The reagents will be assumed to be in a certain physical vacuum. The state of this vacuum is characterized by a certain ensemble of excitation quanta and adequately represents the real state of the condensed phase surrounding the monomer and the active center. In terms of the quantum field theory, if a chemical reaction is carried out, this implies that a probabilistic event Q takes place. This event involves fortuitous simultaneous encounter of both reagents at the same place and a suitable (in energy, momentum etc.) excitation quantum.

Theoretically, many variants are possible, however, only three main variants will be considered (Scheme 3).

1) $v \sim [A] \cdot [B] \cdot q \cdot \Delta t$

2) $v \sim [A] \cdot [B] \cdot q \cdot \Delta t$

3) $v \sim [A] \cdot [B] \cdot q \cdot \tau$

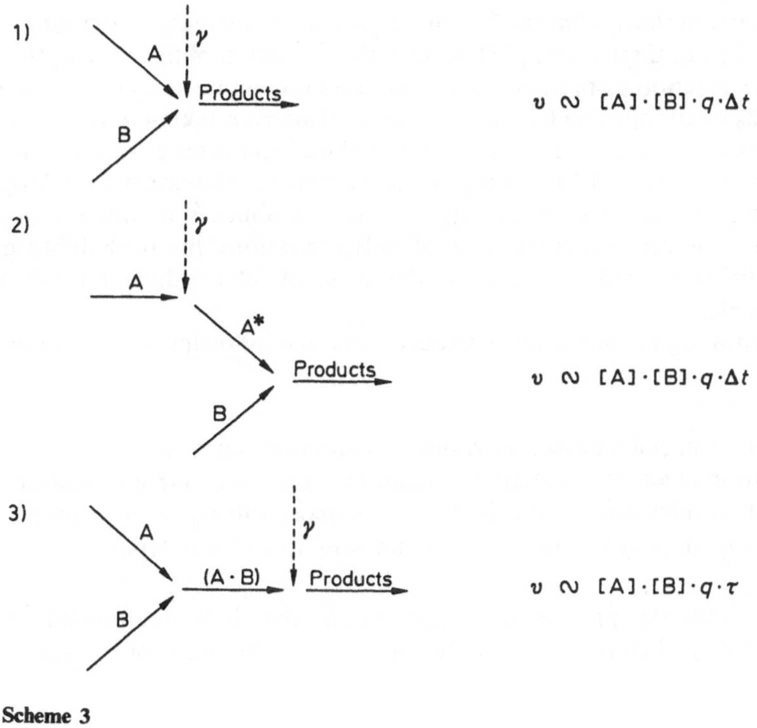

Scheme 3

1) If all the three particles (two reagents and an excitation quantum) meet simultaneously, the probability of the event Q and at the same time the reaction rate v will be proportional to $[A]\cdot[B]\cdot q\cdot\Delta t$ where [A] and [B] are reagent concentrations, q is the probability density of the appearance of the suitable excitation quantum γ in the place of a encounter between A and B at the corresponding time moment t and Δt is the time of the physical contact of reagents A and B in the bimolecular collision (in liquid $\Delta t \sim 10^{-13}$ s [21]).

2) If the event Q takes place stepwise, e.g. by the primary absorption of the excitation quantum γ by the reagent A, A turning to the excited state A*, the reaction rate will still be proportional to $[A]\cdot[B]\cdot q\cdot\Delta t$, since in solution the lifetime of the molecule in the excited state coincides with Δt in the order of magnitude [22].

3) If the event Q is preceeded by the complexation stage between reagents A and B, a considerable acceleration of the reaction should be expected. However, for this acceleration the reaction centers of substrates in the complex should be oriented with respect to each other in a suitable manner. If this condition is obeyed, i.e. if the complex being formed lies at the reaction coordinate and, hence, is an intermediate, it may be expected that the process rate will be proportional to $[A]\cdot[B]\cdot q\cdot\tau$ where τ is the potential lifetime of the complex in the absence of the reaction.

If we speak about polymerization, the effect of its acceleration by a factor of $\tau/\Delta t \approx 10^{13}\tau$ is of a purely entropic nature because it is not due to a change in the energetical levels of the system but is caused only by an increase in the time of residence on each of these levels [23]. In fact, the formation of the intermediate transforms the reaction from inter- to intramolecular conditions and thus leads to the catalysis by the approach considered here. Moreover, taking into account that even weak complexes with $\tau \approx 10^{-8}$ s should lead to the acceleration of the process by about a factor of 10^5, it may be expected that in the case when a large number of degrees of freedom provide a real contribution to the reaction coordinate (the situation characteristic of polymerization), the probability of the process being carried out without the stage of intermediate formation will be negligible.

Hence, summing up the above considerations the following points should be noted:

1. Complexation in polymerization is not an exception but a rule,
2. The formation of an intermediate accelerates the reaction (entropy catalysis),
3. Complexation does not necessarily lead to stereospecificity (in the opposite case, virtually all polymerization processes would be of this type).

It follows from the preceeding considerations that it is not possible to develop the theory of stereospecific polymerization on the basis of traditional concepts.

3.2 Principle of Intermediate Cyclicity

Principle of intermediate cyclicity. Intermediates corresponding to the main reaction channel usually have a cyclic structure.

As has been shown above, the formation of a complex between an active center and a monomer can greatly increase the addition rate of the active center to the monomer via some reaction channels and decrease it via other channels. Everything depends on the structure of the complex: it may be an intermediate for some processes and not an intermediate for other processes.

The postulation of the cyclic structure of the complex is mainly intuitive. The active center exhibits a certain "localized stress" which is expressed mainly in the lability of active bonds or in the existence of non-compensated charge or spin density. In the interaction with the unsaturated π-system of the monomer, the possibility of delocalization of the corresponding functional stress appears on the intermolecular level. This delocalization is probably most effective for the cyclic structure of the complex.

In the framework of the LCFMOF approximation, the polymerization of a vinyl monomer, e.g., by a free-radical or free-ion mechanism, may be shown as follows:

$$----F^* + \; F_1{=}F_2 \longrightarrow ----F \cdots \begin{array}{c} F_1 \\ \| \\ F_2 \end{array} \longrightarrow ----F{-}F_1{-}F_2^*$$

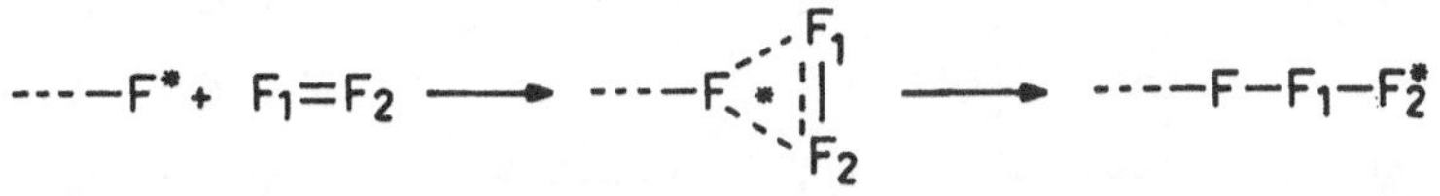

The formulation of the principle of the intermediate cyclicity (for brevity in further discussion it will be called the cyclicity principle) decreases to a considerable extent the quantity of possible (to be considered) variants of monomer insertion into the chain, which makes the problem more definite.

3.3 Principle of Correspondence

Principle of correspondence. Reactions that, other conditions being equal, are in better correspondence with respect to symmetry in the sense of Woodward-Hoffmann proceed at higher rates.

Note. Processes with symmetry correspondence should mostly obey the principle of the intermediate, i.e., they should proceed via the complexation stage.

As can be seen from the formulation, the principle of correspondence is an extension of the Woodward-Hoffmann principle to polymerization. The original Woodward-Hoffmann approach [13] was based on the analysis of correlation diagrams. At the beginning of the 1970s, Pearson [24] greatly simplified the analysis of symmetry by formulating a simple and effective rule: a reaction is allowed if the symmetry of the bonds being formed coincides with that of broken bonds (Scheme 4).

According to Pearson, whether the bonds being broken in the reagents correspond (after symmetrization) to the bonds being formed in the product is an adequate and equivalent criterion for the reaction to be allowed or forbidden in the sense of Woodward-Hoffmann. Apart from simplicity, the efficiency of the rule of bond symmetry is also ensured by the fact that it is equally applicable to π- and σ-bonds. Hence, its predictions are equally applicable to both the "full-face" ethylene dimerization (Scheme 4a) or to the Diels-Alder reaction (Scheme 4b) and to addition reaction, e.g., the addition of molecular hydrogen to ethylene or butadiene.

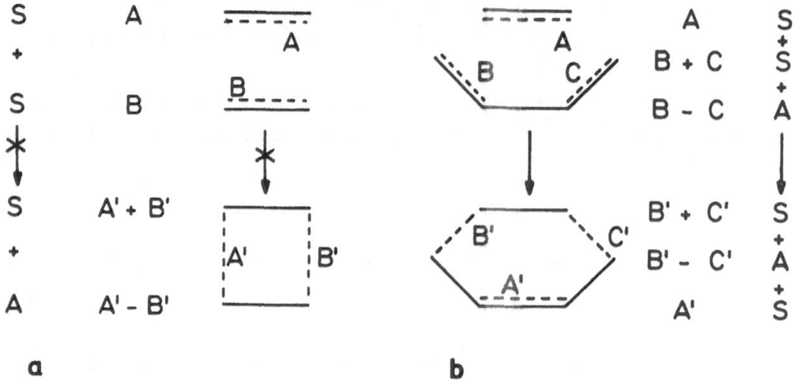

Scheme 4

Furthermore, it is taken into account in the formulation of the principle of correspondence that the experimental confirmation of predictions in the sense of Woodward-Hoffmann extends much further than for unsubstituted molecules [25]. A kind of "symmetry" control without symmetry is observed in nature, e.g., the addition of propylene to ethylene obeys the same limitation with respect to orbital symmetry as the "full-face" ethylene dimerization.

Hence, nature itself suggests the concepts of "pseudosymmetry". The greater the difference between pseudosymmetry and symmetry, the weaker is the role of limitations in the sense of Woodward-Hoffmann and vice versa. The physical meaning of the correspondence principle is the postulation of the existence of not only qualitative but also quantitative correlation between pseudosymmetry as a measure of the "quality" and "quantity" of symmetry on the one hand and the polymerization rate on the other.

In the LCMOF approximation, pseudosymmetry reflects not so much the geometrical symmetry of interacting substrates, as, rather, the topological symmetry of interacting orbitals.

3.4 Principle of Local Symmetry

Principle of local symmetry. In the interaction between a symmetric or a quasisymmetric monomer and its own active center being in the free-radical or free-ionic state, a cyclic adduct of the donor-acceptor type exhibiting the local pseudosymmetry of the third order is formed as one of the intermediates.

Definition. A monomer will be called "quasisymmetric" if the quality of local pseudosymmetry is such that the difference between the two lowest unoccupied or half-occupied (with one electron) orbitals of at least one of the conceivable structures of the intermediate complex does not exceed in value Fermi's correlation energy [26] between these orbitals.

A locally symmetric intermediate may be represented in the LCFMOF approximation in the form of an "equilateral triangle" having a pair of energetically degenerate or quasidegenerate frontier orbitals:

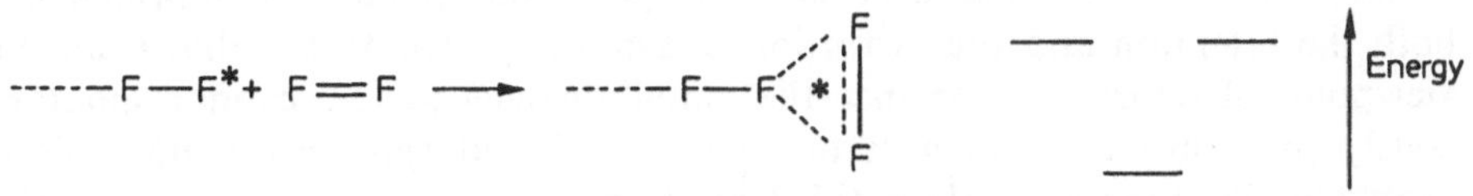

Occupation of frontier orbitals in the intermediate:

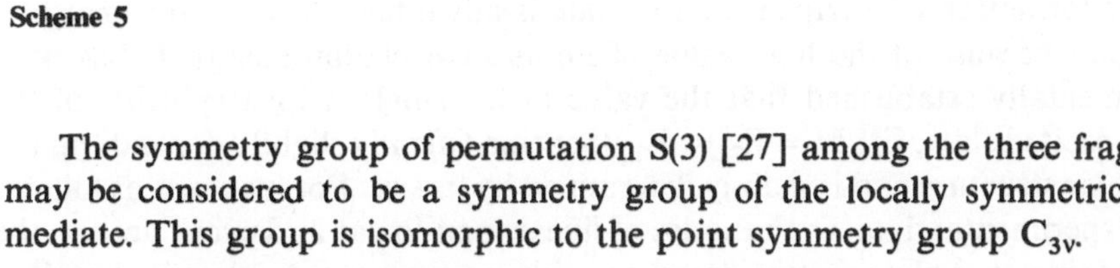

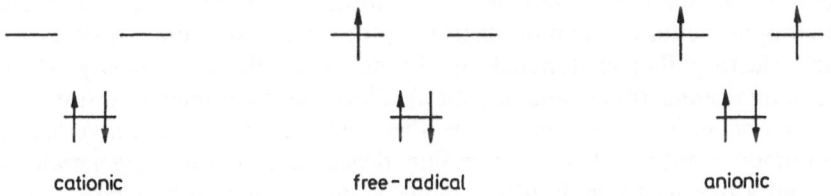

Scheme 5

The symmetry group of permutation S(3) [27] among the three fragments may be considered to be a symmetry group of the locally symmetric intermediate. This group is isomorphic to the point symmetry group C_{3v}.

Note. In the case of free-radical polymerization, the locally symmetric intermediate being formed will be extremely unstable because of the Jahn-Teller effect [28].

Among all five postulates formulated in this work, the principle of local symmetry is of the most "polymerization" character. The validity of its introduction is based on intuition, the complex of experimental data on the investigation of paramagnetic stilbene adducts [29–32] and the predicting ability of the concept as a whole.

3.5 Spin Exclusion Principle

Spin exclusion principle. The reactions the coordinates of which lead to the formation of intermediates being in the triplet state proceed at relatively low rates.

The difference between this principle and the four preceeding ones is that the range of its application is limited to anionic and anionic-coordination polymerization. Only in the anionic propagation mechanism, is the formation of triplet intermediates possible (Scheme 5). This is probably one of the main reasons for the high structure- and stereoregulating ability of anionic and anionic-coordination catalytic systems including Ziegler-Natta catalysts and is one of their distinguishing features.

In contrast to the well-known Wigner-Witmer rule [24] that multiplicity is retained during a chemical reaction, the origin of the spin exclusion principle is both the retention and the distortion of symmetry. The fact is that from the viewpoint of rigorous invariance the initially singlet system cannot spontaneously pass into the triplet state. Spin-orbital and spin-lattice interactions probably play the main role in this transition.

The process is assumed to occur in two stages. At first the singlet π-complex of the monomer with an active center is formed. In the second stage, the symmetric monomer and the active center continue to approach each other. The antibonding π-orbital of the monomer-acceptor (which in anionic processes is naturally electrophilic) is donated by the excessive electron density of the anionic active center (nucleophilic action). Thus, the monomer insertion into the polymer chain is accompanied by further destabilization of the π-system of the monomer, frontier orbitals becoming degenerate or quasidegenerate in energy, and a paramagnetic locally symmetric intermediate is formed.

The formation of a triplet intermediate is advantageous from the energetic viewpoint because of the high value of Fermi's correlation energy. It has been experimentally established that the value of the singlet-triplet splitting of the lowest excited state, $E_{corr}^{Fermi} = E_{S_1} - E_{T_1}$ (Scheme 6a) only slightly depends on the molecular structure and is mainly determined by its size. For instance, as follows from experimental data, in the series of linear, non-linear and cyclic acenes, the value of singlet-triplet splitting decreases with increasing molecular size but does not become negligible and approaches the limiting value $\geqq 1$ eV [26].

Hence, it should be expected a priori that in the triplet locally symmetric intermediate considered here the value of the Fermi correlation energy will not be lower than 1 eV in the order of magnitude. Rigorousness of spin exclusion should probably depend on the value of the singlet-triplet splitting in the intermediate $\Delta E = E_{corr}^{Fermi} - \Delta E$ where ΔE characterizes the degree of degeneration of frontier orbitals (Scheme 6b) and in contrast to E_{corr}^{Fermi} is a structure-

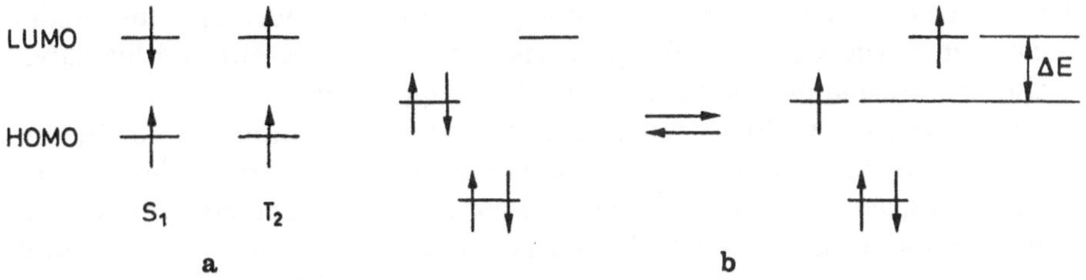

Scheme 6

sensitive parameter, a kind of quantitative measure of pseudosymmetry. The higher ΔE, i.e., the lower the symmetry of the intermediate and, correspondingly, the weaker the degeneration of frontier orbitals, the less rigorously the spin exclusion principle is obeyed.

A symmetric or a quasisymmetric monomer can avoid the "spin trap" and be satisfactorily inserted into the polymer chain in the following cases:

a) in the interaction with the "alien" active center (e.g. in copolymerization when active centers differ by the terminal monomer unit of the chain),

b) in the interaction with the "native" active center, if a system exhibits the degrees of freedom for changing the overlapping topology of molecular orbitals and if this change will lead to a decrease in the local symmetry of the intermediate (e.g., in the case of diene homopolymerization),

c) in the polymerization on ion pairs or on a less polar bond, when the monomer is exposed to the electric and "quantum-chemical" fields of the counterion if the perturbing force of these fields is sufficiently high for effective removal of local symmetry (e.g., in polymerization in hydrocarbon media). In these cases the counterion plays the role of an external "symmetry switch".

On the other hand, in the case of anionic-coordination polymerization on paramagnetic catalytic centers, the existence of free spins in the active center can lead to the stabilization of the triplet intermediate by involving unpaired electrons of the transition metal in the exchange interaction with those of the intermediate. This should increase the effect of spin exclusion. However, the final result will depend on the competition of two groups of factors stabilizing (pseudosymmetry and exchange interaction) and destabilizing (electric and "quantum-chemical" fields of the counterion) a locally symmetric triplet intermediate.

4 Discussion

The aim of the next part of the paper is to demonstrate the validity of the above concept of the elementary act of addition polymerization of non-cyclic monomers by comparing some of its conclusions with the available experimental data.

The choice of specific examples for comparison is both adequate and random because it depends on the subjective tendencies of the author and his natural desire to show mainly the advantages of this approach.

Moreover, it should be noted that, taking into account the euristic and program character of this concept, in the beginning its discussion should not be limited to the consideration of a detailed picture of the interaction between the monomer and the active center. The purpose of the discussion, just as that of the entire concept, is to show the prospects and the trend of development of the future axiomatic microscopic theory of polymerization as a whole on the basis of the phenomenological analysis of "particular" examples.

4.1 Free-Radical Polymerization of Symmetric Vinyl Monomers

In the LCFMOF approach, a symmetric vinyl monomer may be represented in the form of two identical radical fragments interacting with each other. During the process, these fragments interact with a radical fragment of the same type at the end of the polymer chain (the "native" radical center) in agreement with the principle of local symmetry leading to the formation of a free-radical locally symmetric intermediate (Scheme 7). As a result of the Jahn-Teller effect which is assumed to be a first-order effect [24], this intermediate will undergo fast isomerization of the type shown in Scheme 7 and is actually very close in structure to the transition state of the free-radical addition of the "native" active center to the double bond of the symmetric monomer.

The change in the free energy as a function of the reaction coordinate should have approximately the appearance of a plateau (Fig. 2). Note that the free energy profile shown in Fig. 2 does not contradict the Jahn-Teller theorem [33]. This theorem affirms that the adiabatic potential has no minimum near the degeneration point, but this is not the case for the thermodynamic potential, in particular for free energy. However, the absence of the minimum of the adiabatic

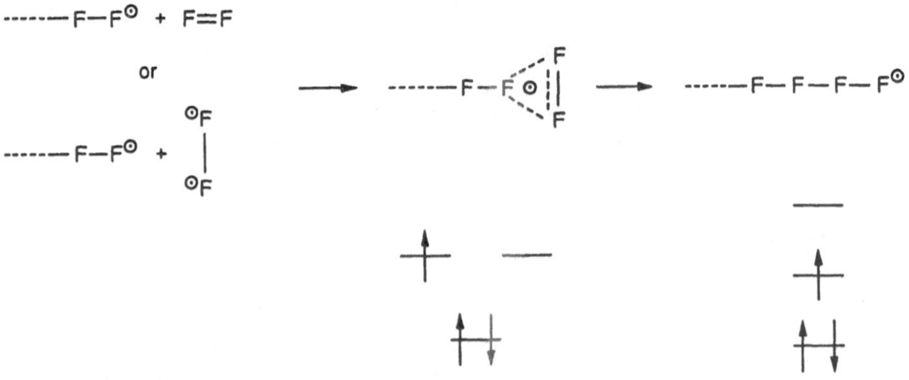

Scheme 7

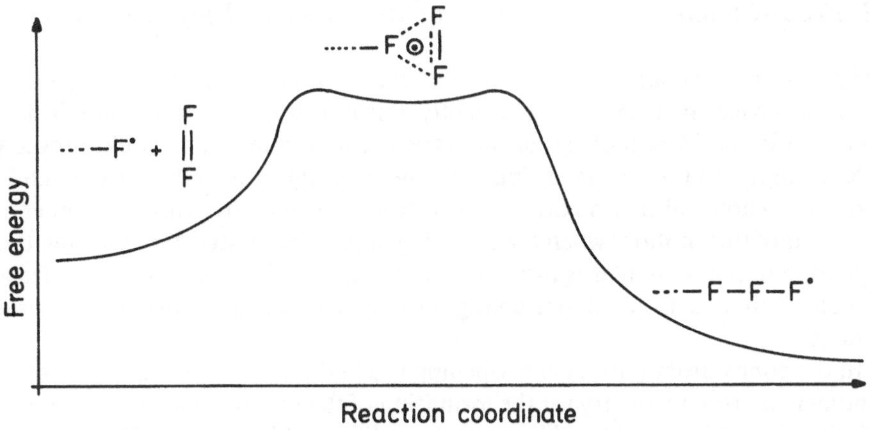

Fig. 2. Suggested free energy profile for the free radical polymerization of symmetric monomers

potential is not an indispensable condition of the immediate and necessary distortion of the initial symmetric nuclear configuration [34].

Moreover, Fig. 2 does not contradict the well-known Murrel-Laidler theorem [35, 36] which forbids taking the locally symmetric intermediate for the transition state because for the reason of symmetry at least two independent paths exist for its isomerization, i.e., for the insertion of the monomer into the polymer chain. In other words, the bifurcation of the reaction coordinate proceeds in the locally symmetric intermediate. Nevertheless, this is forbidden for the transition state by the Murrel-Laidler theorem which asserts that the matrix of force constants in the transition state has a single negative value, i.e., that the transition state corresponds to a single reaction coordinate.

However, it is difficult to say at present to what an extent the picture shown in Scheme 7 and Fig. 2 corresponds to reality. In any case, the free-radical polymerization of symmetric vinyl monomers is not forbidden from the standpoint of symmetry, and ethylene is known to be polymerized by the free-radical mechanism. It should be noted that in the case of quasisymmetric ethylene derivatives (e.g. propylene the quasisymmetric character of which is due to real symmetry of its frontier π-orbitals), free-radical polymerization is sometimes interrupted as early as in the stage of oligomer formation because chain termination reactions proceed intensively [37]. This has naturally no relationship to any exclusion with respect to symmetry.

As a conclusion it should be borne in mind that the simultaneous existence of the bifurcation point on the reaction coordinate and the Jahn-Teller effect imply most probably the impossibility of the formation of even slightly regular polymers in the free-radical polymerization of quasisymmetric monomers, at least at temperatures which are not very low.

4.2 Free-Radical Polymerization of Asymmetric Vinyl Monomers

In this case, a cyclic intermediate does not exhibit high symmetry and degenerate energy levels (Scheme 8). Consequently, it is not subjected to the Jahn-Teller effect. The isomerization of this asymmetric cyclic intermediate should proceed more regularly. In the course of ring opening, the fragments at the system's ends rotate as a whole (all this is naturally also true for symmetric vinyl monomers). As a result of this motion, when the initially planar atomic structure forming the vinyl bond in the monomer is inserted in the polymer chain, it acquires, at least for a short time, a fixed spatial configuration determined by the type of ring opening.

In the conrotatory path of ring opening [13] taking into account the overall monomer rotation necessary for the formation of the new σ-bond, as a result the fragments F and F_2 considered here rotate in different directions with respect to the chain (Scheme 9a). In the disrotatory type of ring opening, fragments F and F_2 rotate in the same direction with respect to the chain (Scheme 9b). Note that if the direction of rotation of fragment F is determined by the structure of the terminal part of the chain, in this case the rotation of fragment F_2 will correspond to that of F.

It is logical to assume that the type of ring opening of the intermediate determines to a certain extent the initial configuration of two end units of the chain. Moreover, it is natural to assume that the conrotatory and disrotatory motion lead to the formation of racemic and meso-diads, respectively.

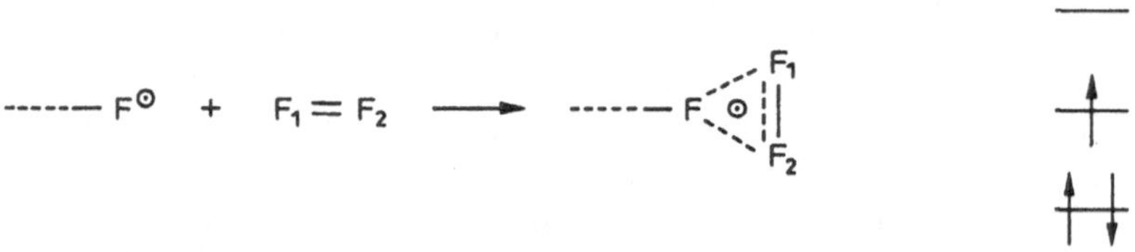

Scheme 8

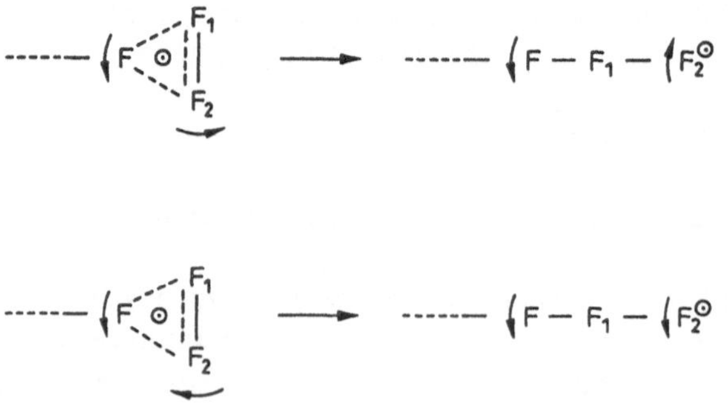

a

b

Scheme 9

From the topological viewpoint, the breaking of intramolecular bonds in a cyclic intermediate should proceed similarly to that in the cyclopropyl radical (or cation, or anion). In both cases the same quantity of labile electrons takes part in the process and the character of their motion changes in the course of ring opening. Proceeding from topological equivalence, it may be expected that the type of ring opening in the intermediate coincides with that for the cyclopropyl analog.

The real process of the formation of an iso- or syndiotactic polymer is without doubt much more complex than is shown in Scheme 9. But the point is that the transition to the LCFMOF approximation was actually caused by the necessity for the topologic simplification of the real situation. This transition makes it possible to single out and elucidate the role of the electronic factor as the driving force of polymerization, in particular, to understand the direction in which it attempts to influence the process in this case.

Moreover, other factors, the most important of which is the steric factor, are not taken into account. It cannot be ruled out that in some cases ring opening and hence, the formation of end units' configuration is controlled by steric interaction rather than by electronic interaction. However, when side substituents at the vinyl bond are flexible and not very bulky, the electronic factor should be of paramount importance. It should be noted that in this case the reasons for the formation of stereoregular vinyl polymers are not yet clear [38].

Taking into account the fact that the cyclopropyl radical is opened mainly by the conrotatory type [39] it should be expected that free-radical polymerization tends mainly to the syndiotactic type of stereoregularity. This conclusion does not seem to contradict reality [40, 41]. As an illustration Table 1 gives the data [43–45] on the content of syndiotactic structures in three polymers.

Actually, the fixation of diad configuration in the polymer chain proceeds only when the next monomer molecule is attached. Until this moment, the end unit can still undergo isomerization becoming either planar or its mirror antipode. In the absence of isomerization, the formation of a stereoregular syndiotactic polymer should be expected. In the case when an end unit has the time (before the attack of the following monomer molecule) to adopt a planar configuration, the distribution of diads in the polymer is described by Bernoulli's statistics (binomial distribution) [42]. If the state to two end units of

Table 1. Content of syndiotactic triads (S) in vinyl polymers obtained by free-radical polymerization [43–45]

Polymer	Temperature, °C	S, %
Poly(methyl methacrylate)	60	75.7
	– 100	90.0
Poly(methacrylic anhydride)	60	65.0
	– 100	73.0
Poly(vinyl chloride)	100	63.0
	– 45	73.5

the chain in the course of the polymerization (and also the formation mechanism of polymer tacticity) may be tentatively represented by the following chain of transformation:

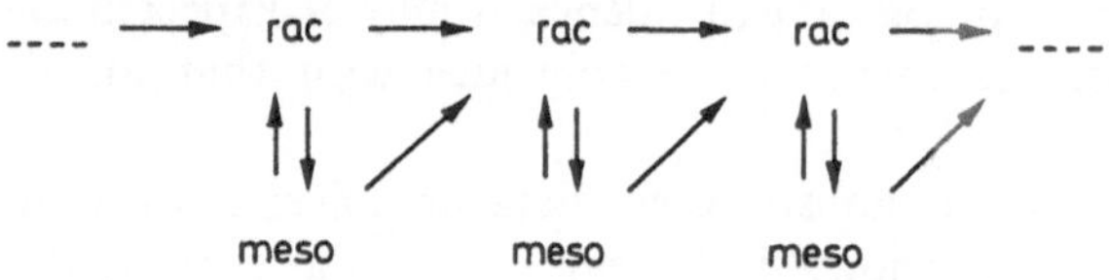

Scheme 10

then the distribution and content of diads in the polymer is described by more complex statistics. In particular, both the distribution and the content of diads may depend on monomer concentration and on conversion. Everything is determined by the ratio of propagation and isomerization rates.

4.3 Free-Radical Polymerization of Symmetric Dienes

Let us assume that the polymerization of quasisymmetric dienes proceeds in accordance with the principle of local symmetry, i.e., via the stage of formation of a locally symmetric intermediate. The structure of the locally symmetric diene intermediate (Scheme 11) is tentative. However, this is not very important since the train of thoughts is based only on the hypothesis of the existence of the Jahn-Teller effect in this case.

The modern theory of chemical reaction is based on the concept of the potential energy surface, which assumes that the Born-Oppenheimer adiabatic approximation [16] is obeyed. However, in systems subjected to the Jahn-Teller effect, adiabatic potentials have the physical meaning of the potential energy of nuclei only under the condition that non-adiabatic corrections are small [28]. In the vicinity of the locally symmetric intermediate, these corrections will be very large. The complete description of nuclear motion, i.e. of the mechanism of the chemical reaction, can be obtained only from Schroedinger's equation without applying the Born-Oppenheimer approximation in the vicinity of the locally

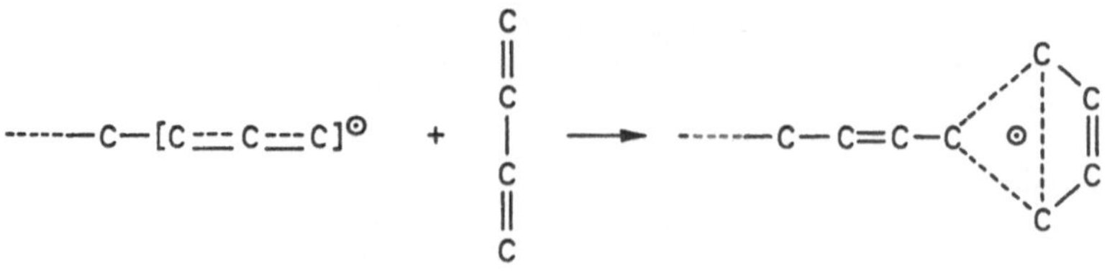

Scheme 11

symmetric intermediate and thus without applying the concept of the potential energy surface.

Although Fig. 2 is not quite correct in the above sense, the conclusion is probably physically justified that the free energies of the locally symmetric intermediate and the nearest transition state are very close to each other. First, the Jahn-Teller system, not being at the minimum of the adiabatic potential, is hardly long-lived in the condensed liquid phase, in other words it will hardly be at any considerable minimum of free energy. Secondly, this system cannot be at the maximum of free energy (in this case it would be the transition state) because of the Murrel-Laidler theorem [36]. Hence, the concept of a plateau in Fig. 2 is actually physically meaningful.

As already mentioned, bifurcation of the reaction coordinate takes place in the locally symmetric Jahn-Teller intermediate. Let us assume that one branch of the reaction coordinate leads to the fixation of the 1,4 (or 4,1)-structure in the penultimate chain unit (Scheme 12a) and the other branch causes the formation of the vinyl structure of this unit (Scheme 12b). Then it should be expected that the relative content of these structures in the polymer will not depend on the conditions of the process, in particular on the polymerization temperature (because a plateau exists in Fig. 2).

Table 2 gives the structural composition of polybutadiene and polyisoprene obtained by free-radical polymerization [46–48]. The fact that the content of the vinyl structure units in polydienes is independent of the polymerization temperature over a wide range and that their content is much higher than the

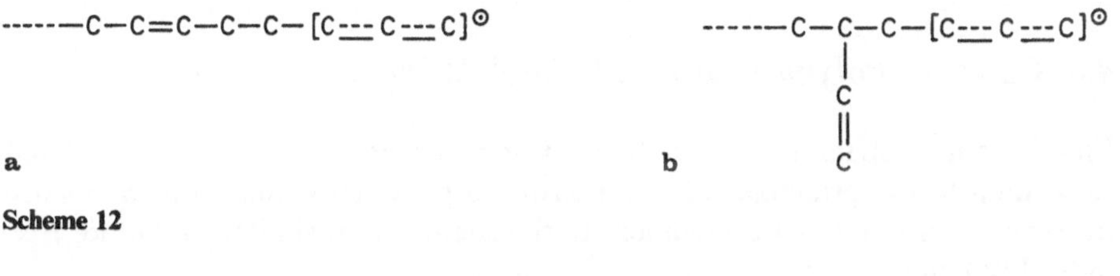

Scheme 12

Table 2. Effect of polymerization temperature on the microstructure of polydienes obtained by free-radical polymerization [46–48]

Polymer	t, °C	Content of units, %		
		1,4	1,2	3,4
Polybutadiene	– 35	83	17	
	5	82	18	
	50	81	19	
	100	78	22	
	223	82	18	
Polyisoprene	– 47	86	8	6
	50	88	7	5
	100	89	5	6
	257	89	2	9

equilibrium content [49] is so unusual that it is probably difficult to find an alternative explanation outside the framework of the present concept (at least, the author does not know of any attempt to interpret the data in Table 2).

4.4 Free-Radical Copolymerization of Symmetric Dienes

Let us assume that we are dealing with an alternating copolymerization of butadiene (or isoprene) with a monomer drastically differing from butadiene. In this case the symmetric diene will be attached all the time to an "alien" radical active center and, vice versa, an "alien" monomer will be bonded to a diene radical center. The principle of local symmetry is no longer valid, and in the absence of the Jahn-Teller effect the formation of only the 1,4 (4,1)-structure of diene units in the copolymer should be expected.

It is interesting to note that in the free-radical copolymerization of butadiene and 1,1-diphenylethylene an alternating copolymer is formed, all its diene units having 1,4-structure [49].

4.5 Free-Radical Polymerization of Asymmetric Diene

In this case, since the principle of local symmetry is weakened, it should be expected that the content of vinyl structure in polydiene is lower than in polybutadiene. In fact, e.g., in polychloroprene, the fraction of vinyl units in the temperature range from -40 to $+40\,°C$ varies from 1 to 3% [49].

4.6 Cationic Polymerization of Vinyl Monomers

First, chain propagation by a free-cationic mechanism will be considered. According to the principles of the intermediate, cyclicity and correspondence, the scheme for monomer attachment to the chain will be similar to that for free-radical polymerization:

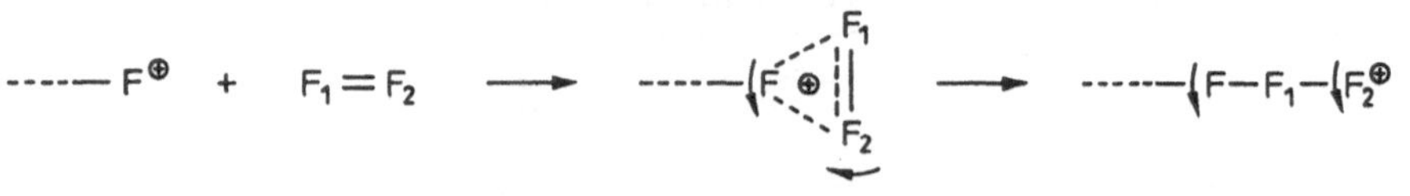

Scheme 13

However, in contrast to the free-radical process, ring opening in the cationic intermediate will proceed mainly by the disrotatory type [39]. This should lead (Scheme 9b) to the predominant formation of the mesodiad at the chain end.

Consequently, in the free-cation propagation mechanism, conditions exist for the predominant formation of the isotactic chain structure. In order that these conditions can actually be fulfilled, the following monomer molecule should, naturally, interact (be complexed) with an active center before its configuration changes (is transformed into a planar structure or into the mirror-image of the initial structure). As in free-radical polymerization (Scheme 10), the formation of the polymer microstructure by the free-cationic mechanism can be formally represented in the most general form by the following chain of transformations (it is evident that the planar structure of the end unit is statistically equivalent to "infinitely" rapid isomerization):

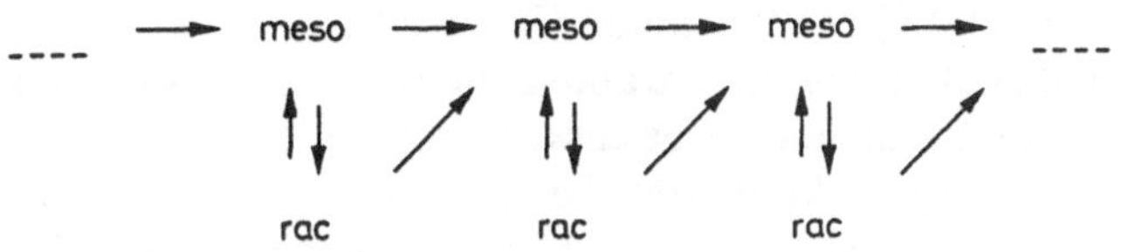

Scheme 14

If isomerization is absent or slow, an isotactic polymer is obtained. In the case of fast isomerization or a planar structure of the end unit, diad distribution in the polymer is described by Bernoulli's statistics. If the rates of isomerization and propagation are comparable, the precise form of the distribution function may be determined on the basis of a method of solution of multicenter problems developed by the author [6].

It should be noted that, in cationic polymerization, complexation between a free cation and a monomer has been reported in recent Plesch papers [50]. Moreover, the following hypothesis of cationic polymerization has been discussed: the propagation on an ion pair also proceeds via the formation of a complex the structure of which coincides with that of the ion pair separated by the monomer: $M_n^{\oplus}$, M, $A^{\ominus}$ [51]. In both cases, according to the general concept of the structure of π-complexes [52], it is certain that the cyclic structure is present both in the complex of the monomer with a free cation and in the ion pair separated by the monomer. However, in the latter case the presence of the counterion can drastically (in the sense of symmetry) change the electronic characteristics of the complex [25]. This, in turn, can lead to a change in the type of ring opening in the intermediate from the disrotatory to the conrotatory type [52] and, correspondingly, to the reversion of stereospecificity.

On the other hand, the counterion can greatly slow down the isomerization of the end unit and thus favour the formation of a stereoregular polymer. However, these very important problems require further development and a more detailed discussion.

In conclusion of this section it will be noted that one of the first examples of homogenous polymerization in which isotactic polymers with the content of isotactic triads higher than 80% have been obtained is the cationic polymerization of vinyl ethers [53, 54].

4.7 Anionic Polymerization of Symmetric Vinyl Monomers

In contrast to free-radical polymerization, in anionic polymerization it is possible to change gradually the structure of the active center passing from the ion-covalent carbon-metal bond to the ion pair and, in the limit, to a free anion by increasing the polarity and solvating ability of the solvent. For the simplification of the interaction picture without reducing the consideration area five basic states of the carbon-metal bond in an anionic active center will be considered: 1) the slightly polar carbon-metal bond (R-X), 2) the polarised carbon-metal bond ($R^{\ominus} \cdots X^{\oplus}$), 3) the contact ion pair ($R^{\ominus}, X^{\oplus}$), 4) the solvent-seperated ion pair ($R^{\ominus}|X^{\oplus}$) and 5) the free anion ($R^{\ominus} + X^{\oplus}$).

The problem of the real existence of the five above states of the carbon–metal bond in any specific situation will not be discussed because it is considered comprehensively in Refs. [55–57]. It should only be noted that in the case of the carbon–lithium bond all five bond states appear to exist, whereas other alkali metals cannot form the slightly polar carbon–metal bond. We mean by the slightly polar bond the bond in which the electron density distribution may be assumed to be virtually symmetric (pseudosymmetric) from the viewpoint of the correspondence principle.

Let us consider the interaction between a symmetric vinyl monomer and a free anion. According to the local symmetry principle, in this case a locally symmetric intermediate is formed being in the ground triplet state (Schemes 5 and 15). In accordance with the spin exclusion principle, it should be expected that the polymerization of symmetric vinyl monomers on free anions proceeds very slowly.

The attachment of an active center with a slightly polar carbon–metal bond to a symmetric monomer is in poor correlation with respect to symmetry (Scheme 16) and is "forbidden" because of the correspondence principle.

Hence, the symmetric vinyl monomer should polymerize most actively on contact ion pairs. First, the presence of a counterion should eliminate or at least weaken the effect of the local symmetry principle and at the same time that of the spin exclusion principle. Secondly, on the base of Epiotis' results [58], it may be expected that the polar character of the carbon–metal bond should favor cycloaddition. However, this also follows from the formulation of the correspondence principle if the transition to a more polar bond is considered to be a transition to a more asymmetric bond.

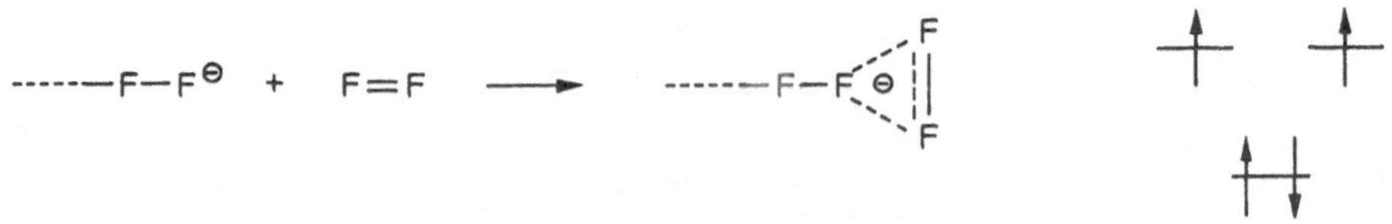

Scheme 15

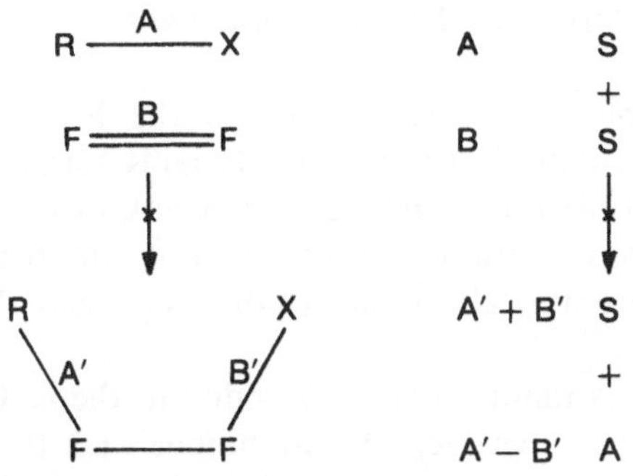

Scheme 16

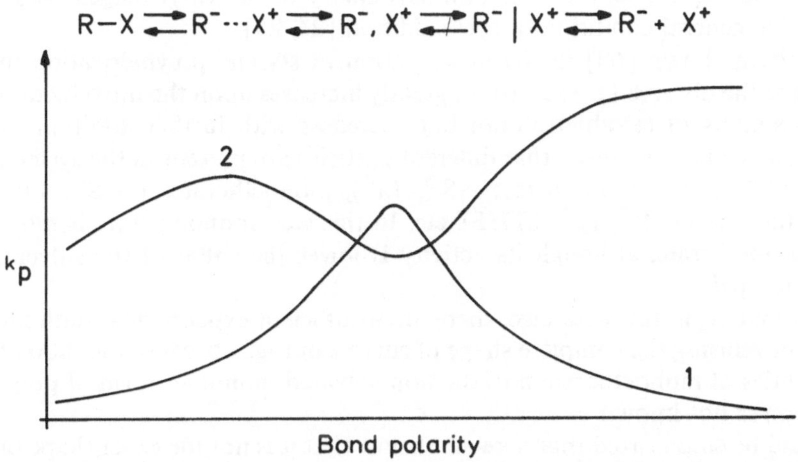

Fig. 3. Approximate character of the dependence of propagation rate constant k_p on the polarity of the carbon–metal bond in the active center for the polymerization of symmetric (*curve 1*) and asymmetric (*curve 2*) vinyl monomers

Summing up all the above considerations it should be expected that the dependence of the propagation rate constant k_p (not the rate itself) on bond polarity in the active center should be a bell-like curve schematically shown in Fig. 3 (curve 1).

It may be tentatively assumed that curve 1 in Fig. 3 does not contradict the very scarce experimental data on the anionic polymerization of symmetric vinyl monomers. For instance, it is known (bibliography see Ref. [59]) that ethylene is anionically polymerized on the polarized carbon–lithium bond or the corresponding contact ion pair. However, additional experimental investigations are needed for drawing a more definite conclusion about the validity of curve 1 in Fig. 3.

4.8 Anionic Polymerization of Asymmetric Vinyl Monomers

In this case the spin exclusion principle is not valid because the locally symmetric intermediate (there are intermediates but not of this type) is absent. The correspondence principle is also weakened to a considerable extent. For all these reasons, the dependence of the propagation rate constant k_p on the polarity of the carbon–metal bond is approximately similar to that represented qualitatively by curve 2 in Fig. 3.

The existence of the assumed local maximum on curve 2 is due to the fact that the mutual donor-acceptor interaction between the monomer and the active center is probably most effective when the asymmetry of the former and the carbon–metal bond polarity (pseudosymmetry) in the latter supplement each other to a "certain asymmetry" which is attained, e.g. in the interaction between the symmetric vinyl monomer and the contact ion pair. It should be noted that the requirement of a certain asymmetry of the vinyl reagent is an indispensable general condition of cycloaddition [24, 25].

It has been shown [60] in the investigation of styrene polymerization in benzene that the observed rate constant greatly increases upon the introduction of small amounts of tetrahydrofuran but decreases with further addition. A detailed kinetic analysis shows that different particles are present in the system: dimers $(\cdots S^{\ominus}, Li^{\oplus})_2$, monomers $(\cdots S^{\ominus}, Li^{\oplus})$, monoetherates $(\cdots S^{\ominus}, Li^{\oplus}$, THF), dietherates $(\cdots S^{\ominus}, Li^{\oplus}$, 2THF), etc. In this case monoetherate is more active than dietherate, although its activity is lower than that of the solvent separated ion pair.

Unfortunately, in this case also, there are insufficient experimental data for accepting or refusing the complete shape of curve 2 in Fig. 3 because the ratio of the reactivities of monoetherate and the non-solvated monomer form of polystyryllithium is not known.

It should be emphasized that here and below that it is not the exact shape of the curves which is significant but only the tendency represented by these curves towards an increase or decrease in mutual reactivity of monomers and active centers depending on the carbon–metal bond polarity and hence depending on the process conditions.

In the polymerization of vinyl monomers on free anions or solvent separated ion pairs, ring opening in a cyclic intermediate occurs in the conrotatory manner [39]. Hence, anionic polymerization in polar media should tend to form syndiotactic polymers and should be slightly similar in this respect to free-

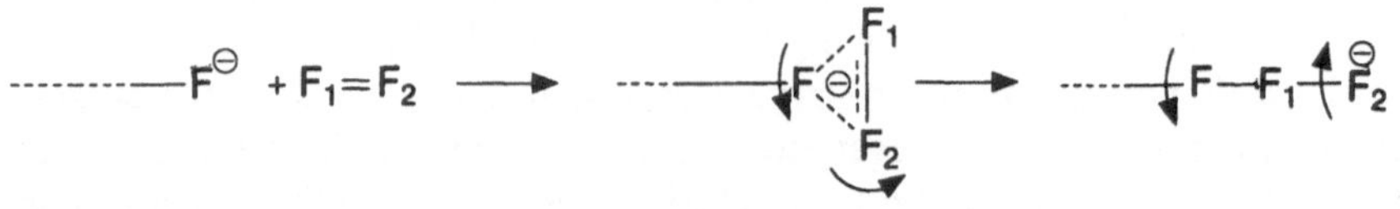

Scheme 17

Table 3. Content of racemic diads (r) in anionic polystyrene [62, 63]

Initiator	Solvent	t, °C	r, %
n-Butyllithium	Toluene	− 20	70
Naphthalene sodium	Toluene	30	66
	Heptane	30	71
	THF	30	67
	THF	− 78	71
	Dimethoxyethane	30	68
Naphthalene potassium	Toluene	30	46
	Toluene	− 78	38
	Heptane	30	41
	THF	30	67
	THF	− 78	67
	Dimethoxyethane	30	68
Naphthalenep caesium	Toluene	30	68
	Toluene	0	37
	THF	30	58

radical polymerization. Formally, in this case chain propagation may be illustrated by Schemes 10 and 17.

As in the case of cationic polymerization, the presence of a metal atom can drastically change the electronic parameters of ring opening. In other words, the counterion may play the role of a symmetry "switch", i.e. it can induce the reversion of stereospecificity of the active center. The possibility of anionic chain propagation on contact ion pairs via on intermediate stage of the formation of monomer-separated ion pairs was considered by Erusalimsky as far back as 1970 [61]. However, even 20 years later the author of the present paper does not attempt to discuss this problem in detail.

In order to compare the above conclusions about the actual situation, Table 3 gives the data on the microstructure of polystyrene obtained by anionic polymerization [62, 63]. In spite of the complex character of the data in Table 3, the general trend towards the formation of racemic diads in polar media is quite evident.

4.9 Anionic Polymerization of Symmetric Dienes

The attachment of a slightly polar carbon–metal bond to a symmetric or quasisymmetric diene is in correspondence with respect to symmetry (pseudosymmetry) and, according to the correspondence principle, should proceed relatively rapidly. It is interesting to note that as far back as 30 years ago Stearns and Forman [64] and Szwarc [65] drew an analogy between the elementary act of diene insertion on the carbon–metal bond and the Diels-Alder reaction. Moreover, taking into account the fact that dienes can take part in the diene synthesis only in the *cisoid* conformation [66] it should be assumed by analogy

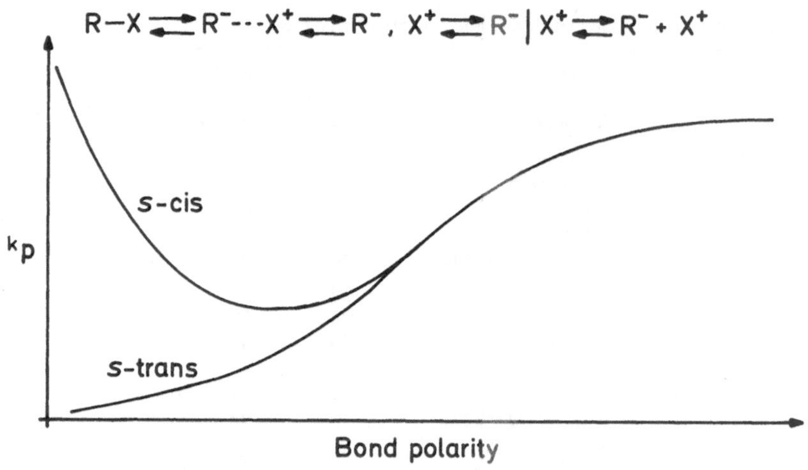

Fig. 4. Approximate character of the dependence of the polymerization rate constant k_p for *s-cis-* and *s-trans*-conformation of quasisymmetric dienes on the polarity of the carbon–metal bond in the active center

that in the polymerization on an active center with a slightly polar carbon–metal bond, the *s-cis*-conformation of the symmetric diene is much more active than the *s-trans*-conformation. When the polarity of the carbon–metal bond increases, the correspondence with respect to symmetry decreases, which should lead to a gradual equalization of the reactivities of *s-cis-* and *s-trans*-conformations of the quasisymmetric diene. This situation is shown qualitatively in Fig. 4 (the polymerization of symmetric dienes on a free anion will be discussed below).

It is known [66] that in *trans*-piperylene the methyl group does not prevent the molecule from adopting the cisoid conformation:

The formation of the *cisoid* conformation is very difficult for *cis*-piperylene because of the steric repulsion between the hydrogen atom and the methyl group [66]:

Proceeding from the above considerations it should be assumed that *trans*-piperylene behaves in polymerization according to the curve for the *s-cis*-conformation in Fig. 4, whereas the behavior of *cis*-piperylene corresponds to the curve of the *s-trans*-conformation. The available literature data reviewed in detail in Ref. [67] (see also Refs. [49, 68]) confirm this assumption. Moreover, this assumption is in an agreement with the fact that in isoprene polymerization the monomer form complexed with tetramethylene diamine is less reactive than the non-solvated form [69].

4.10 Stereospecificity of the Anionic Polymerization of Symmetric Dienes

It follows from the existence of "conformational scissors" (Fig. 4) that in the polymerization of symmetric or quasisymmetric dienes in hydrocarbon media on an active center with a slightly polar carbon–metal bond the primary acts of monomer attachment lead to the *cis*-conformation of the end unit. This conclusion is in good agreement with the modern concepts of the formation mechanism of the *cis*-structure of anionic polydienes [70]. According to these concepts this is followed by either the attachment of the next monomer molecule or by the *cis-trans* isomerization of the end unit. The microstructure is fixed at the moment of the attachment of a new monomer unit to the active center, the configuration (*cis*- or *trans*-) of the end unit being retained in the polymer chain:

$$\cdots \text{cis}^* + \text{monomer} \rightarrow \cdots \text{cis—cis}^*$$

$$\cdots \text{trans}^* + \text{monomer} \rightarrow \cdots \text{trans—cis}^*$$

When the polarity of the carbon–metal bond increases, the primary conformation of the end unit will gradually increasingly correspond to the thermodynamical equilibrium conformation of the initial diene monomer (Fig. 4). As a result, the stereostructure of the growing chain is controlled by the kinetic but

Table 4. Structure of polybutadiene obtained in tetrahydrofuran at 0 °C in the presence of naphthalene complexes of alkali metals as catalysts [68]

Alkali metal	Polybutadiene structure, % of units		
	cis-1,4	*trans*-1,4	1,2
Li	0	3.6	96.4
Na	0	9.2	90.8
K	0	17.5	82.5
Rb	0	24.7	75.3
Cs	0	25.5	74.5

Table 5. Structure of polybutadiene obtained in pentane at 0 °C in the presence of alkali metals as catalysts [68]

Alkali metal	Polybutadiene structure, % of units		
	cis-1,4	*trans*-1,4	1,2
Li	35	52	13
Na	10	25	65
K	15	40	45
Rb	7	31	62
Cs	6	35	59

also by the thermodynamic factors even in the initial stage of the process. This conclusion contradicts neither the existing theoretical concepts [70] nor the available experimental data (Tables 4 and 5), however, it requires a more rigorous verification.

4.11 Microstructure of Anionic Polydienes

In the interaction of the symmetric diene with its own free anion, in case of an attack at the α-carbon atom of the chain, according to the local symmetry principle, a triplet locally symmetric intermediate should be formed (Scheme 18a). In accordance with the spin exclusion principle, this reaction path, i.e., the attachment of the molecule of a quasisymmetric diene to the α-carbon atom, is suppressed. It should be noted that it is the assumption of the triplet character of the locally symmetric intermediate that has a physical meaning rather than its specific structure proposed in Scheme 18a (it should be established experimentally).

In the case of the vinyl monomer, its polymerization would stop at the stage of the formation of the locally symmetric intermediate. In contrast, for dienes the attack at the γ-carbon atom is possible and results in the formation of an intermediate of lower symmetry. Hence, it becomes possible to eliminate or at least to weaken the spin exclusion principle. Consequently, the polymerization at the γ-carbon atom of a free anion should proceed more intensively than at the α-carbon atom. This should lead to the predominant formation of the vinyl structure of chain units.

Upon the transition to the polymerization on ion pairs and, further, on the carbon–metal bond, the presence of the counterion in the active center will distort and decrease the local symmetry of the intermediate, thus weakening the spin exclusion principle and gradually leading to the predominant formation of the 1,4(4,1)-structure of diene units in the polymer. This situation is qualitatively shown in Fig. 5.

Table 4 gives the data on the microstructure of polybutadiene obtained in an electron donor medium [68].

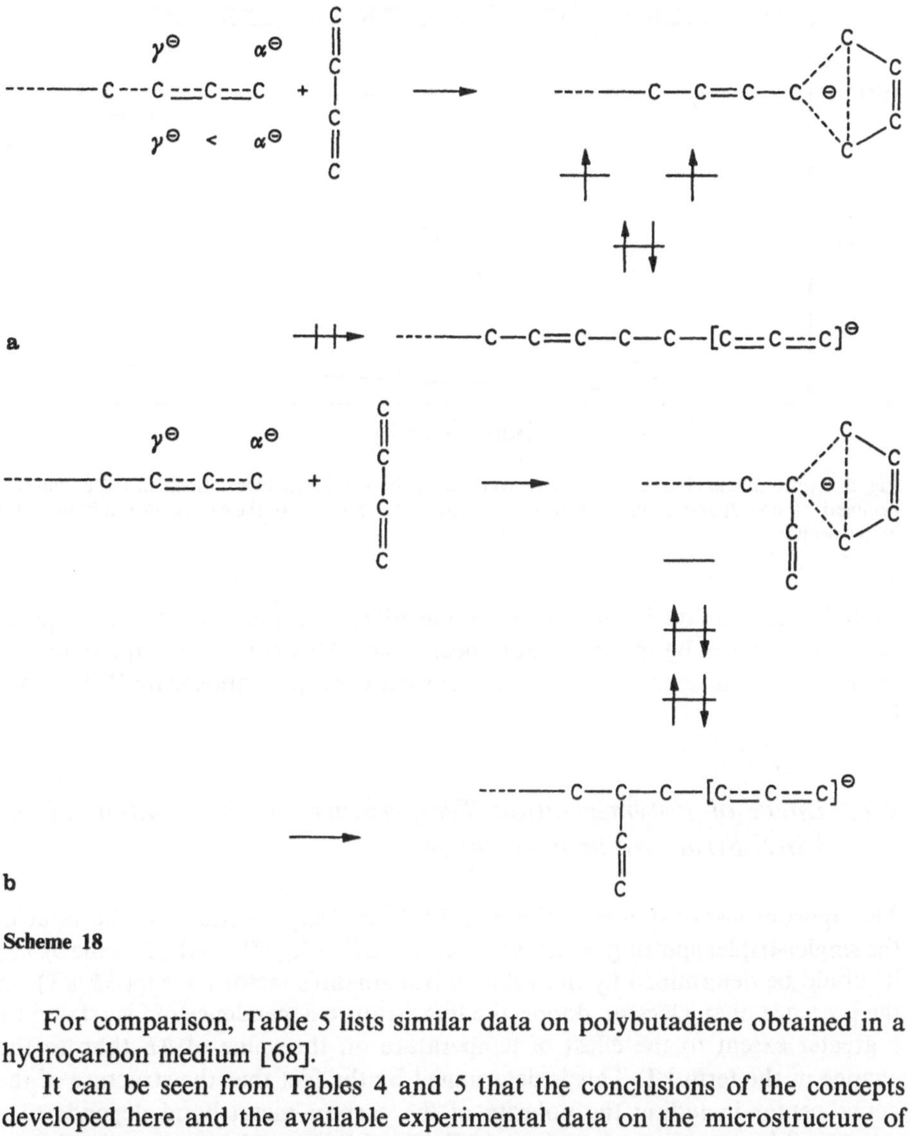

Scheme 18

For comparison, Table 5 lists similar data on polybutadiene obtained in a hydrocarbon medium [68].

It can be seen from Tables 4 and 5 that the conclusions of the concepts developed here and the available experimental data on the microstructure of polydienes are in good agreement. The greatest changes in the microstructure of polybutadiene on passing from the electron donor to the hydrocarbon medium are observed for the lithium catalysts, and the smallest changes are observed for rubidium and cesium catalysts. This agrees with the fact that if all alkali metals are considered, on passing from tetrahydrofuran to pentane the polarity of the carbon-rubidium or carbon-caesium bond changes to the lowest extent and that of carbon-lithium bond to the highest extent [56].

Hence, although even in strongly solvating media the main fraction of electron density is on the α-carbon atom [71, 72], according to the local symmetry and spin exclusion principles, the vinyl structure of the polymer chain

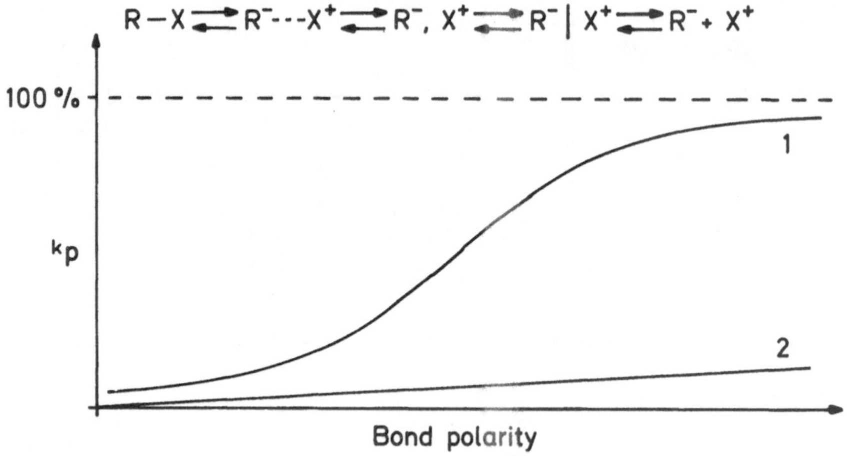

Fig. 5. Approximate character of the dependence of the vinyl structure content in the polymer of symmetric (*curve 1*) and asymmetric (*curve 2*) diene on the polarity of the carbon–metal bond in the active center

should predominate in polydienes obtained by the polymerization of quasi-symmetric dienes by the free-anion mechanism. This is probably the nature of the monomer-induced reaction center transfer concept proposed by Dolgoplosk [73].

4.12 Effect of Polymerization Temperature on the Content of the Vinyl Structure in Polydienes

The rigorousness of spin exclusion should depend exponentially on the value of the singlet-triplet splitting in the intermediate $\Delta E = E^{Fermi}_{corr} - \Delta E$ (Scheme 6), i.e., it should be determined by the value of Boltzmann's factor $f = \exp(\Delta E/kT)$. In the presence of an electron donor, the temperature dependence of f is related by a greater extent to the effect of temperature on the value of ΔE than to the change in the term kT. This is determined by the fact that the structure of the active center including the polarity of the carbon–metal bond depends to a considerable extent on temperature [56]. The solvation of the counterion by the electron donor increases with decreasing temperature, thus increasing its protection and decreasing the distorting action of the counterion on the local symmetry. Formally this corresponds to the lower value of ΔE and, hence, to the higher values of ΔE and f.

Hence, in the presence of or, to an even greater extent, in a medium of an electron donor the decrease in polymerization temperature should lead to an increase in vinyl structure content in polydienes. The same conclusion also follows directly from Fig. 5 if it is taken into account that the polarity of the carbon–metal bond decreases with increasing temperature. Figure 6 shows the

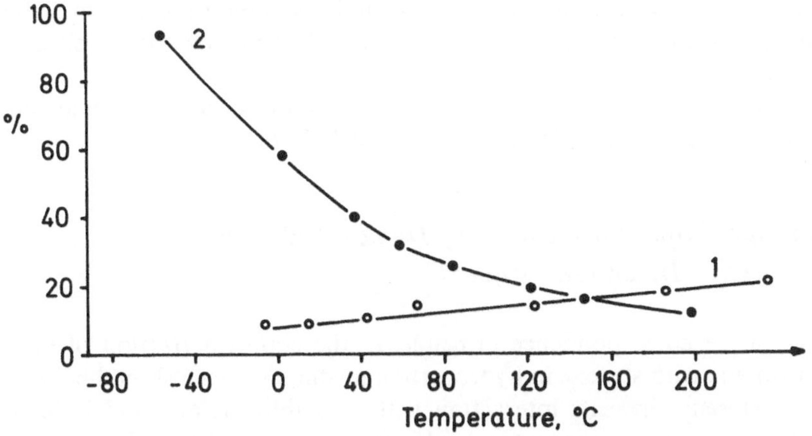

Fig. 6. Vinyl structure content in polybutadiene obtained in cyclohexane (*curve 1*) and in cyclohexane with THF addition (*curve 2*) vs polymerization temperature

data on the content of 1,2-structure units in polybutadiene synthesized in cyclohexane [14] and in cyclohexane with the addition of tetrahydrofuran [75, 76].

In a hydrocarbon medium, the formation of a triplet intermediate is not possible since $\Delta E < 0$ because of the perturbing action of the counterion. The value of f increases with temperature and the content of the vinyl structure also increases. This change is not so marked as in the case of the electron donor presence because in the hydrocarbon medium, ΔE depends slightly on temperature.

4.13 *Anionic Copolymerization of Symmetric Dienes*

Just as in the free-radical copolymerization of dienes considered above, in this case also, the action of the local symmetry principle is weakened and hence, the spin exclusion principle is no longer valid. Consequently, the predominant

Table 6. Structure of butadiene units in homo- and copolymers obtained in solvating media under identical conditions [77, 78]

Structure of units, %		Structure of units, %	
1,4	1,2	1,4	1,2
alternating copolymer with stilbene		homopolymer	
88	12	18	82
alternating copolymer with 1,1-diphenylethylene		homopolymer	
70	30	18	82
copolymer with styrene		homopolymer	
83	17	20	80

formation of the 1,4(4,1)-structure of diene units should be expected in an alternating anionic copolymer including the case of polymerization in electron donor media.

Table 6 gives the microstructure content of butadiene units in homo- and copolymers obtained under identical conditions [77, 78].

4.14 Anionic Copolymerization of Dienes and Styrene ("Activity Inversion" Effect)

According to the correspondence principle, in the copolymerization of butadiene or isoprene with styrene in hydrocarbon media, the diene should be more active than styrene since its interaction with a slightly polar bond is better correlated with respect to symmetry. In this case, other conditions being equal, the overall rate of diene homopolymerization may be lower than that of styrene since it is determined not only by the reactivity of the monomer but also by the effective concentration of active centers.

On passing to more polar media, the situation should drastically change: styrene should gradually become more reactive than butadiene or isoprene. It is probably the correspondence principle that provides the explanation of an experimental phenomenon which obtained a special title, the "activity inversion" effect, because of its apparent anomality.

It has been established [79, 80] that the initial stage of copolymerization in hydrocarbon media is determined only by the influence of diene. Styrene virtually does not take part in the reaction. As a result, the chain propagation rate is close to that in the polymerization of diene alone. Only after almost all the diene has been consumed, does the insertion of styrene into the growing chain begin. In this case the polymerization rate is higher than in styrene homopolymerization (presumably, because of the higher concentration of the monomer form of active centers). Finally, instead of the expected random copolymer, a block copolymer is formed in this system. Its formation mechanism has not yet been completely explained in the literature.

The addition of small amounts of ethers or amines changes the situation to such an extent that the effect of styrene begins to dominate. This behaviour also indicates that the electron factor plays a non-trivial role in the polymerization process.

4.15 Anionic Polymerization of Asymmetric Dienes

In this case there are no symmetry restrictions, and regardless of the conditions of the process mainly the formation of the 1,4(4,1)-structure of diene units should be expected both in homo- and copolymers. The content of *cis*- or *trans*-units in the polymer is entirely determined by the thermodynamical equilibrium conformation of the initial monomer.

Table 7. Effect of monomer symmetry on the structure of homopolydienes obtained by anionic polymerization in tetrahydrofuran at – 78 °C [76, 81–83]

Diene	Polydiene microstructure, % of units			
	cis-1,4	*trans*-1,4	1,2	3,4
$CH_2{=}CH{-}CH{=}CH_2$	0	9	91	
$CH_2{=}CH{-}C(Ph){=}CH_2$	74	0	26	0
$CH_2{=}CH{-}CH{=}CH{-}Ph$	11	79	0	10
$CH_2{=}CH{-}CH{=}CH{-}C({=}O){-}OCH_3$	0	100	0	0

Table 7 lists for comparison selected data on the structure of homopolymers obtained by the anionic polymerization of dienes in a solvating medium [76, 81–83].

4.16 Anionic-Coordination Diene Polymerization

According to the opinion of the author, there are no fundamental differences between anionic and anionic-coordination polymerization. Moreover, the former should be regarded as an adequate simplified model for the latter. From this viewpoint, the effect of the principles considered above should also be extended to the range of anionic-coordination processes and, possibly, to Ziegler-Natta heterogeneous catalysis. However, although these types of polymerization are similar, they naturally should exhibit great differences.

First, the nature of the bonds, say, carbon-alkali metal and carbon-transition metal bond is slightly different. However, as regards interaction with the monomer, it is not the detailed shape of the orbital that is of the greatest importance, but the distribution of the electron density between the hydrocarbon part of the chain and the counterion. The problem is whether the monomer during the interaction draws the electron density from one (the case of the ionic bond) or two (the case of the covalent bond) centers. This alternative should be closely related to the concept of the symmetry or asymmetry of the carbon-transition metal bond. This can be seen even from the fact that, as known from quantum mechanics, it is not the molecular orbital itself as a wave function that has a physical meaning, but only its square describing the electron density distribution in the corresponding bond. Hence, the problem of the symmetry or asymmetry of electron density distribution of any bond is physically equivalent to that of the symmetry or asymmetry of electron density distribution in this bond. In other words, a covalent bond should be considered to be symmetric (with respect to a plane passing approximately through the center of the bond normal to it), whereas the ionic bond should be regarded as asymmetric in complete agreement with all the above considerations.

Secondly, in contrast to alkali metals, transition metals can complex with both electron donor and electron acceptor ligands. Complexation with the former leads to increasing electron density on the transition metal, to the resulting decrease in electronegativity and, correspondingly, to an increase in the ionic character (asymmetry) of the carbon-transition metal bond. This should inevitably affect the microstructure of the resulting polydiene. The expected tendency of the change in this structure is as follows: from the 1,4-*cis* to 1,4-*trans* and further to an increase in vinyl structure content. In contrast, complexation with electron acceptor ligands leads to an increasing deficit of electron density on the transition metal, to an increase in its electronegativity and the covalent character (symmetry) of the carbon-transition metal bond. Correspondingly, the predicted tendency of the change in the microstructure is opposite to the above case: a gradual transition from the vinyl structure to 1,4-*trans*- and, further, to 1,4-*cis*-structure should be observed. Moreover, the better the correlation of the reaction with respect to symmetry, the higher the propagation rate constant (Fig. 4). All these conclusions are in relatively good agreement with experimental data [49].

Third, a characteristic feature of transition metals greatly distinguishing them from alkali or alkaline-earth metals is the fact that in some cases unpaired spins exist on atomic orbitals or the molecular orbitals of complexes formed by transition metals with a very wide range of ligands. These spins can increase the efficiency of spin exclusion for at least two reasons. First, when the active center contains a heavy atom, moreover in the paramagnetic state, the multiplet transitions are virtually no longer forbidden [26]. This facilitates the singlet-triplet conversion in the intermediate. Secondly, the "paramagnetic" electrons of the transition metal participating in exchange interaction with unpaired electrons of the triplet intermediate lead to its considerable stabilization and, correspondingly, to the extension of the applicability of the spin exclusion principle.

Hence, proceeding from the above considerations it is possible to introduce the following problem: to attempt to predict a certain experimentally checked result which would be difficult to explain from the standpoint of other concepts or, to put it simply, it would be virtually impossible. This test would evidently be a very rigorous verification of the entire ideological basis of the proposed approach.

Consequently, it is clear from the above considerations that the polymerization of quasisymmetric dienes on paramagnetic complexes of transition metals should be accompanied by the struggle between two trends. On the one hand, the existence of the electric field and specific quantum-chemical interactions ("quantum-chemical" field) should decrease the symmetry of the intermediate and the validity of the spin exclusion principle. On the other hand, as already mentioned, the intrinsic paramagnetism of the transition metal should increase the validity of the spin exclusion principle. Which of the two trends prevails, is difficult to predict. In any case it may be expected that the polymerization of symmetric dienes, such as butadiene or quasisymmetric isoprene, on paramagnetic anionic-coordination centers is accompanied by a considerable formation of the vinyl microstructure of the polymer.

The polymerization on purely π-allyl complexes of transition metals is of the greatest interest for checking these predictions. Allyl itself is a good low molecular weight model for the end unit of the growing polymer chain. In this connection it may be expected that the symmetry of the ligand environment of the transition metal actually remains invariable during polymerization. This fact, in turn, ensures the retention and constancy of the spin state of the central atom.

Only four purely π-allyl complexes of transition metals capable of initiating the polymerization of dienes yielding high molecular weight polymers have been mentioned in the literature [49, 68]. All of them are listed in Table 8 which shows that each of them contains three allyl (crotyl) ligands. Proceeding from general considerations [84] it may be assumed that the symmetry of each of these complexes is close to octahedric symmetry (each allyl is a bidentate ligand [49]).

In agreement with the ligand field theory [85], Scheme 19 shows the qualitative energetic diagram of molecular orbitals of the valence shell of the expected octahedric complex ML_6. These orbitals are constructed from the valence orbitals of the metal and the group orbitals of ligands. The splitting of levels of d-atomic orbitals of the metal and the group orbitals of ligands in the octahedric crystalline field are shown separately.

Table 8 gives the data on the electron structure of valence shells of atoms of the corresponding transition metals [86]. Taking into account the fact that the total number of electrons provided by three allyl (crotyl) ligands to the molecular orbitals of the complex is nine, it may be concluded that the complexes of Cr (chromium), Nb (niobium) and Ti (titanium) should be paramagnetic and those of Rh (rhodium) should be diamagnetic. In this case the total

Table 8. Catalysts of anionic-coordinative polymerization based on π-allyl complexes of transition metals [49, 68]

Metal	Structure of the valence shell of the metal	π-allyl complex	Polybutadiene structure, % of units	
			1,4	1,2
Cr	$3d^5 4s^1$	$(C_3H_5)_3Cr$ $(C_4H_7)_3Cr$	10–19	81–90
Nb	$4d^4 5s^1$	$(C_4H_7)_3Nb$	0	100
Ti	$3d^2 4s^2$	$(C_4H_7)_3Ti$	17	83
Rh	$4d^8 5s^1$	$(C_4H_7)_3Rh$	94	6
Nb		$(C_4H_7)_2NbCl$	96.5	3.5
Ni	$3d^8 4s^2$	(C_3H_5) NiX	97–99	1–3

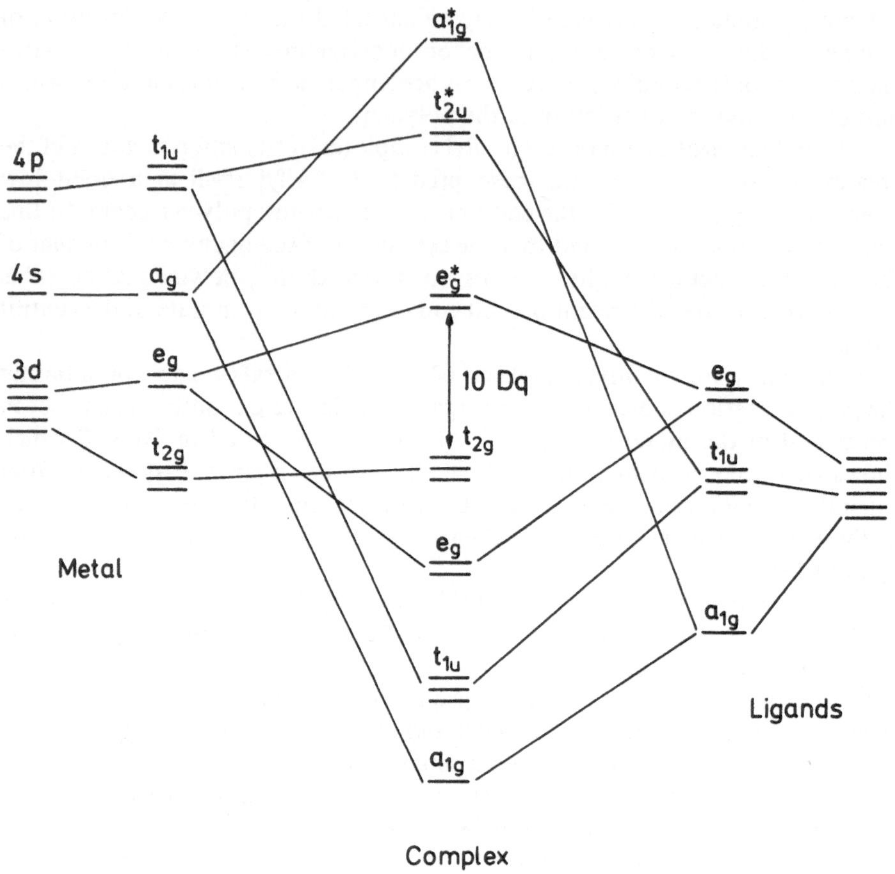

Scheme 19

spin of transition metals should be

Cr – 3/2; Nb – 1; Ti – 1/2

Although the precise value of total spins has not been reported in the literature, it is known that the complexes of Cr, Nb and Ti are actually paramagnetic [87] and that of Rh is diamagnetic [88].

Hence, it should be expected that the content of the vinyl structure should be higher in polybutadiene obtained on π-allyl complexes of Cr, Nb and Ti than in that polymerized with the aid of $(C_4H_3)_3Rh$. When the data given in Table 8 are compared to the predicted results, the exact coincidence between them seems even surprising. It is hardly possible that this agreement is fortuitous.

Moreover, if one crotyl group in the π-crotyl complex of Nb is replaced by chlorine, the complete inversion of the structure-controlling ability of the catalyst is observed. Furthermore, nickel catalysts most widely used in industry for the preparation of *cis*-1,4-polydienes are also diamagnetic [87] because of the low symmetry of the ligand environment and the even number of total electrons.

4.17 Mechanism for the Formation of Alternating Copolymers

Alternating copolymers can probably be formed by two paths. First, an alternating copolymer is formed when one of the monomers participating in the reaction is very reactive but is unable to undergo homopolymerization because of unfavourable steric factors (1,1-diphenylethylene is a good example of such a monomer).

Secondly, proceeding from the concept of the elementary polymerization act developed here, in the case when two quasisymmetric monomers with quite different structures undergo copolymerization, the formation of an alternating copolymer may be expected in accordance with the local symmetry and spin exclusion principles. This tendency should be particularly pronounced in copolymerization on paramagnetic anionic-coordination centers.

In fact, it has been established [89] in the investigation of factors determining the structure of alternating copolymers of butadiene and propylene by the $TiCl_4$–$C_6H_5COCH_3$–$Al(i\text{-}Bu)_3$ catalyst system that an alternating copolymer is formed on paramagnetic binuclear complexes of dialkylated derivatives of Ti^{+3}.

5 Conclusions

The axiomatic approach to the development of the theory of addition polymerization presented here is the first consequent attempt to interpret as many various experimental facts as possible on the basis of as few assumptions as possible. It should be noted that at present this concept has neither a wide recognition nor has been the subject of profound criticism.. However, the author hopes that this is due to the fact that these concepts are not very well known. In this connection, the author should like to draw the attention of the reader to the following points:

First, these concepts should not be considered as settled. Many of the points are, no doubt, open to discussion and some of them may even be erroneous. However, this is not essential. The main purpose of the author was to apply a new ideology to the already settled and relatively conservative field of chemical science, polymer chemistry. The reader can judge to what an extent this attempt was a success.

Secondly, the possibilities of using the axiomatic approach are not exhausted by the present paper. From the methodological standpoint, this is the most flexible and universal method of the analysis of natural phenomena. As the range of experimental facts becomes wider, the primary formulation of the basic postulates may, if necessary, either be made more precise or have new postulates added.

Thirdly, the established correlation between the microstructure of the polymer and the paramagnetism of the active center in anionic-coordination polymerization is both unexpected and not trivial. The predictions of results of this kind cannot be fortuitous. The correct predictions suggest that these postulates actually anticipate the real physical bases of the elementary act of catalytic polymerization.

Acknowledgements: The author should like to express his gratitude to Prof. S.S. Skorokhodov and Prof. B.L. Erussalimsky for moral support and attention to his work, to Prof. V.D. Shteingarts and Dr. I.I. Bilkis for their help and the informal interest in the author's ideas, to his colleagues Sapurina I.Yu. and Shubin N.V. for the discussion of the paper and to Koroleva E.A. for the translation of the paper into English.

References

1. Shamanin VV (1984) 4th International Symposium on Homogeneous Catalysis, Leningrad, prepr. vol 1, p 201
2. Shamanin VV (1990) Vysokomol Soedin ser A 32: 2283
3. Reid C (1970) Hilbert. Springer, Berlin Heidelberg New York
4. Messia A (1978) Quantum mechanics. Naŭka, Moscow
5. Orear J (1979) Physics. Collier Macmillan, London
6. Shamanin VV (1982) Dissertation, Leningrad
7. Elliott JP, Dawber PG (1979) Symmetry in physics. The Macmillan, London
8. Struik DJ (1963) Abriss der Geschichte der Mathematik. VEB Deutscher Verlag der Wissensc-haften, Berlin
9. Poincare A (1983) On Science, Moscow
10. Born M (1962) Einstein's Theory of Relativity. Dover, New York
11. Wigner EP (1970) Symmetries and reflections. Indiana University Press, Bloomington
12. Kokkedee JJJ (1969) The Quark Model. WA Benjamin, Amsterdam
13. Woodward RB, Hoffmann R (1970) Die Erhaltung der Orbitalsymmetrie. Verlag Chemie, Weinheim
14. Dirac PAM (1958) The Principles of quantum mechanics. Clarendon, Oxford
15. Eyring H, Lin SH, Lin SM (1980) Basic chemical kinetics. John Wiley, New York
16. Minkin VI, Simkin B Ya, Minyaev RM (1986) Quantum chemistry of organic reactions, Moscow
17. Dewar MJS, Dougherty RC (1975) The PMO theory of organic chemistry. Plenum, New York
18. Gilchrist TL, Storr RC (1972) Organic reaction and orbital symmetry. The University Press, Cambridge
19. Robinson PJ, Holbrook KA (1972) Unimolecular reactions. Wiley-Interscience, London
20. Physics of Microcosm. Edited by Shirkov DV (1980) Moscow
21. Shachparonov MI (1980) The mechanisms of rapid processes in liquids. Visshaja shkola, Moscow
22. Barltrop JA, Coyle JD (1975) Excited states in organic chemistry. John Wiley, London
23. Bender ML, Bergeron RJ, Komiyama M (1984) The bioorganic chemistry of enzymatic catalysis. John Wiley, New York
24. Pearson RG (1976) Symmetry rules for chemical reactions. Orbital topology and elementary processes. John Wiley, New York
25. Salem L (1982) Electrons in chemical reactions: First principles. John Wiley, New York

26. McGlynn SP, Azumi T, Kinoshita M (1969) Molecular spectroscopy of the triplet state. Prentice-Hall, New Jersey
27. Flurry RL (1980) Symmetry groups. Theory and chemical applications. Prentice-Hall, New York
28. Bersuker IB, Polinger VZ (1983) The vibratory interactions in molecules and crystals. Naŭka, Moscow
29. Podolsky AF, Boldyrev AG, Sapurina I Yu, Shamanin VV, Ushakova IL, Orlova NG (1982) Acta Polymerica 33: 181
30. Boldyrev AG, Podolsky AF, Sapurina I Yu, Ushakova IL, Shamanin VV (1985) Acta Polymerica 36: 208
31. Sapurina I Yu (1983) Dissertation, Leningrad
32. Shamanin VV, Podolsky AF, Boldyrev AG, Sapurina I Yu, Ushakova IL (1984) 4th International Symposium on Homogeneous Catalysis, Leningrad, prepr. vol 2, p 96
33. Jahn HA, Teller E (1937) Proc Roy Soc 161A: 220
34. Dmitriyev IS (1986) The electron in the eyes of the chemist. Khimia Leningrad
35. Murrell JN, Laidler KJ (1968) Trans Faraday Soc 64: 371
36. Murrell JN (1972) Chem Commun 1044
37. Bresler SE, Erussalimsky BL (1965) Physics and chemistry of macromolecules. Naŭka, Moscow
38. Pino P (1985) Stereoregulation in polymer synthesis. In: Vogl O (ed) Polymer science in the next decade. John Wiley, New York
39. Woodward RB, Hoffmann R (1965) J Am Chem Soc 87: 395
40. Bagdasaryan HS (1966) Theory of radical polymerization, Moscow
41. Erussalimsky BL (1977) Radical polymerization. In: Encyclopaedia of polymers. Sovetskaja enciklopedia, Moscow, vol 3, p 267
42. Hudson DJ (1964) Statistics, Geneva
43. Bovey F (1960) J Polym Sci 46: 59
44. Tiers G, Bovey F (1960) J Polym Sci 47: 479
45. Johnsen U (1961) J Polym Sci 54: 86
46. Belonovskaya GP, Dolgoplosk BA, Tinyakova EI (1957) Izvestiya AN SSSR 1: 65
47. Condon F (1953) J Polym Sci 11: 139
48. Richardson WS (1954) J Polym Sci 13: 229
49. Dolgoplosk BA, Tinyakova EI (1985) Metalloorganic catalysis in processes of polymerization, Moscow
50. Plesch PH (1989) Intern Symp on Ionic Polymerization, Strasbourg
51. Sauvet G, Moreau M, Sigwald P (1986) Macromol Chem, Macromol Symp 3: 33
52. Dewar MJS (1969) The molecular orbital theory of organic chemistry. McGraw-Hill, New York
53. Matsuzaki K, Hamada M, Arita K (1967) J Polym Sci Part A-1, 5: 1233
54. Yuki H, Hatada K, Ota K, Kinoshita I, Murahasi S, Ono K, Ito Y (1969) J Polym Sci Part A-1, 7: 1517
55. Szwarc M (1968) Carbanions living polymers and electron transfer processes. John Wiley, New York
56. Szwarc M (ed) (1972) Ions and ion pairs in organic reactions. John Wiley, New York
57. Bates RB, Ogle CA (1983) Carbanion chemistry. Springer, Berlin Heidelberg New York
58. Epiotis ND (1978) Theory of organic reactions. Springer, Berlin Heidelberg New York
59. Young RN, Quirk RP, Fetters LJ (1984) Adv Polym Sci 56: 1
60. Worsfold DJ, Bywater S (1962) Can J Chem 40: 1564
61. Erussalimsky BL (1970) Ionic polymerization of polar monomers. Naŭka Leningrad
62. Uryu T, Seki T, Kawamura T, Funamoto A, Matsuzaki K (1976) J Polym Sci, Polym Chem Ed 14: 3035
63. Uryu T, Kawamura T, Matsuzaki K (1979) J Polym Sci, Polym Chem Ed 17: 2019
64. Stearns RS, Forman LE (1959) J Polym Sci 41: 381
65. Szwarc M (1959) J Polym Sci 40: 583
66. Dneprovsky AS, Temnikova TI (1979) The theoretical basisses of organic chemistry. Khimija, Leningrad
67. Zgonnik VN (1982) Dissertation, Leningrad
68. Saltman WM (ed) (1977) The stereo rubbers. John Wiley, New York
69. Davidjan A, Nikolaew N, Sgonnik V, Belenkii B, Nesterow V, Krasikow V, Erussalimsky B (1978) Macromol Chem 179: 2155
70. Erussalimsky BL (1986) Mechanisms of ionic polymerization. Plenum, New York
71. Worsfold DJ, Bywater S, Hollingsworth G (1972) Macromolecules 5: 389

72. Shue F, Worsfold DJ, Bywater S (1970) Macromolecules 3: 509
73. Dolgoplosk BA, Golshtein SB, Volershtein EA (1990) Dokl AN SSSR 252: 880
74. Hsieh HL (1965) J Polym Sci 3A: 181
75. Kropatshev VA, Dolgoplosk BA, Nikolaev NI (1957) Dokl AN SSSR 115: 516
76. Basova RV, Arest-Yakubovitsh AA, Solovich DA, Desyatova NV, Gantmacher AR, Medvedev SS (1963) Dokl AN SSSR 149: 1067
77. Yuki H, Okamoto Y, Tsubota K, Kasai K (1970) Polymer J 1: 147
78. Melenevskaya E, Sgonnik V, Dolinskaya E, Erussalimsky B (1978) Macromol Chem 179: 2759
79. Korotkov AA (1958) Angew Chem 70: 85
80. Rakova GV, Korotkov AA (1958) Dokl AN SSSR 119: 982
81. Suzuki T, Tsuji Y (1978) Macromolecules 11: 639
82. Suzuki T, Tsuji Y, Takegami Y, Harwood HJ (1979) Macromolecules 12: 234
83. Podolsky AF, Unpublished results
84. Cotton FA, Wilkinson G (1966) Advanced inorganic chemistry. John Wiley, New York
85. Minkin VI, Simkin B Ya, Minyaev RM (1979) The theory of structure of molecules. Visshaja skhola, Moscow
86. Vonsovsky SV (1973) Magnetism of microparticles. Naŭka, Moscow
87. Wilke G, Bogdanovic B, Hardt P, Heimbach P, Keim W, Kröner M, Oberkirch W, Tanaka K, Steinrücke E, Walter D, Zimmermann H (1966) Angew Chem 78: 157
88. Powell J, Shaw BL (1966) Chem Comm 323
89. Yacenko LA (1983) Dissertation, Leningrad

Editor: Prof. Dr. H. Höcker
Received November 1992

Phase Transitions in Polymer Blends and Block Copolymer Melts: Some Recent Developments

K. Binder
Institut für Physik, Johannes Gutenberg Universität Mainz, D-55099 Mainz, Staudinger Weg 7, FRG

The classical concepts about unmixing of polymer blends (Flory-Huggins theory) and about mesophase ordering in block copolymers (Leibler's theory) are briefly reviewed and their validity is discussed in the light of recent experiments, computer simulations and other theoretical concepts. It is emphasized that close to the critical point of unmixing non-classical critical exponents of the Ising universality class are observed, in contrast to the classical mean-field exponents implied by the Flory-Huggins theory. The temperature range of this non-mean-field behavior can be understood by Ginzburg criteria. The latter are also useful to discuss the conditions under which the linearized (Cahn-like) theory of spinodal decomposition holds. While Flory-Huggins theory predicts correctly that the critical value of the Flory χ-parameter scales with chain length N (for symmetrical mixtures) $\chi_c \propto 1/N$, it strongly overestimates the prefactor and its use for fitting experimental data yields spurious concentration dependence. Also the chain radii depend on both χ and the composition of the mixture, thus invalidating the random phase approximation (RPA). Particular strong deviations from the RPA are predicted for block copolymer melts, where chains may stretch out in a dumbbell-like shape even in the disordered phase, before the microphase separation transition is approached. This review concludes with an outlook on interfacial phenomena and surface effects on these systems and other open problems in this field.

Advances in Polymer Science, Vol. 112

1 Introduction

The phase behavior of binary (AB) and ternary polymer mixtures is of long-standing interest [1–22] for several reasons: (1) one hopes that by "alloying" various polymers, one may be able to produce materials with more favorable properties than the pure constituents, in the same way it is possible for metallic alloys [23–25]; (2) due to the interplay of the random coil structure [1, 2, 26] of long flexible macromolecules with their enthalpic interactions, the behavior of polymer mixtures poses challenging problems to a treatment in the framework of statistical thermodynamics; (3) the large molecular weight of the polymer chains not only provides a convenient control parameter whose variation allows a more stringent test of theories than is possible for mixtures consisting of small molecules, but is also responsible for making many phenomena much slower and occurring on larger wavelengths than in the small molecule systems, and hence often more easily observable (e.g. the dynamics of early stages of unmixing [27–36], or the concentration profile of wetting layers [37, 38], etc.).

In this review we shall disregard any viscoelastic and mechanical properties of polymer blends and hence disregard their applications; we will rather focus on the above two points (2) and (3), to bring out that polymer mixtures are *model systems* by which theoretical concepts which have a much broader application can be tested. For this reason, we also include diblock copolymers (A-B) into the discussion, since their treatment in many respects can be viewed as a suitable generalization [39–43] of concepts developed for polymer blends. For both types of system, we shall give details only for the simplest theories, such as the Flory-Huggins theory [1, 2, 13–22] of blends and the Leibler [43] theory of block copolymer mesophase formation, emphasizing the drastic approximations that are made and discussing their validity or their limitations. The more recent and often very sophisticated theories (e.g. [44–64]) will be quoted only rather briefly, first because a detailed treatment would fill at least one complete book, secondly because there is no general agreement on which of these theories constitutes the most significant progress. Rather we will focus on the interplay between theory, computer simulation, and experiment, since we feel that the insight gained from simulations has stimulated recent theoretical developments and more careful experiments – and will continue to do so, since not all the issues seem to be settled!

At this point it should be clear that the present review does not aim at a complete coverage of the field, but necessarily gives only a selective treatment which is biased by the limited knowledge and restricted interests of the author. But nevertheless it is to be hoped that the present facet is an instructive view of the field and a useful guide to the literature, and will stimulate further work, both theoretical, simulational and – last but not least – experimental.

As a final disclaimer, we mention that we concentrate here on static structure and thermodynamic properties, extension to dynamic phenomena is not treated in full but only in a limited way and occasionally. Similarly, emphasis is on the bulk properties of materials, not on interfacial effects; and finally, linear flexible polymers are assumed throughout, and extensions to semiflexible polymers, where unmixing may compete with liquid-crystalline order [65], is not considered, nor the extension to branched polymer architecture (such as the question of compatibility of interpenetrating binary (AB) polymer networks [66, 67]), etc.

2 Flory–Huggins–Theory of Polymer Mixtures and Its Extensions

In this section, we discuss the phase behavior of binary (AB) polymer mixtures. For large molecular weight, the configurational entropy of mixing contribution to the Gibbs free energy is strongly reduced in comparison with mixtures of small molecules; therefore many polymer mixtures are strongly incompatible and hence practically entirely segregated over the experimentally accessible "temperature window" (from the glass transition temperature, where the melt is frozen into an amorphous structure, to the temperature where chemical disintegration or crosslinking of one species of the mixture sets in). Such strongly incompatible situations will not be considered here, rather we are interested in partially compatible mixtures which, in some temperature regions are miscible over the entire concentration range. The miscibility gap then ends either in an upper critical consolute temperature (UCST) or in a lower critical consolute temperature (LCST). An example for UCST-behavior is provided by mixtures of deuterated and protonated polymers of otherwise the same chemical species [68] and the polyisoprene (PI)-poly(ethylene-propylene) (PEP) system [69–71], an example for LCST-behavior by mixtures of polystyrene (PS) and polyvinyl-methylether [27, 29, 32–34], see Fig. 1. This phase diagram shows a typical complication of real polymer mixtures, namely due to the molecular weight distribution ("polydispersity") the critical point {where one gets from the one-phase region directly into the unstable (spinodal) region bounded by the "spinodal curve"} is not an extremum temperature (minimum for LCST, maximum for UCST) of the coexistence curve, but somewhat offset [72]. Such complications will be disregarded here, mostly we shall be concerned with strictly monodisperse mixtures. Also, we are not aiming at a theoretical explanation why some systems show UCST and others show LCST behavior, nor are we trying any explicit connections between chemical structure of the polymers and their compatibility: rather we would like to focus on the general statistical thermodynamics of such systems. Then the phase diagram of Fig. 1 is interpreted by the generic phase diagram of Fig. 2, where the miscibility is described in terms of a competition between entropy of mixing and an effective enthalpy contribution driving the two species of the mixture apart. This effective enthalpy is parametrized in terms of the Flory-Huggins χ-parameter [1, 2, 13–22] which may depend on both temperature T and volume fraction ϕ (of, say, the B species) in the mixture [6–8].

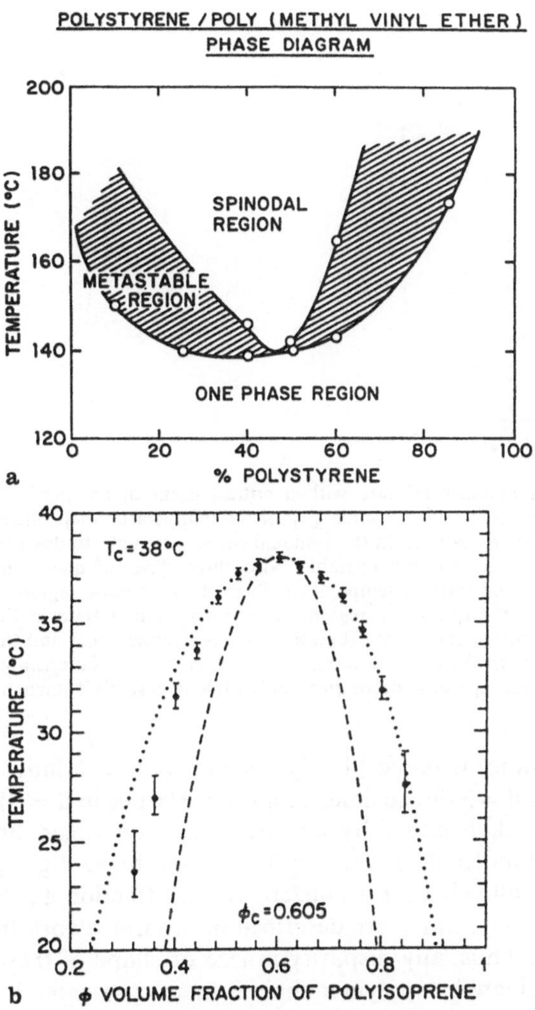

Fig. 1a. Phase diagram of the mixture polystyrene and polyvinyl methyl ether, molecular weights being $\bar{M}_W = 62700$ (PVME) and $\bar{M}_W = 60000$ (PS), as obtained from light scattering. The *lower curve* describes the miscibility gap ("binodal", "coexistence curve"), the *upper curve* describes the spinodal curve, which touches the coexistence curve in the critical point. The *shaded region* in between binodal and spinodal is believed to describe homogeneously mixed metastable one-phase states. From Snyder et al. [27]. **b** Phase diagram for polyisoprene-poly(ethylene-propylene) with molecular weights of 2000 and 5000, respectively. From Cumming et al. [70]

2.1 The Flory-Huggins Lattice Model and the Mean-Field Approximation

More than 50 years ago, Flory and Huggins [13–17] formulated a lattice model which captures the essential features of this competition between configurational entropy of mixing and enthalpy contributions, and even today this extremely simplified model is the basic ground on which most of the discussion

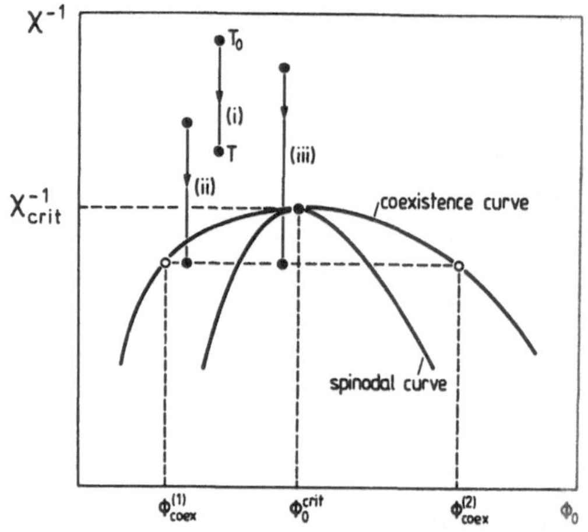

Fig. 2. Schematic phase diagram of a polymer mixture with a critical point of an unmixing transition occuring at a volume fraction ϕ_0^{crit} and the value χ_{crit} of the (temperature-dependent) effective interaction parameter χ. The region underneath the spinodal curve is thought to describe a region where homogeneous states are intrinsically unstable. Also three types of quenching experiments are indicated: (i) from an initial state at temperature T_0 in the one-phase region to another state at temperature T, also in the one-phase region. (ii) From an initial state in the one-phase region to a final state in the metastable region (between the coexistence curve and the spinodal curve). (iii) From an initial state in the one-phase region to a final state in the unstable region: phase separation then proceeds via "spinodal decomposition", while in case (ii) it starts by nucleation and growth

on polymer mixing thermodynamics is based [1–12]. One introduces a lattice (usually taken as simple cubic, in d = 3 dimensions, or a square lattice in d = 2), where each lattice site can be taken either by an effective A-monomer, an effective B-monomer or a solvent molecule or vacancy V, Fig. 3 (here we slightly generalize the original model by allowing for a nonzero volume fraction ϕ_V of "free volume" or solvent, respectively, which are described in this framework to a large extent in a similar way). Thus, any disparity in size or shape of these subunits of the chains is not considered. A polymer chain is then represented by a self-avoiding walk on the lattice (which must not intersect any other chain, of course, since each lattice site can be occupied at most once). Thus, the chains are allowed to differ at most by their index of polymerization (expressed here in terms of the number N_A, N_B of the subunits), but different chain stiffness (expressed via variable persistence lengths of the chains [1, 2, 26]) is also disregarded, and thus the model focuses on the "universal" aspects of the phase behavior of polymer mixtures. While modern theories on phase transitions [73, 74] indeed imply that near the critical temperature T_c certain aspects are "universal" (i.e., material-independent), such as critical exponents [73, 74], ratios between critical amplitudes [75], etc., we shall see that this truly universal region is very narrow in polymer mixtures with high molecular weight [76–79], so there is a great deal of interest in understanding the region with non-universal, material-dependent behavior, too. Only recently have attempts being made considering effective monomers blocking several lattice sites from further occupation which thus allow to some extent the modelling of a different size and

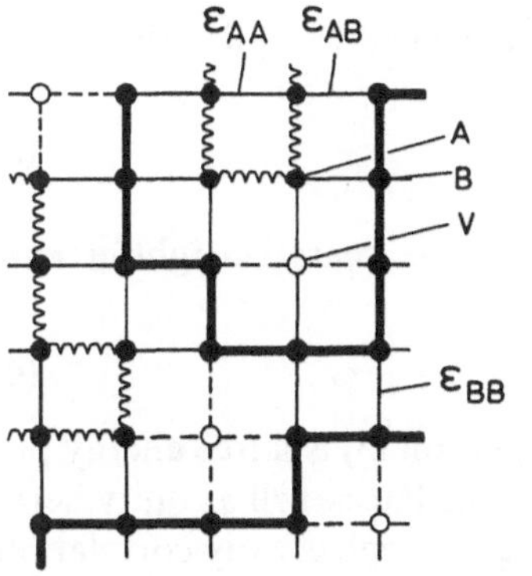

Fig. 3. Schematic illustration of the Flory-Huggins lattice model for a polymer mixture. Lattice sites taken by (effective) monomers are indicated by *full dots*; lattice sites taken by vacancies are denoted by *empty circles*. Chains of type A are indicated by *thick bonds* between the monomers, and B chains by *wavy bonds*. Nearest neighbor nonbonded interactions between monomers of the same kind (ε_{AA} or ε_{BB}) are shown as *full straight lines* and those between monomers of a different kind (ε_{AB}) by *broken lines*. Interactions between monomers and vacancies (or solvent molecules, respectively), ε_{AV} and ε_{BV}, could be introduced as well but will be assumed here to be zero throughout

shape of subunits [51, 80–99]. In particular, the bond fluctuation model [84] allows the efficient use of potentials for bond angles [94–98] and thus modelling of chemical detail: using such an approach in the context of mixing thermodynamics will be a promising direction of future research, but stays outside our consideration here; we shall be concerned with the simple model of Fig. 3 almost exclusively.

According to statistical thermodynamics [100] the task now would be to enumerate all configurations of A-chains (containing N_A effective monomers) and B-chains (containing N_B effective monomers) compatible with the given volume fractions ϕ_A, ϕ_B ($\phi_A + \phi_B + \phi_V = 1$) and the constraints that no chain intersections occur. For each configuration ($\mathscr{C}$) one needs to evaluate the internal energy $E(\mathscr{C}) = \varepsilon_{AA}n_{AA} + \varepsilon_{BB}n_{BB} + \varepsilon_{AB}n_{AB}$, where n_{AA}, n_{BB} and n_{AB} are the numbers of (nearest-neighbor) AA, BB and AB pairs in this configuration. Then the partition function Z and free energy F of the system follow from weighting each configuration with the appropriate Boltzmann factor $\exp[-E(\mathscr{C})/k_BT]$,

$$Z = \sum_{\mathscr{C}} \exp[-E(\mathscr{C})/k_BT], \quad F = -k_BT \ln Z\,. \tag{1}$$

Other quantities can be obtained from suitable derivatives of the free energy or direct statistical mechanics averages; e.g. in the semi-grandcanonical ensemble of a symmetrical mixture ($N_A = N_B = N$) (where the chemical potential difference $\Delta\mu = \mu_A - \mu_B$ per effective monomer is a given independent thermodynamical variable) the order parameter $\langle M \rangle$ defined as the relative excess in the number of A-chains (n_A) over the number of B-chains (n_B) in the system

$$M = \frac{n_A - n_B}{n_A + n_B}, \quad \langle M \rangle = \frac{1}{Z}\sum_{\mathscr{C}} M \exp[-E(\mathscr{C})/k_BT]\,. \tag{2}$$

Obviously, the problem posed by Eqs. (1), (2) is very difficult and an exact analytic solution does not exist. So Flory and Huggins and others [1, 2, 13–22] resorted to drastic approximations. Rather than presenting the mathematical details of their calculations, we shall only discuss the physical content of their final result here. Their excess contribution to the free energy of mixing becomes

(relative to the free energy of a pure one-component melt, for $\phi_V = 0$)

$$\frac{\Delta F_{mix}}{k_B T} = \frac{\phi_A \ln \phi_A}{N_A} + \frac{\phi_B \ln \phi_B}{N_B} + \chi \phi_A \phi_B, \tag{3}$$

where the Flory-Huggins parameter χ is related to the nearest-neighbor pair energy defined above

$$\chi = z[\varepsilon_{AB} - (\varepsilon_{AA} + \varepsilon_{BB})/2]/k_B T, \tag{4}$$

z being the coordination number of the lattice. Equation (3) is a free energy per lattice site and for simplicity we take the volume of one lattice cell as unity here.

Obviously the enthalpy expression $\chi \phi_A \phi_B$ in Eq. (3) neglects any correlation effects in the occupancy of lattice sites: the probability that on neighboring lattice sites A-B-pairs occur is simply taken as the product $\phi_A \phi_B$ of the respective volume fractions. This is a special case of a mean-field approximation (MFA), which is known to yield a critical behavior described by the Landau theory of phase transitions [100], which differs from the correct critical behavior expected [73, 74] in the "universal" regime close to the critical point χ_{crit}, ϕ^{crit} in Fig. 2. We shall discuss these various types of critical behavior in Sect. 2.2.

A second approximation concerns the number of monomers (from other chains!) that a considered monomer interacts with. In Eq. (4), this is taken to be equal to the coordination number of the lattice, while it is clear that a monomer not at the end of the chain has at most z-2 neighboring monomers from other chains, two sites being taken by its neighbors along the chain, and there are at most $z - 1$ neighbors along the chain. This overestimation in the number of neighbors is to some extent, corrected in Guggenheim's approximation [20, 21], which (for $N_A = N_B = N$ large) amounts to replacing z in Eq. (4) by $z_{eff} = z - 2 + 2/N$. However, even then, χ is still overestimated: interactions of monomers with other monomers of the same chain, as they occur e.g. in crankshaft-type configurations (Fig. 3), do not contribute to unmixing of different chains: depending on the signs of ε_{AA} and ε_{BB}, such intrachain contacts may rather lead to coil contraction [101–103] or expansion. As we shall see in Sect. 4, the Guggenheim [20, 21] z_{eff} rather than z in Eq. (4) still significantly overestimates the actual number of effective neighbor contacts.

A third approximation is involved in the entropy of mixing terms in Eq. (3). While for $N_A = N_B = 1$ the lattice sites can be occupied independently from each other, this is no longer true for chains which must be both selfavoiding and mutually avoiding. Eq. (3) largely neglects such constraints. Given all these approximations whose accuracy is rather uncertain, it is not clear to what extent the result Eq. (3) accounts for the basic model, described in Fig. 3. As a consequence, it is not clear to what extent the shortcomings that comparison of Eq. (3), (4) to experiments reveal should be attributed to the model used (Fig. 3) or to the approximate statistical mechanics by which it is treated. This question can be clarified, however, by Monte Carlo simulations (Sect. 4) which yield essentially the exact statistical mechanics of the model in Fig. 3 – apart from systematic errors due to the finite size of the simulated box and statistical errors due to the finite time of the simulation run, which are both well under control [104–106].

As an example for the predictions which follow from Eq. (3), we note the condition for the spinodal curve, which becomes using $\phi_A = \phi$, $\phi_B = 1 - \phi$ and neglecting any possible ϕ-dependence of χ [1, 2]

$$\left.\frac{\partial^2(\Delta F_{mix}/k_B T)}{\partial\phi^2}\right|_{\phi=\phi_s(\chi)} = 0 \Rightarrow \frac{1}{\phi N_A} + \frac{1}{(1-\phi)N_B} = 2\chi_s(\phi)\,. \tag{5}$$

Here we have denoted the equation of the spinodal curve as $\phi = \phi_s(\chi)$ and its inverse function as $\chi_s(\phi)$. For a mixture of monodispersive polymers, the critical point is at the maximum of $\chi_s(\phi)$, which yields

$$\phi^{crit} = (\sqrt{N_A/N_B} + 1)^{-1}, \quad \chi_{crit}^{-1} = 2(N_A^{-1/2} + N_B^{-1/2})^{-2}\,. \tag{6}$$

For the simplest case of a symmetric mixture ($N_A = N_B = N$) this reduces to

$$\phi^{crit} = 1/2, \quad \chi_{crit} = 2/N\,. \tag{7}$$

As a last point of this subsection, we briefly discuss the extension [22] of the theory to ternary systems (including vacancies V or solvent, respectively), treating only the fully symmetrical case further as a simple example. Then Eq. (3) is replaced by

$$\frac{\Delta F}{k_B T} = (\phi_A \ln\phi_A + \phi_B \ln\phi_B)/N + \phi_V \ln\phi_V + \chi_{AB}\phi_A\phi_B + \tfrac{1}{2}\chi_{AA}\phi_A^2 + \tfrac{1}{2}\chi_{BB}\phi_B^2\,. \tag{8}$$

Here χ_{AB}, χ_{AA} and χ_{BB} are the phenomenological counterparts of the pair energies ε_{AB}, ε_{AA} and ε_{BB} in Fig. 3. Writing now (remember $\phi_A + \phi_B = 1 - \phi_V$ and $\phi_A^{crit} = \phi_B^{crit} = (1 - \phi_V)/2$)

$$\phi_A = \phi_A^{crit}(1 + m), \quad \phi_B = \phi_B^{crit}(1 - m) \tag{9}$$

one finds for the part $\Delta F'$ of ΔF that depends on m

$$\begin{aligned}\frac{\Delta F'(\phi_V)}{k_B T(1-\phi_V)} &\equiv \frac{\Delta F/k_B T - \phi_V \ln\phi_V}{(1-\phi_V)} - \frac{1}{N}\ln(1-\phi_V) - \frac{1-\phi_V}{8}(\chi_{AA} + 2\chi_{AB} + \chi_{BB}) \\ &= \frac{1}{N}\left\{\frac{1+m}{2}\ln\left(\frac{1+m}{2}\right) + \frac{1-m}{2}\ln\left(\frac{1-m}{2}\right)\right\} + \frac{(1-\phi_V)(\chi_{AA} - \chi_{BB})}{4}m \\ &\quad - \frac{1-\phi_V}{4}[\chi_{AB} - (\chi_{AA} + \chi_{BB})/2]\,m^2\,. \end{aligned} \tag{10}$$

It can be seen that $\Delta F'(\phi_V)/(1-\phi_V)$ has the same structure as $\Delta F'(\phi_V=0)$, if one introduces the renormalized Flory parameters $\tilde{\chi}_{AA}=(1-\phi_V)\chi_{AA}$, $\tilde{\chi}_{AB}=(1-\phi_V)\chi_{AB}$, and $\tilde{\chi}_{BB}=(1-\phi_V)\chi_{BB}$. Thus the condition for the critical point simply becomes [22, 107]

$$\chi_{crit} \equiv [\chi_{AB}-(\chi_{AA}+\chi_{BB})/2]_{T=T_c} = 2/[(1-\phi_V)N] \,. \tag{11}$$

Since $\mu_A = \partial(\Delta F)/\partial\phi_A$, $\mu_B=\partial(\Delta F)/\partial\phi_B$, we obtain from Eq. (8) for $\Delta\mu \equiv \mu_A-\mu_B$

$$\begin{aligned}\frac{\mu_A-\mu_B}{k_BT} &= (\ln\phi_A - \ln\phi_B)/N + \chi_{AB}(\phi_B-\phi_A)+\chi_{AA}\phi_A-\chi_{BB}\phi_B \\ &= \left[\ln\frac{1+m}{1-m}\right]\Big/ N - [\tilde{\chi}_{AB}-(\tilde{\chi}_{AA}+\tilde{\chi}_{BB})/2]m \\ &\quad + (\tilde{\chi}_{AA}-\tilde{\chi}_{BB})/2\,.\end{aligned} \tag{12}$$

Phase coexistence for $\chi > \chi_{crit}$ occurs for

$$(\mu_A-\mu_B)/k_BT = (\tilde{\chi}_{AA}-\tilde{\chi}_{BB})/2\,. \tag{13}$$

The equation for the coexistence curve then is given by $\{\tilde{\chi}=\tilde{\chi}_{AB}-(\tilde{\chi}_{AA}+\tilde{\chi}_{BB})/2\}$

$$\ln\frac{1+m_{coex}}{1-m_{coex}} = N\tilde{\chi}m_{coex}, \quad m_{coex} = \tanh\tfrac{1}{2}N\tilde{\chi}m_{coex}\,. \tag{14}$$

Expanding the tanh near the critical point as $\tanh x \approx x - x^3/3$ yields the critical behavior of the coexistence curve (interpreting ϕ_0 in Fig. 2 as ϕ_A).

$$\begin{aligned} m_{coex} &= \pm\sqrt{3}\left(\frac{\chi}{\chi_{crit}}-1\right)^{1/2}, \\ \phi^{(1)}_{A,coex} &= \frac{1}{2}(1-\phi_V)\left[1-\sqrt{3}\left(\frac{\chi}{\chi_{crit}}-1\right)^{1/2}\right] = \phi^{(2)}_{B,coex}, \\ \phi^{(2)}_{A,coex} &= \frac{1}{2}(1-\phi_V)\left[1+\sqrt{3}\left(\frac{\chi}{\chi_{crit}}-1\right)^{1/2}\right] = \phi^{(1)}_{B,coex}\,. \end{aligned} \tag{15}$$

Thus the two signs of the order parameter $m_{coex}(=\langle M\rangle$ for $\Delta\mu = \mu_A-\mu_B=\Delta\mu_{coex})$ simply correspond to A-rich $(+)$ and A-poor $(-)$ branches of the coexistence curve in Fig. 2, which develops mirror symmetry for $N_A=N_B$ around $\phi_A^{crit}=\phi_B^{crit}=(1-\phi_V)/2$. A possible asymmetry in the pair interactions ($\varepsilon_{AA} \neq \varepsilon_{BB}$ and hence $\tilde{\chi}_{AA}\neq\tilde{\chi}_{BB}$) only has the effect that phase coexistence is shifted away from $\Delta\mu=0$ to $\Delta\mu = k_BT(\tilde{\chi}_{AA}-\tilde{\chi}_{BB})/2$. The strictly binary case (no vacancies, $\phi_V=0$) is of course included as simple limiting case in Eqs. (12)–(15).

Equation (15) exhibits the characteristic square root shape of the coexisting curve, corresponding to the Landau [100] value for the critical exponent [73, 74] of the order parameter, $\beta = 1/2$. It turns out that $m_{coex} = \pm\sqrt{3}(\chi/\chi_{crit} - 1)^{1/2}$ remains true also for asymmetric chain lengths ($N_A \neq N_B$), but the regime in the (χ, ϕ) plane where this equation is a good approximation to the actual coexistence curve resulting from Eq. (3) gets smaller the larger the asymmetry becomes.

While so far we have considered the volume fraction ϕ_V of vacancies as fixed (as done in many computer simulations [84–99, 101–103, 107], it also is useful to consider the melt as compressible and then the pressure controls the vacancy concentration. This interpretation is the starting point of the "lattice fluid" model [108–110] of polymer mixtures and related models [111, 112], but will not be pursued further here.

2.2 Scattering from Concentration Fluctuations and the Random Phase Approximation (RPA)

Now we return to the incompressible binary mixture where $\phi_A + \phi_B = 1$ and hence concentration variations must satisfy $\delta\phi_B = -\delta\phi_A$. In order to describe scattering of neutrons, X rays or light [113], we must extend the treatment to include the wavevector dependence of concentration variations. The convenient framework for this problem equivalent to the Flory-Huggins level of description in the long wavelength limit is the RPA introduced by deGennes [2, 114]. We now sketch this approach and introduce first the concept of linear response to a change δU of an external potential that a chain feels [115]. A single chain with $N + 1$ monomers on a lattice in an (inhomogeneous) potential $U(\vec{r})$ has a statistical weight proportional to $\exp\{-[U(\vec{r}_0) + U(\vec{r}_1) + \ldots + U(\vec{r}_N)]\}$. Let us define the sum of all statistical weights with fixed $\vec{r}_0 = \vec{r}\,''$ and $\vec{r}_N = \vec{r}\,'$, which we define as $z^N G_N(\vec{r}\,'', \vec{r}\,')$ (z is the coordination number of the lattice – the partition function of an ordinary random walk in the absence of potentials thus is taken out as a normalization factor). Obviously $G_N(\vec{r}\,'', \vec{r}\,')$ then satisfies the recursion formula

$$G_N(\vec{r}\,'', \vec{r}\,') = \sum_{\vec{r}} G_{N'}(\vec{r}\,'', \vec{r}\,')\, G_{N-N'}(\vec{r}, \vec{r}\,'), \quad 0 \leq N' \leq N \tag{16}$$

because a monomer (with label N′) must be at some site $\vec{r}$ of the lattice, and the statistical weight factorizes in the contribution of the chain from $\vec{r}\,''$ to $\vec{r}\,'$ and the remaining part from $\vec{r}$ to $\vec{r}\,'$. If we consider the total partition function $Z(\{U\})$ as a functional of the potentials $\{U(\vec{r}_i)\}$, we also have

$$Z(\{U\}) = \sum_{\vec{r}\,'', \vec{r}\,'} G_N(\vec{r}\,'', \vec{r}\,') . \tag{17}$$

Now we observe that the average monomer concentration $\langle\varphi(\vec{r})\rangle$ can also be written as the fraction of the sum of all weights of configurations which have a monomer at $\vec{r}$ relative to $Z(\{U\})$, the unrestricted sum of all weights,

$$\langle\varphi(\vec{r})\rangle = \frac{1}{Z(\{U\})} \sum_{\vec{r}'',\vec{r}'} \sum_{N'=0}^{N} G_{N'}(\vec{r}'',\vec{r}')\, G_{N-N'}(\vec{r},\vec{r}') \,. \tag{18}$$

Here $\varphi(\vec{r}) = \sum_i \delta_{\vec{r}=\vec{r}_i}$ as long as one considers the Flory-Huggins lattice, but ultimately one wishes to consider some coarse-graining over many cells of the lattice to construct a smoothly varying field $\varphi(\vec{r})$ defined in continuous space.

Now we consider the change of $G_N(\vec{r}'',\vec{r}')$ due to a change of the potential $U(\vec{r}) \to U(\vec{r}) + \delta U(\vec{r})$. Since $\exp[-(U+\delta U)/k_BT] - \exp(-U/k_BT) \approx -(\delta U/k_BT)\exp(-U/k_BT)$, we can write, using Eq. (16) and summing over all chain configuration that contain $\vec{r}$,

$$\delta G_N(\vec{r}'',\vec{r}') = -\frac{1}{k_BT}\sum_{N'=0}^{N} G_{N'}(\vec{r}'',\vec{r}')\, G_N(\vec{r}',\vec{r})\, \delta U(\vec{r}) \,, \tag{19}$$

which implies $\delta Z(\{U\}) = \sum_{\vec{r}',\vec{r}''} \delta G_N(\vec{r}'',\vec{r}') = -(1/k_BT)\, Z(\{U\}) \langle\varphi(\vec{r})\rangle \delta U(\vec{r})$, or

$$\langle\varphi(r)\rangle = -k_B T\, \delta \ln Z(\{U\})/\delta U(\vec{r}) \,. \tag{20}$$

On the other hand we may also consider the linear response of $\langle\varphi(\vec{r})\rangle$ in Eq. (18) to the change $\delta U(\vec{r})$ of the potential as well, which yields $[\delta\varphi(\vec{r}) \equiv \varphi(\vec{r}) - \langle\varphi(\vec{r})\rangle]$

$$\delta\langle\varphi(\vec{r})\rangle = -\frac{1}{k_BT}\langle\delta\varphi(\vec{r})\,\delta\varphi(\vec{r}')\rangle\delta U(\vec{r}') \,, \tag{21}$$

or alternatively

$$\langle\delta\varphi(\vec{r})\,\delta\varphi(\vec{r}')\rangle = (k_BT)^2 \frac{\delta^2 \ln Z(\{U\})}{\delta U(\vec{r})\,\delta U(\vec{r}')} \,. \tag{22}$$

Now in thermal equilibrium, the correlation function $\langle\delta\varphi(\vec{r})\,\delta\varphi(\vec{r}')\rangle$ is translationally invariant, it does not depend on the two points $\vec{r}$, $\vec{r}'$ separately, but only on the relative distance. This consideration is useful when we generalize Eq. (21) to the case that the change of potential is not restricted to a single site $\vec{r}'$, but an (inhomogeneous) change occurs in the whole system.

$$\delta\langle\varphi(\vec{r})\rangle = -\frac{1}{k_BT}\sum_{\vec{r}'}\langle\delta\varphi(\vec{r})\,\delta\varphi(\vec{r}')\rangle\delta U(\vec{r}') \,, \tag{23}$$

which upon Fourier transformation reduces to

$$\delta\langle\varphi(\vec{q})\rangle = -\frac{1}{k_BT} S_N(\vec{q})\, \delta U(\vec{q}) \,, \tag{24}$$

where $\langle \varphi(\vec{q}) \rangle$ is the Fourier transform of $\langle \varphi(\vec{r}) \rangle$, $U(\vec{q})$ the Fourier transform of $U(\vec{r})$, and $S_N(\vec{q})$ is the single-chain structure factor,

$$S_N(\vec{q}) = \sum_{\vec{r}} \exp(i\vec{q}\cdot\vec{r}) \langle \varphi(\vec{r})\, \varphi(0) \rangle$$

$$\underset{(\vec{q} \neq 0)}{=} \sum_{\vec{r}} \exp(i\vec{q}\cdot\vec{r}) \langle \delta\varphi(\vec{r})\, \delta\varphi(0) \rangle . \tag{25}$$

Equation (24) shows that the response of the Fourier components of the density of a chain to a change of potential acting on that chain can be expressed in terms of the static structure factor $S_N(\vec{q})$ of the chain: for gaussian chains this is simply the well-known Debye function [2].

$$S_N(\vec{q}) = N f_D(x), \quad f_D(x) = \frac{2}{x^2}[\exp(-x) - 1 + x],$$

$$x = R_g^2 q^2 = N\sigma q^2/6 , \tag{26}$$

σ being the statistical segment length and R_g the gyration radius of the chain.

This approach is next generalized to an ideal solution containing two species A, B at volume fractions ϕ_A, ϕ_B. Then the relation between ϕ_A, ϕ_B and the respective chemical potentials $\mu_{A',B'}$ is $\phi_A = C_A \exp(\mu'_A/k_BT)$, $\phi_B = C_B \exp(\mu'_B/k_BT)$, where C_A, C_B are constants. Response functions κ_A, κ_B are then defined as [78]

$$\kappa_A \equiv (\partial\phi_A/\partial\mu'_A) = \phi_A/k_BT, \quad \kappa_B \equiv (\partial\phi_B/\partial\mu'_B) = \phi_B/k_BT . \tag{27}$$

We now turn to the case where A is a polymer consisting of N_A subunits and B of N_B subunits. Introducing chemical potentials μ_A, μ_B referring to a subunit each, we have $\mu'_A = N_A\mu_A$, $\mu'_B = N_B\mu_B$. Hence

$$\delta\phi_A = (\phi_A/k_BT)\,\delta\mu'_A = (\phi_A N_A/k_BT)\,\delta\mu_A , \tag{28a}$$

$$\delta\phi_B = (\phi_B/k_BT)\,\delta\mu'_B = (\phi_B N_B/k_BT)\,\delta\mu_B . \tag{28b}$$

Considering now the chemical potential difference

$$\delta(\Delta\mu) \equiv \delta\mu_A - \delta\mu_B = k_BT\left\{\frac{\delta\phi_A}{\phi_A N_A} - \frac{\delta\phi_B}{\phi_B N_B}\right\} . \tag{29}$$

and making use of the "incompressibility condition" $\phi_A + \phi_B = 1$, $\delta\phi_B = -\delta\phi_A$, $\phi = \phi_A/(\phi_A + \phi_B)$, we can write Eq. (29) as

$$\delta(\Delta\mu) = k_BT\{(\phi N_A)^{-1} + [(1-\phi)N_B]^{-1}\} \equiv \kappa^{-1}\delta\phi , \tag{30}$$

where we thus have defined as inverse response function $(k_BT\kappa)^{-1} = (\phi N_A)^{-1} + [(1-\phi)N_B]^{-1}$, which agrees, of course, with $\partial^2(\Delta F_{mix}/k_BT)/\partial\phi^2)_T$ obtained from Eq. (3) for $\chi = 0$. We now generalize these results, introducing q-dependence: for an inhomogeneous chemical potential acting on one chain apart from a change of sign since μ enters the statistical weight with the opposite sign as U, we can use Eq. (23), (24). In the present normalization of volume fractions (which differs from the single-chain problem in Eqs. (16–25)), we have

$$\kappa_A(\vec{q}) \equiv \partial\phi_A(\vec{q})/\partial\mu_A(\vec{q}) = \frac{\phi_A}{k_B T} S_A(\vec{q}) \,, \tag{31a}$$

$$\kappa_B(\vec{q}) \equiv \partial\phi_B(\vec{q})/\partial\mu_B(\vec{q}) = \frac{\phi_B}{k_B T} S_B(\vec{q}) \,, \tag{31b}$$

where we have denoted the single chain structure factors of A and B chains as $S_A(\vec{q})$, $S_B(\vec{q})$. For $q = 0$ we have $f_D(0) = 1$, $S_N(q \to 0) \to N$, see Eq. (26), and hence Eq. (31a, b) reduce to Eqs. (28a, b), as it should be. From Eq. (29), we can similarly conclude

$$\delta[\Delta\mu(\vec{q})] \equiv \delta\mu_A(\vec{q}) - \delta\mu_B(\vec{q}) = k_B T \left\{ \frac{\delta\phi_A(\vec{q})}{\phi_A S_A(\vec{q})} - \frac{\delta\phi_B(\vec{q})}{\phi_B S_B(\vec{q})} \right\}, \tag{32}$$

and using the local incompressibility condition $\delta\phi_A(\vec{q}) = -\delta\phi_B(\vec{q}) \equiv \delta\phi(\vec{q})$ one finds the inverse collective response function of the noninteracting ideal mixture

$$[k_B T \kappa_{ni}(\vec{q})]^{-1} = \frac{\partial[\Delta\mu(\vec{q})/k_B T]}{\partial\phi(\vec{q})} = \frac{1}{\phi S_A(\vec{q})} + \frac{1}{(1-\phi) S_B(\vec{q})} \,. \tag{33}$$

At this point we emphasize that Eqs. (16)–(33) should not be understood as a formally rigorous derivation (such derivations by various techniques can be found in the literature [114, 116, 117]) but rather the present treatment (which follows Ref. 78) is a plausibility argument. In this spirit, one can also extend the theory to the interacting case, within the framework of mean field theory: the inverse collective response function of the interacting system within RPA is always found from that of the noninteracting system by subtracting the Fourier transform of the interaction [118]. In our case we have a collective structure factor

$$\begin{aligned} S_{coll}^{-1}(\vec{q}) &= [\kappa_B T \kappa(\vec{q})]^{-1} = [\kappa_B T \kappa_{ni}(\vec{q})]^{-1} - 2\chi_{eff}(\vec{q}) \\ &= [\phi S_A(\vec{q})]^{-1} + [(1-\phi) S_B(\vec{q})]^{-1} - 2\chi_{eff}(\vec{q}) \,. \end{aligned} \tag{34}$$

Here $\chi_{eff}(\vec{q})$ is a wavevector-dependent generalization of the Flory-Huggins χ parameter [78]. For $q \to 0$, Eq. (34) yields

$$S_{coll}^{-1}(\vec{q} \to 0) = [\phi N_A]^{-1} + [(1-\phi) N_B]^{-1} - 2\chi_{eff}(0) \,, \tag{35}$$

which is the same as $\partial^2(\Delta F_{mix}/k_B T)/\partial\phi_A^2)$ calculated from Eq. (3) (note $\phi_A = \phi$, $\phi_B = 1 - \phi$ in the present case) and $\chi_{eff}(0) = \chi$ if χ is independent of

volume fraction ϕ. If χ depends on ϕ, we have to make the identification [78]
$-2\chi_{eff}(0) = \frac{\partial^2}{\partial\phi^2}[\phi(1-\phi)\chi(\phi)]$.

We now discuss the consequences of Eq. (34) and we are particularly interested in the behavior at small q, where we may use the expansions [78]

$$f_D(x) \cong 1 - x/3, \quad \chi_{eff}(\vec{q}) = \chi_{eff}(1 - \frac{1}{6}q^2 r_0^2)\,, \tag{36}$$

where r_0 plays the role of an effective interaction range. This yields, for $qR_g \ll 1$,

$$S_{coll}^{-1}(\vec{q}) \cong \frac{1}{\phi N_A} + \frac{1}{(1-\phi)N_B} + \frac{1}{18}\left(\frac{\sigma_A^2 q^2}{\phi} + \frac{\sigma_B^2 q^2}{(1-\phi)}\right) - 2\chi_{eff} + \frac{\chi_{eff}}{3} q^2 r_0^2\,. \tag{37}$$

Being interested in values of the χ-parameter comparable to χ_{crit}, which is of order $1/N_A$ (or $1/N_B$, respectively, whichever is larger, cf. Eq. (6)), and noting that for short range interactions r_0 is of the same order as the size of the effective subunits of the chains σ_A, σ_B, it is clear that the last term on the rhs of Eq. (37) can be neglected: the basic q-dependence in Eq. (37) thus results from the expansion of the Debye function, i.e. it is entropic in origin, not from the spatial dependence of the interaction strength. So Eq. (37) can be rewritten as

$$S_{coll}(\vec{q}) = S_{coll}(0)/[1 + q^2\xi^2]\,, \tag{38}$$

where ξ can be interpreted as the *correlation length of concentration fluctuations* and is given by

$$\xi = (a/6)\,/\sqrt{(1-\phi)\,/(2N_A) + \phi/(2N_B) - \chi_{eff}\phi(1-\phi)}\,, \tag{39}$$

where we have defined an "effective lattice spacing" a of the Flory-Huggins lattice as follows

$$a^2/[\phi(1-\phi)] = \sigma_A^2/\phi + \sigma_B^2/(1-\phi)\,. \tag{40}$$

Of course both $S_{coll}(0)$ – Eq. (35), and ξ – Eq. (39) diverge when χ_{eff} reaches the value $\chi_s(\phi)$ of χ_{eff} at the spinodal curve {Eq. (5)}. For simplifying the notation, we will henceforth disregard the distinction between χ and χ_{eff}. In the one-phase region ($\chi < \chi_{crit}$), we can read off from Eqs. (35), (39) the Landau-like critical behavior for $\phi = \phi^{crit}$,

$$S_{coll}(0) = \frac{1}{2\chi_{crit}}(1 - \chi/\chi_{crit})^{-1}\,, \tag{41}$$

$$\xi = \frac{a}{6}(4N_A N_B)^{1/4}(1 - \chi/\chi_{crit})^{-1/2}\,. \tag{42}$$

This implies, the critical exponents γ and ν of the scattering function $\{S_{coll}(0)\} \propto (1-\chi/\chi_{crit})^{-\gamma}\}$ and correlation length $\{\xi = \hat{\xi}(1-\chi/\chi_{crit})^{-\nu}\}$ have the values $\gamma = 1$ and $\nu = 1/2$, respectively [73]. The same is true also for $\chi > \chi_{crit}$ (where one is particularly interested in the approach to χ_{crit} along the coexistence curve). Here we consider only the fully symmetric case $N_A = N_B = N$ for simplicity. For this case the coexistence curve has already been derived {see Eq. (15), for $\phi_V = 0$}. Inserting $\phi^{(1)}_{coex}$ {or $\phi^{(2)}_{coex}$, respectively} in Eqs. (35), (39) simply yields

$$S_{coll}(0) \approx \frac{1}{4\chi_{crit}}(1-\chi/\chi_{crit})^{-1}, \quad \chi \to \chi_{crit}, \tag{43}$$

$$\xi \approx \frac{a}{6}\sqrt{N}(\chi/\chi_{crit}-1)^{-1/2}, \quad \chi \to \chi_{crit}. \tag{44}$$

Equations (42), (44) display one very remarkable feature: the prefactor $\hat{\xi}$ of the power law for the correlation length is *not* a microscopic length (of the order of the effective lattice spacing a), but it is much larger, namely of the order of the gyration radius of the coils. This larger prefactor of the correlation length is responsible for the fact that polymer mixtures behave over a wide range of χ mean-field like, except in the close vicinity of the critical point, as a Ginzburg criterion shows (see Sect. 2.5). At this point we note, that the static collective structure factor considered in Eqs. (34), (38) cannot be solely derived as a response function (Eq. (33)) but rather we have a fluctuation relation linking it to a Fourier transform of a correlation function (just as has been derived in Eqs. (22)–(25) for the single-chain case). Thus we can write

$$S_{coll}(\vec{q}) = \langle \phi_{\vec{q}}\phi_{-\vec{q}} \rangle = \langle |\phi_{\vec{q}}|^2 \rangle, \tag{45}$$

where $\phi_{\vec{q}}$ is the Fourier transform of the local concentration. Eq. (45) is important, because it constitutes the link to the small angle X-ray or neutron scattering (as well as light scattering) from concentration fluctuations in polymer blends. An analysis of such scattering data in terms of Eqs. (34), (37)–(39) and in particular the location of the spinodal temperature $T_s(\phi)$ {which corresponds to a Flory-Huggins parameter χ_s as considered in Eq. (5)} by extrapolating data for $S^{-1}_{coll}(0)$ linearly to zero as a function of temperature (see Fig. 32b below) is a common experimental practice. However, as early as at this point we emphasize that the existence of a spinodal curve is well-defined only within mean-field theory; we shall critically examine the validity of mean-field theory and the significance of the spinodal curve in Sect. 2.5.

Finally we mention that the RPA result, Eq. (37), can also be extended to the case of polydisperse polymer mixtures [119]. The result is

$$S^{-1}_{coll}(\vec{q}) = [\phi\overline{N^w_A}]^{-1} + [(1-\phi)\overline{N^w_B}]^{-1} - 2\chi + (q^2/18)\,[(\overline{N^z_A}/\overline{N^w_A}) \times(\sigma^2_A/\phi) + (\overline{N^z_B}/\overline{N^w_B})\,\sigma^2_B/(1-\phi)] \tag{46}$$

where $\overline{N^z}$, $\overline{N^w}$, are "number average" or "weight average" of the molecular weight distribution defined in terms of moments as $\overline{N^z} = \overline{N^3}/\overline{N^2}$, $\overline{N^w} = \overline{N^2}/\overline{N^1}$.

2.3 The deGennes-Flory-Huggins-Free-Energy-Functional

So far we have restricted the treatment of unmixing of polymers to a mean-field level and have also restricted attention to the fully uniform situation (e.g. computation of the coexistence curve) or small nonuniform concentration deviations from equilibrium, as described by $S_{coll}(\vec{q})$ discussed in the previous section. However, there is also interest in situations where strong concentration inhomogeneities occur, and thus one is beyond the regime where linear response theory is valid: formation of unmixed regions after quenching a mixture from the one phase region to the two-phase region ("spinodal decomposition" [120, 121] see Sect. 2.4), formation of interfaces between coexisting phases and the related question of the formation free energy of B-rich droplets in a supersaturated A-rich phase (nucleation [122, 123]). Such interface formation also plays a role in surface enrichment, wetting phenomena [124–127] and related effects (see Sect. 6).

In addition, the nonlinear effects of statistical fluctuations need to be considered when one wishes to go beyond mean field theory. Following the practice with other phase transition phenomena [73, 74, 128], one considers the mean-field free energy, Eq. (3), as a density of a free energy functional $\Delta\mathcal{F}$, where ϕ no longer is assumed homogeneous but rather is treated as a concentration field $\phi(\vec{r})$, and thus one also has to augment the expression by a "gradient energy" contribution describing the free energy cost of the concentration inhomogeneities [129],

$$\frac{\Delta\mathcal{F}}{k_BT} = \int d^3\vec{r}\left\{f_{FH}[\phi(\vec{r})] + \frac{a^2}{36\phi(1-\phi)}[\nabla\phi(\vec{r})]^2\right\}, \tag{47}$$

where $f_{FH}[\phi(\vec{r})]$ is identical to $\Delta F_{mix}/k_BT$ as given by Flory-Huggins (FH) theory in Eq. (3). The choice of the coefficient in front of the $[\nabla\phi]^2$ term will be justified below. We emphasize that the $[\nabla\phi]^2$ term is only the first term of an expansion valid for slow spatial variations, e.g. terms such as $[\nabla^2\phi]^2$ etc. are neglected. By slow spatial variation it is meant that the gradient term is small in comparison with the other term in the free-energy expression $f_{FH}(\phi)$, i.e. [78]

$$a^2(\nabla\phi)^2 \ll N_A^{-1}, \quad a^2(\nabla\phi)^2 \ll N_B^{-1}. \tag{48}$$

This implies that only spatial variations on scales large in comparison to the coil radii can be considered, as will be discussed below.

It is useful to derive a local chemical potential difference $\mu(\vec{r})$ between A- and B-monomers. This is given by the functional derivative of Eq. (47), namely [78]

$$\mu(\vec{r}) = \frac{\delta(\Delta\mathscr{F})}{\delta\phi(\vec{r})} = k_B T\left\{\frac{\ln\phi + 1}{N_A} - \frac{\ln(1-\phi)+1}{N_B} + \chi(1-2\phi) - \frac{a^2}{18\phi(1-\phi)}\nabla^2\phi + \left[\frac{\sigma_A^2}{36\phi^2} - \frac{\sigma_B^2}{36(1-\phi)^2}\right](\nabla\phi)^2\right\}, \quad (49)$$

where in the last term the ϕ-dependence of the effective lattice spacing a (Eq. (40)) was used. Considering now a case where $\phi(\vec{r})$ is only very weakly inhomogeneous, $\phi(\vec{r}) = \phi_0 + \delta\phi(\vec{r})$, one may linearize Eq. (49) in $\delta\phi(\vec{r})$ to find

$$\frac{\mu(\vec{r})}{k_B T} = \text{const.} + \left\{\frac{1}{N_A\phi_0} + \frac{1}{N_B(1-\phi_0)} - 2\chi - \frac{a^2}{18\phi_0(1-\phi_0)}\nabla^2\right\}\delta\phi(\vec{r}). \quad (50)$$

Introducing Fourier transforms (V = volume)

$$\phi_q = \frac{(2\pi)^3}{V}\int d^3\vec{r}\,\phi(\vec{r})\exp(i\vec{q}\cdot\vec{r}),$$

$$\mu_q = \frac{(2\pi)^3}{V}\int d^3\vec{r}\,\mu(\vec{r})\exp(i\vec{q}\cdot\vec{r}) \quad (51)$$

one immediately finds from Eq. (50)

$$\frac{\mu_{\vec{q}}}{k_B T} = \left\{\frac{1}{N_A\phi_0} + \frac{1}{N_B(1-\phi_0)} - 2\chi + \frac{a^2q^2}{18\phi_0(1-\phi_0)}\right\}\delta\phi_{\vec{q}} = [k_B T\kappa(\vec{q})]^{-1}\delta\phi_{\vec{q}}, \quad (52)$$

where in the last step we have recalled (cf. Eq. (33)) that the linear response of ϕ_q with respect to $\mu_{\vec{q}}$ defines the collective response function $S_{coll}(\vec{q}) = k_B T\kappa(\vec{q})$ of the mixture. Comparing Eqs. (37) and (52) then we recover our previous result exactly (where the correction $\chi_{eff}q^2r_0^2/3$ in Eq. (37) is henceforth neglected). This derivation of $S_{coll}^{-1}(\vec{q})$ from Eq. (47) shows that the coefficient of the gradient term in Eq. (47) has been adjusted so that the long wave length limit of RPA treatment is recovered. The RPA also clarifies the nature of the correction terms to the long wavelength limit: the expansion of the Debye function (Eq. (26)) yields terms of order $x^2 = N_{A,B}^2\sigma_{A,B}^4 q^4/36$ as higher order corrections, which are negligible against the terms of order x (that were kept in Eq. (37), see Eq. (36)) only if $x \ll 1$ i.e. $\sigma_{A,B}^2 q^2 \ll N_{A,B}^{-1}$, which is equivalent to the condition mentioned in Eq. (48).

Obviously, on such large length scales where Eq. (48) holds, polymers look like small molecules, and this is the physical content of the simple $(\nabla\phi)^2$ free energy functional, Eq. (47), that is indeed very reminiscent of a much broader class of materials. Only the particular ϕ-dependence of the coefficient of this "gradient energy" reminds us that we are dealing with polymers, where this coefficient is mainly of entropic origin, rather than with small molecule mixtures, where it is of mostly enthalpic origin (for a small molecule mixture there would only be the term $\chi_{eff}\, q^2 r_0^2/3$ in Eq. (37) and then this term would yield the coefficient of the gradient energy).

It sometimes is useful to consider a slight generalization so that one not only has segment *lengths* σ_A, σ_B different from each other, but one also introduces different monomeric *volumes* v_A, v_B. Then the Flory-Huggins free energy density is written as (130)

$$f(\phi) = \frac{\phi \ln \phi}{z_A N_A} + \frac{(1-\phi)\ln(1-\phi)}{z_B N_B} + \chi\phi(1-\phi)\,, \tag{53}$$

where z_A, z_B denote these monomeric volumes normalized to some reference volume v_0 i.r. $z_A = v_A/v_0$, $z_B = v_B/v_0$. The free energy functional then takes the form

$$\frac{\Delta F}{k_B T} = \int d^3\vec{r}\, \frac{1}{v_0}\Bigg\{ f[\phi(\vec{r})] + \frac{1}{36}\left[\frac{\sigma_A^2}{z_A \phi(\vec{r})} + \frac{\sigma_B^2}{z_B[1-\phi(\vec{r})]}\right] |\nabla\phi(\vec{r})\,|^2 \Bigg\}\,. \tag{54}$$

Equation (47), is of course, recovered for $z_A = z_B = 1$ and choosing $v_0 = 1$. The choice of the reference volume is arbitrary. In the asymmetric case a convenient choice is $v_0 = \sqrt{v_A v_B}$ since then $z_A z_B = 1$ and thus the quantity in the braces in Eq. (54) is a function of v_A/v_B only [130]. We also note the corresponding generalization of the collective structure factor,

$$[S_{coll}(q)]^{-1} = \frac{1}{N_A z_A \phi} + \frac{1}{N_B z_B (1-\phi)} - 2\chi + \frac{q^2}{18}\left[\frac{\sigma_A^2}{z_A \phi} + \frac{\sigma_B^2}{z_B(1-\phi)}\right]. \tag{55}$$

It is desirable to modify the expression for the free energy such that it is valid not just in the long wavelength limit but yields the full expression of the RPA (Eq. (34)) at all wave numbers. This is achieved for [131]

$$\frac{\Delta \mathcal{F}}{k_B T} = \int d^3\vec{r}\, \frac{1}{v_0}\left\{ \frac{\ln \phi_A}{z_A N_A} F_A(\nabla^2)\phi_A + \frac{\ln \phi_B}{z_B N_B} F_B(\nabla^2)\phi_B + \chi\phi_A\phi_B \right\}, \tag{56}$$

with $\phi_A = \phi(\vec{r})$, $\phi_B = 1 - \phi(\vec{r})$, and the operator $F_\alpha(\nabla^2)$ is defined as

$$F_\alpha(\nabla^2) = \frac{1}{2}\left[1 + \frac{1}{f_D(-R_{g\alpha}^2 \nabla^2)}\right], \tag{57}$$

$f_D(x)$ being the Debye function, Eq. (26). In the remainder of this article, however, we shall restrict ourselves to the simplest case, Eq. (47).

As a first application for the use of the free energy functional, we will discuss the calculation of interfacial concentration profiles $\phi(x)$ between coexisting unmixed phases and interfacial tension [132–134]. For a symmetric mixture ($N_A = N_B = N$) phase coexistence occurs for $\mu = 0$, and since the interfacial profile $\phi(x)$ also must be found by minimizing Eq. (47), we look for a solution of ($\sigma_A = \sigma_B = \sigma$)

$$\frac{\delta(\Delta\mathcal{F}/k_BT)}{\delta\phi(x)} = 0 = \frac{1}{N}\{\ln\phi - \ln(1-\phi)\} + \chi(1-2\phi) - \frac{\sigma^2}{18\phi(1-\phi)}\frac{d^2\phi}{dx^2} + \frac{\sigma^2}{36}[\phi^{-2} - (1-\phi)^{-2}](d\phi/dx)^2 \tag{58}$$

which satisfies boundary conditions $\phi(x \to -\infty) = \phi^{(1)}_{A,coex} \equiv \phi_1$, $\phi(x \to +\infty) = \phi^{(2)}_{A,coex} = 1 - \phi_1$. Using these boundary conditions one can integrate this "Euler-Lagrange"-equation to find [134]

$$\xi^2_{coex}\left(\frac{d\phi}{dx}\right)^2 = 2\phi(1-\phi)\{\phi\ln\phi - \phi_1\ln\phi_1 + (1-\phi)\ln(1-\phi) - (1-\phi_1)\ln(1-\phi_1) + 2\left(\frac{\chi}{\chi_{crit}}\right)[\phi(1-\phi) - \phi_1(1-\phi_1)]\}\Big/ \left[1 - \frac{2\phi_1(1-\phi_1)}{1-2\phi_1}\ln\frac{1-\phi_1}{\phi_1}\right], \tag{59}$$

where the correlation length at coexistence has been used [78]

$$\xi_{coex} = \frac{\sigma}{6}(2N)^{1/2}\left[1 - \frac{2\phi_1(1-\phi_1)}{1-2\phi_1}\ln\frac{1-\phi_1}{\phi_1}\right]^{-1/2} \tag{60}$$

For ϕ near ϕ_1 an expansion of Eq. (59) to second order yields

$$\xi_{coex}\frac{d\phi}{dx} \approx \phi - \phi_1, \quad \phi(x) - \phi_1 \approx \exp(x/\xi_{coex}), \quad x \to -\infty\,. \tag{61}$$

Thus it follows that far away from the center of the profile (the center is fixed at $\phi = \phi_{crit} = 1/2$ for $x = 0$ per definition by the choice of the coordinate system) the deviation from the bulk value is exponentially small, and the decay constant is always given by ξ_{coex}. On the other hand, inspection of Eq. (59) shows that a characteristic length L defined from the maximum slope of the profile as $L^{-1} \equiv 2\{\phi^{(2)}_{A,coex} - \phi^{(1)}_{A,coex}\}^{-1}/(d\phi/dx)|_{x=0}$ agrees with ξ_{coex} only in the critical region, but in general differs from ξ_{coex}. In particular, for $\chi/\chi_{crit} \gg 1$ this length becomes much smaller than ξ_{coex}. Within this limit $\phi_1 \to 0$ and Eq. (59)

simplifies to

$$\xi_{coex}^2 (d\phi/dx)^2 = \frac{\chi}{\chi_{crit}} [2\phi(1-\phi)]^2 \tag{62}$$

by keeping only the leading terms. Eq. (62) is solved by the familiar tanh profile

$$\phi(x) = \phi_{crit}[1 + \tanh(x/\ell)], \ell = \xi_{coex}(\chi_{crit}/\chi)^{1/2} = \frac{\sigma}{3}\chi^{-1/2}\,. \tag{63}$$

A more elaborate theory [132] confirms the tanh profile but yields a different prefactor in the relation for the characteristic length ℓ, namely $\ell = \sigma(6\chi)^{-1/2}$. The reason for this discrepancy is due to a limitation of the gradient square theory, which can only describe accurately slow concentration variations (Eq. (48)), and describes hence the much more rapid variation of volume fraction occurring in the center of the interfacial profile for $\chi \gg \chi_{crit}$ only qualitatively. For $\chi/\chi_{crit} \gg 1$ the profile is hence governed by two different lengths: in the center the length ℓ (independent of molecular weight) applies, and in the wings the correlation length $\xi_{coex} \approx \sigma/6(2N)^{1/2}$ which is then of the same order as the gyration radius [134].

Near the critical point the situation simplifies as both lengths ξ_{coex}, ℓ merge into one length again. In fact, Eq. (58) then simplifies as $\{m(x) = 2\phi(x) - 1\}$

$$2m(x) + \frac{2}{3}m^3(x) - N\chi m(x) - \frac{N\sigma^2}{9}\frac{d^2m}{dx^2} = 0\,, \tag{64}$$

which is again solved by a tanh profile [133, 78]

$$m(x) = m_{coex}\tanh(x/\ell), \quad \ell = \frac{\sigma}{3}\sqrt{N}\left(\frac{\chi}{\chi_{crit}} - 1\right)^{-1/2} = 2\xi_{coex}\,. \tag{65}$$

From Eqs. (47, 65) one then obtains the interfacial excess free energy, by subtracting the bulk free energy and normalizing the result per unit interface area,

$$\frac{\Delta F_{int}}{k_B T} = \frac{9}{\sigma^2\sqrt{N}}(1 - \chi_{crit}/\chi)^{3/2}, \quad \chi \to \chi_{crit}\,, \tag{66}$$

which explicitly displays the mean-field exponent of the interface tension $\bar{\mu} = 3/2$ {in general $\Delta F_{int} \propto (1 - T/T_c)^{\bar{\mu}}$ [73–75]}. For $\chi \gg \chi_{crit}$, on the other hand, the interfacial tension is [132]

$$\frac{\Delta F_{int}}{k_B T} = \frac{1}{\sigma^2}\left(\frac{\chi}{6}\right)^{1/2}\,. \tag{67}$$

For further discussion of the point that the square gradient functional, Eq. (47), holds only in the "weak segregation limit" but not in the "strong segregation limit", see Refs. [135–137].

2.4 Dynamic Phenomena: Interdiffusion, Spinodal Decomposition

For the moment, let us assume that the local volume fraction is the only slow variable of interest in the system, $\phi(\vec{r}, t)$. Since the overall composition of the blend is conserved, we have a continuity equation

$$\frac{\partial \phi(\vec{r}, t)}{\partial t} + \nabla \cdot J(\vec{r}, t) = 0 , \tag{68}$$

where the current $\vec{J}(\vec{r}, t)$ can be identified with the local current of species A treating the mixture as incompressible. Near equilibrium, one then postulates a linear relation between J and the gradient of the local chemical potential $\mu(r)$, that was introduced in the last subsection [129]

$$J(\vec{r}, t) = -\int \frac{\Lambda(\vec{r} - \vec{r}\,')}{k_B T} \nabla' \mu(\vec{r}\,', t)\, d^3\vec{r}\,' , \tag{69}$$

where $\Lambda(\vec{r} - \vec{r}\,')$ is a nonlocal generalization of an Onsager coefficient [138]. This nonlocality reflects the connectivity of the polymer chains: here the local dynamics of the chains (e.g. described by the Rouse [139, 140] or reputation models [140, 141], depending whether one deals with short chains, that are not entangled, or with long, mutually entangled chains) enters. However, the explicit construction of $\Lambda(\vec{r} - \vec{r}\,')$ still seems to be a problem not yet fully solved [78, 129, 142]. Combining then Eqs. (68), (69) yields, since

$$\nabla \cdot \vec{J}(\vec{r}, t) = \int \frac{\Lambda(\vec{r} - \vec{r}\,')}{k_B T} \nabla'^2 \mu(\vec{r}\,', t)\, d^3\vec{r}\,' , \tag{70}$$

and introducing Fourier transforms as in Eq. (51),

$$\left\{ \Lambda(\vec{q}) = \frac{(2\pi)^3}{V} \int d^3\vec{r} \exp[i\vec{q} \cdot (\vec{r} - \vec{r}\,')] \Lambda(\vec{r} - \vec{r}\,') \right\} ,$$

$$\frac{\partial \phi_q(t)}{\partial t} = -\frac{\Lambda(\vec{q})\, q^2}{k_B T} \mu_q(t) . \tag{71}$$

It turns out that Eq. (71) is still somewhat incomplete: since it is a deterministic equation, it neglects statistical fluctuations. The latter are incorporated, however, if we add a random force on the right hand side,

$$\frac{\partial \phi_q(t)}{\partial t} = -\frac{\Lambda(\vec{q})\, q^2}{k_B T} \mu_{\vec{q}}(t) + \eta_q(t) . \tag{72}$$

The strength of this random force is related to the Onsager coefficient $\Lambda(\vec{q})$ via a (generalized) fluctuation-dissipation theorem:

$$\langle \eta_{\vec{q}}(t) \eta_{-\vec{q}}(t') \rangle_T = \langle |\eta_{\vec{q}}|^2 \rangle_T \delta(t - t') = 2\Lambda(\vec{q})\, q^2 \delta(t - t') . \tag{73}$$

As usual, it is assumed that the fluctuating force is uncorrelated in time, which implies that one considers times much larger than the characteristic times during which internal polymer modes equilibrate.

Since $\mu_{\vec{q}}(t)$ (Eqs. (50), (51)) is a highly nonlinear function of the $\phi_{\vec{q}}$'s, no general solutions are known. Assuming that the deviations from equilibrium are very small, $\phi(\vec{r}, t) = \phi_0 + \delta\phi(\vec{r}, t)$, we may linearize $\mu_{\vec{q}}(t)$ in terms of $\delta_{\vec{q}}(t)$, as done in Eqs. (50)–(52). This yields (denoting the Fourier transform of $\delta\phi(\vec{r}, t)$ as $\delta\phi_{\vec{q}}(t)$ now)

$$\begin{aligned}\frac{\partial}{\partial t}\delta\phi_{\vec{q}}(t) &= -\Lambda(\vec{q})q^2\left\{\frac{1}{N_A\phi_0} + \frac{1}{N_B(1-\phi_0)} - 2\chi \right.\\ &\quad \left. + \frac{a^2q^2}{18\phi_0(1-\phi_0)}\right\}\delta\phi_{\vec{q}}(t) + \eta_{\vec{q}}(t)\\ &\equiv -\frac{1}{\tau_{\vec{q}}}\delta\phi_{\vec{q}}(t) + \eta_{\vec{q}}(t)\,, \end{aligned} \tag{74}$$

where in the last step we have defined a relaxation time $\tau_{\vec{q}}$ for the concentration fluctuations. This relaxation time can be studied, for instance, in quenching experiments (Fig. 2) where a step-like change of temperature is performed. In scattering experiments one then can follow the evolution of the collective structure functions $S(q, t) \equiv \langle|\phi_{\vec{q}}(t)|^2\rangle_T$ as a function of time t after the quench. The result is [78, 143]

$$\begin{aligned}\frac{d}{dt}\langle|\phi_{\vec{q}}(t)|^2\rangle_T &= -2\Lambda(\vec{q})q^2\left\{\left[\frac{1}{N_A\phi_0} + \frac{1}{N_B(1-\phi_0)} - 2\chi \right.\right.\\ &\quad \left.\left. + \frac{a^2q^2}{18\phi_0(1-\phi_0)}\right]\langle|\phi_{\vec{q}}(t)|^2\rangle_T - 1\right\}, \end{aligned} \tag{75}$$

which is solved by

$$S(q, t) = S_{coll}(q, T_0)\exp(-2t/\tau_{\vec{q}}) + S_{coll}(\vec{q}, T)[1 - \exp(-2t/\tau_{\vec{q}})]. \tag{76}$$

For a quench from one state to another in the one phase region (case (i) in Fig. 2) or for a quench that leads to a metastable state (case (ii) in Fig. 2), we thus expect that the structure factor relaxes exponentially from its initial shape (described by $S_{coll}(q, T_0)$) to its final shape (described by $S_{coll}(\vec{q}, T)$). The relaxation time which describes how such fluctuations having a wavelength $\lambda = 2\pi/q$ die out (or are built up, respectively) is described by a diffusive law, which we write as

$$\tau_{\vec{q}}^{-1} = \Lambda(q)q^2/S_{coll}(\vec{q})\,, \tag{77}$$

or

$$\tau_{\vec{q}}^{-1} \sim \Lambda(0)\, q^2[2(\chi_s(\phi_0) - \chi) + \frac{a^2q^2}{18\phi_0(1-\phi_0)}]\,. \tag{78}$$

This result clearly displays the mean-field type critical slowing down [144] of the interdiffusion constant $D_{int} = \lim(\tau_{\vec{q}}^{-1}/q^2) = 2\Lambda(0)\ [\chi_s(\phi_0) - \chi]$ at the spinodal curve $\chi_s(\phi_0)$. In particular, at the critical point we have [78]

$$D_{int} = 4\Lambda(0)\,\frac{1}{N}[1 - \chi/\chi_{crit}]\,, \tag{79}$$

where $\Lambda(0)$ is estimated to be of the order Wa^2, where W is a rate for the reorientation of monomers, in the case of short chains that satisfy the Rouse model (self-diffusion constant of the chains $D_{self} \propto Wa^2/N$), while $\Lambda(0) \propto W(d^2/N)$, d being the tube diameter, in the case of entangled chains being described by the reptation model [140, 141]. Equivalently, we can write $\Lambda(0) = \lambda_0 N_e/N$, N_e being the entanglement chain length, and λ_0 an Onsager coefficient for a segment. In any case we hence have $D_{int} \propto D_{self}(1 - \chi/\chi_{crit})$, expressing the critical slowing down of interdiffusion via the factor $(1 - \chi/\chi_{crit})$.

Right at criticality, Eq. (78) predicts an anomalous slowing down described as

$$\tau_{\vec{q}}^{-1}|_{T_c} \propto q^4\,, \tag{80}$$

displaying a "dynamic critical exponent" [144] of mean-field type, $z = 4$.

It must be emphasized, however, that the treatment presented in Eqs. (68)–(80) is too restrictive since it considers a simple diffusive relaxation only: such a treatment applies to solid binary mixtures ("model B" in the Hohenberg-Halperin [144] classification), while in a fluid binary mixture it is necessary to include the long range order parameter fluctuations that are transmitted by velocity fluctuations [145]. The resulting model {"model H" in the Hohenberg-Halperin classification} leads to a "renormalization" of the Onsager coefficient $\Lambda(q)$ due to mode-coupling effects [146]. Equation [77] remains valid but $\Lambda(q)$ is replaced by

$$\Lambda(q) = \Lambda_0(q) + k_B T \int d^3\vec{k}\, S_{coll}^{-1}(|\vec{q} - \vec{k}|)\ \hat{q}\cdot O(\vec{k})\cdot\hat{q}\,, \tag{81}$$

where $\Lambda_0(q)$ is the previous bare Onsager coefficient, $\hat{q}$ is a unit vector in the direction of $\vec{q}$, and the hydrodynamic interactions are described by the Oseen tensor $O(\vec{k})$

$$O_{\alpha\beta}(\vec{k}) = (\delta_{\alpha\beta} - k_\alpha k_\beta/k^2)\,/(8\pi^3\eta k^2)\,, \tag{82}$$

η being the shear viscosity of the polymer blend. In the hydrodynamic regime $qR_q \ll 1$ the result for $\tau_{\vec{q}}^{-1}$ can be expressed in terms of the Kawasaki function [146] K(x) as [145, 147]

$$\Gamma(q) \equiv 6\pi\eta\xi^3/(k_B T\tau_{\vec{q}}) = (q\xi)^2[1 + (q\xi)^2]\,[K(q\xi) + CR_g N^{3/2}/\xi N_e]\,, \tag{83}$$

$$K(x) = \tfrac{3}{4}[(x^{-1} - x^{-3})\tan^{-1}x + x^{-2}]\,, \tag{84}$$

C being a constant of order unity. The previous result, Eq. (80), only applies for $qR \gg N_e/N^{3/2}$: then the mode coupling contribution can be neglected.

However, if one has $q\xi \gg 1$ but $qR \ll N_e/N^{3/2}$, one has $\Gamma(q) \propto (q\xi)^3$, the "universal" behavior for all critical binary fluids.

Also in the long wavelength regime $q\xi \ll 1$ there are two possible behaviors of the rate $\Gamma(q)$: if $\xi/R \ll N^{3/2}/N_e$, the mode coupling corrections are unimportant and $\Gamma(q) \approx CN^{3/2}N_e^{-1}Rq^2\xi$ as in Eqs. (78), (79), while in the opposite case $\Gamma(q) \approx q^2\xi^2$.

Note that all these formulas also contain the result for the limiting case of short chains dynamics described by the Rouse model [139, 140] if we formally put $N_e \approx N$ in these equations. As will be discussed later (Sect. 2.5), there occurs a crossover in the static critical behavior from mean-field-like behavior {where $\xi \propto R_g\varepsilon^{-1/2}$ with $\varepsilon = 1 - \chi/\chi_{crit}$, $S_{coll}^{-1}(0) \propto N^{-1}\varepsilon$} to the nonclassical critical behavior with Ising model [73, 74] critical exponents $\xi \propto \varepsilon^{-\nu}$, $S_{coll}^{-1}(0) \propto \varepsilon^{\gamma}$, $\nu \approx 0.63$, $\gamma \approx 1.24$. This crossover occurs, as predicted by the Ginzburg criterion [76–79, 148] for $\varepsilon_c \propto 1/N$ or [78, 9], equivalently for $\xi \propto R_g\sqrt{N}$. It thus is seen that for $N_e = N$ the crossover from non-mode-coupled dynamics (for $\xi/R_g \ll \sqrt{N}$) to mode-coupled dynamics (for $\xi/R_g \gg \sqrt{N}$) and the crossover from Landau-like to Ising-like critical behavior both occur in the *same* regime ($\xi \propto R_g\sqrt{N}$), see also [147]. This point was confused by Stepanek et al. [71] who erroneously stated that the Ginzburg criterion means $\xi \propto R_g$. As a consequence, they predicted for the reduced rate $\Gamma_r = \Gamma(q)/\xi^3 \propto \eta/(k_BT\tau_q)$ a non-mean field non-mode coupled regime (their regime I) $\Gamma_r \propto \eta q^2\xi^{-2} \propto q^2\varepsilon^{2\nu}$, which in fact should not exist. Thus it is really gratifying, that the excellent experimental data of Stepanek et al. [71] failed to give any evidence for this regime I. In fact, for the sake of clarity in Fig. 4 the reduced rate Γ_r is plotted in the various regimes in the plane of variables ξ/R_g and λ/R_g, where $\lambda = 2\pi/q$ (this is our revision of Fig. 1 of Stepanek et al. [71]). Note that $\eta \propto N$ in the Rouse model and that $\xi \propto N^{1-\nu}\varepsilon^{-\nu}$ in the Ising regime. Using $R_g \propto N^{1/2}$ it is easy to see that, at the boundaries of the regimes shown in Fig. 4, smooth crossover occurs throughout.

We now return to the mean field result for $\tau_q^{\ddagger}$ {Eqs. (77), (78)} but consider the case of quenching experiments that lead into the unstable regime of the polymer blend phase diagram {case (iii) in Fig. 2}. Then $\chi_s(\phi_0) - \chi$ is negative, and this implies also that $\tau_q^{\ddagger}$ is negative for $q < q_c$; using Eq. (78) one finds {denoting $\tau_q^{-1} = R(q)$, the relaxation rate}

$$R(q)/q^2 = -2\Lambda(0)\,[\chi_s(\phi_0) - \chi](1 - q^2/q_c), \tag{85}$$

$$q_c = \frac{2\pi}{\lambda_c} = \frac{6}{a}\sqrt{\phi_0(1-\phi_0)}\,[\chi - \chi_s(\phi_0)]^{1/2}$$

$$= \frac{\sqrt{18}}{a^2}\left(\frac{1-\phi_0}{N_A} + \frac{\phi_0}{N_B}\right)^{1/2}[\chi/\chi_s(\phi_0) - 1]^{1/2}. \tag{86}$$

Equation (86) implies an exponential growth of fluctuations in the range $0 < q < q_c$, with maximum growth rate for $q_m = q_c/\sqrt{2}$. This exponential growth of concentration fluctuations is called "spinodal decomposition" [9–12,

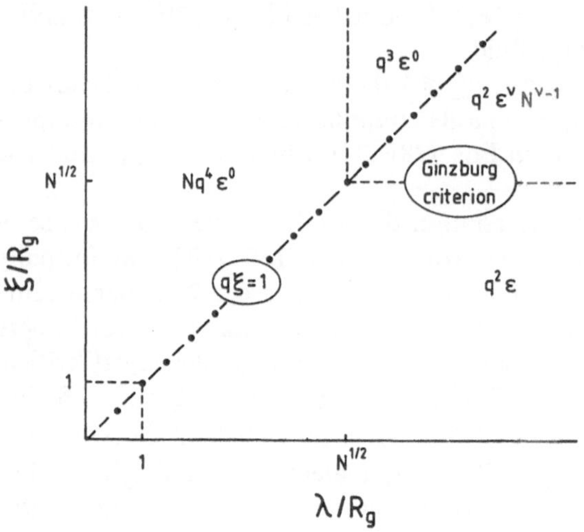

Fig. 4. Schematic illustration of the four different regimes for the reduced relaxation rate $\Gamma_r \propto \eta/(k_B T \tau_q)$ in the different regimes of wavelength $\lambda = 2\pi/q$ and correlation length ξ. Both in the rates and in the crossover lines prefactors of order unity are omitted throughout. The chain curve $q\xi = 1$ separates the hydrodynamic regime, a crossover occurs at $\xi/R_g \propto N^{1/2}$ from the mean-field regime (where also mode-coupling effects can be neglected) to the regime with nonclassical exponents and mode coupling. For small wavelengths in the critical regime, mode coupling causes a crossover from a relaxation rate Nq^4 to q^3 at about $\lambda/R_g \approx N^{1/2}$

78, 120, 121, 129, 140–143, 149–154] Due to the restriction Eq. (48) which leads to $q^2a^2 \ll N_A^{-1}, N_B^{-1}$ (cf. Eq. (52)), Eqs. (85), (86) hold for shallow quenches only, i.e. $\chi/\chi_s(\phi_0) - 1 \ll 1$. In this limit, the characteristic wavelength λ_c is much larger than the gyration radii of the polymer coils, i.e. in this regime the behavior of the polymer mixture is qualitatively the same as that of a mixture of small molecules. Using the full q-dependence of $\Lambda(q)$ and $S_{coll}(q)$ in Eq. (77) this linearized theory of spinodal decomposition can be extended to deep quenches as well [78]. Here we quote the result for symmetrical mixtures ($N_A = N_B = N$, $\sigma_A = \sigma_B = a$) only. Then [78]

$$\tau_q^{-1}/q^2 \propto \frac{\phi_0(1-\phi_0)\ Wa^2\chi}{x}\left\{\frac{1}{2}\frac{\chi_s(\phi_0)}{\chi}x - [1-(1-e^{-x})/x]\right\}, \tag{87}$$

where $x = Na^2q^2/6$. It is seen that in this general case the "Cahn plot" $\{\tau_q^{-1}/q^2$ is plotted vs. q^2 [120, 149–151]$\}$ no longer is linear, unlike Eq. (85), see Fig. 5a). The critical wave-number q_c where τ_q^{-1} vanishes now is found as a solution of the equation $\{\chi_s(\phi_0)/2\chi\}x^2 - x + 1 - e^{-x} = 0$. For deep quenches far underneath the spinodal curve in Fig. 2 {case (iii)} we have $\chi/\chi_s(\phi_0) \gg 1$ and then [78]

$$q_c \cong \left(\frac{6}{Na^2}\right)^{1/2}\left[2\frac{\chi}{\chi_s(\phi_0)} - 1\right]^{1/2}, \quad q_m \cong \left(\frac{6}{Na^2}\right)^{1/2}\left[\frac{2\chi}{\chi_s(\phi_0)}\right]^{1/4}. \tag{88}$$

In this limit the wavenumber of maximum growth q_m is much smaller than the

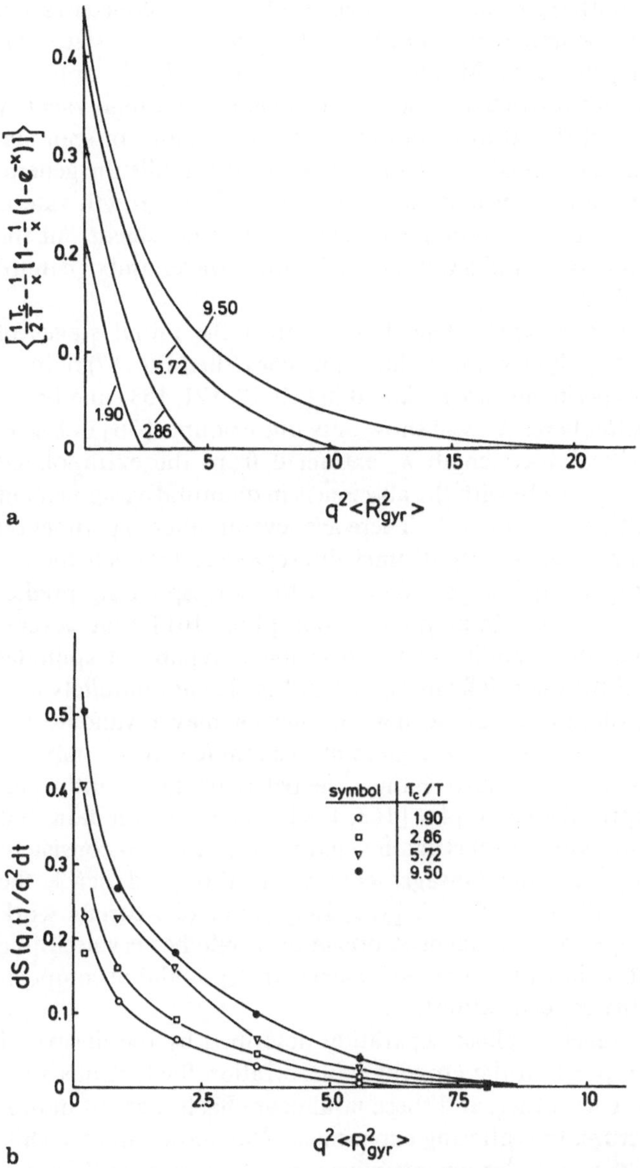

a

b

Fig. 5a. Plot of the normalized relaxation rate τ_q^{-1}/q^2 vs $x = R_g^2 q^2$, for quenches that lead through the critical point such that $\chi_s(\phi_0)/\chi = T/T_c$. Prefactors in Eq. (87) are absorbed in the abscissa scale, and thus only the function $\left\{\frac{1}{2}\frac{T}{T_c} - \frac{1}{x}[1-(1-e^{-x})/x]\right\}$ is shown, for four choices of T_c/T as indicated in the figure. In all cases deep quenches are considered where $q_c R_g > 1$. **b** Corresponding Monte Carlo data for quenching experiments carried out in the three dimensional version of the model shown in Fig. 3, with $\varepsilon_{AB} = \varepsilon$, $\varepsilon_{AA} = \varepsilon_{BB} = 0$, volume fraction of vacancies $\phi_v = 0.6$, chain lengths $N_A = N_B = N = 32$. Dynamics is associated into the Monte Carlo model by a combination of kink-jump and crankshaft motions [104]. The initial growth rate τ_q^{-1} is defined here from the initial derivative of S(q,t), namely $q^{-1}\tau_q \equiv q^{-2}\, d/dt\, S(q,t)|_{t=0}$. From Sariban and Binder [155]

critical wavenumber q_c, both q_m, q_c are in the regime $qR_g \gg 1$, i.e. concentration inhomogeneities grow on length scales much smaller than the coil size. This regime so far has been probed by Monte Carlo simulations [155] only, see Fig. 5b. Since this work was restricted to rather short chains and high vacancy content, rather pronounced deviations from the linearized theory of spinodal decomposition of polymers outlined above were found [155]. While the general shape of the temperature and wavenumber dependence of the growth rate is correctly predicted by Eq. (87), the nonlinear effects completely smear out the sharp vanishing of τ_q^{-1} at $q = q_c$, and a well-defined critical wavenumber cannot be extracted from such data.

There exist several experiments which have probed the initial stages of spinodal decomposition of polymers for shallow quenches (then Eq. (87) reduces to Eq. (85) again). Such experiments are reviewed in [10–12, 121, 153] and hence shall not be treated in detail here. We will show only one example [36] in Fig. 6. In this example, the critical wavelength λ_c extracted from the extrapolated intercept of the Cahn plot (Fig. 6b) with the abscissa is in quantitative agreement with the prediction resulting from Eq. (86). There are several other experimental reports [34, 156–160], however, noting distinct discrepancies between the experimental estimate for q_c (or q_m, respectively) and the corresponding predictions, Eq. (86). However, it has been pointed out [120, 161] that several problems may make the observability of the true linear regime of spinodal decomposition difficult: if the quench from T_0 to T in Fig. 2 is not infinitely fast, the relaxation of the structure factor during the quench may invalidate the analysis [162]; a coupling to other slow degrees of freedom (e.g. if the polymer blend is close to its glass transition) also changes the behavior significantly, and one easily may misinterpret the Cahn plot [163, 164]. Some neutron data that probe the large q behavior where R(q) is again negative seem to yield consistent q_c but several problems with the Onsager coefficient $\Lambda(q)$ {and hence the relaxation rate τ_q^{-1} in Eq. (87)} at $qR_g > 1$ [165, 166]. So further careful work probing the very early stages of spinodal decomposition would be very desirable {e.g. studying the critical behavior as one approaches the spinodal decomposition upon variation of the concentration}.

After the very early stages of phase separation described by the linearized theory of spinodal decomposition the growing concentration fluctuations start to interact strongly with each other, and these nonlinear effects start to induce a coarsening of the structure. In scattering experiments this shows up by a shift of the peak position $q_m(t)$ to smaller wavenumbers as time proceeds (Fig. 7a). Similar behavior is seen in corresponding Monte Carlo simulations (Fig. 7b); however, since these simulations only study rather short chains (N = 32) nonlinear phenomena are important even during the earliest stages of the quenching experiment.

During the later stages one observes that the coarsening behavior changes gradually – an apparent temperature-dependent exponent x in the relation $q_m \propto t^{-x}$ is found in the "intermediate stage", see Fig. 8a, while the behavior in the "transition stage" [36] is similar to the behavior of small molecule mixtures

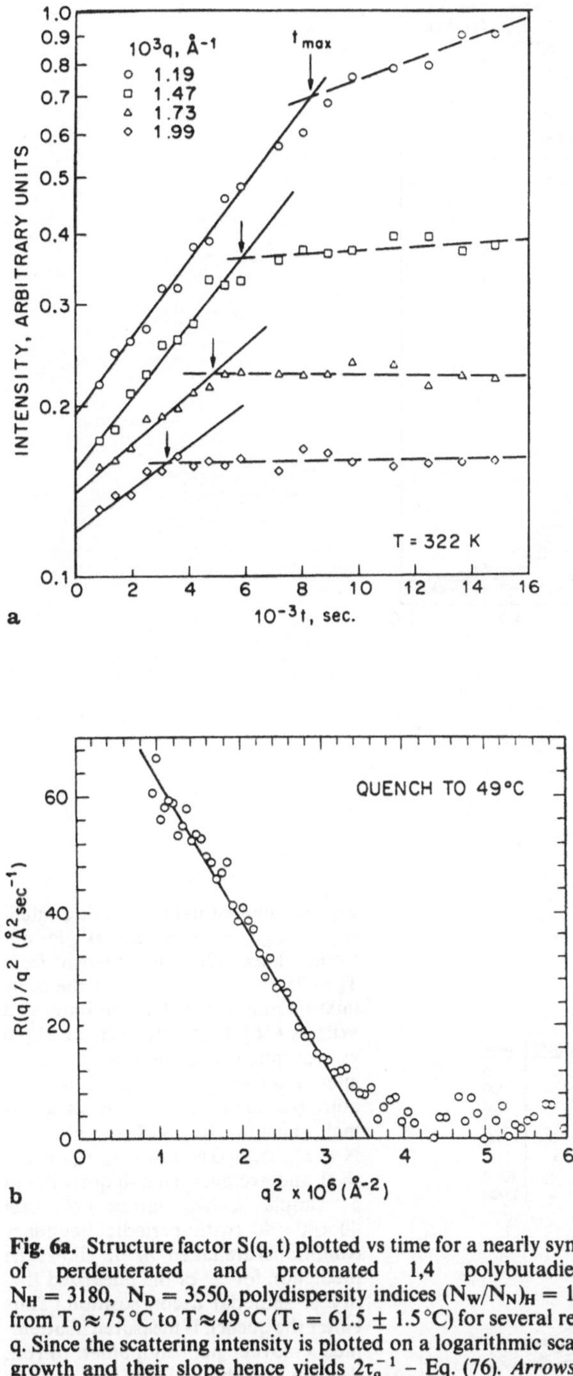

Fig. 6a. Structure factor S(q, t) plotted vs time for a nearly symmetrical critical mixture ($\phi_c = 0.486$) of perdeuterated and protonated 1,4 polybutadiene {degrees of polymerization $N_H = 3180$, $N_D = 3550$, polydispersity indices $(N_W/N_N)_H = 1.03$ and $((N_W/N_N)_D = 1.07$} quenched from $T_0 \approx 75\,°C$ to $T \approx 49\,°C$ ($T_c = 61.5 \pm 1.5\,°C$) for several representative scattering wave numbers q. Since the scattering intensity is plotted on a logarithmic scale, straight lines imply an exponential growth and their slope hence yields $2\tau_q^{-1}$ – Eq. (76). *Arrows* show the time t_{max} where nonlinear effects start to limit the growth. **b** Cahn plot $R(q)/q^2$ vs q^2, for the quenching experiment of a), cf. Eq. (85). Deviations from linearity here are attributed to the neglect of thermal noise – i.e. only the first term on the right hand side of Eq. (76) is kept. From Bates and Wiltzius [36]

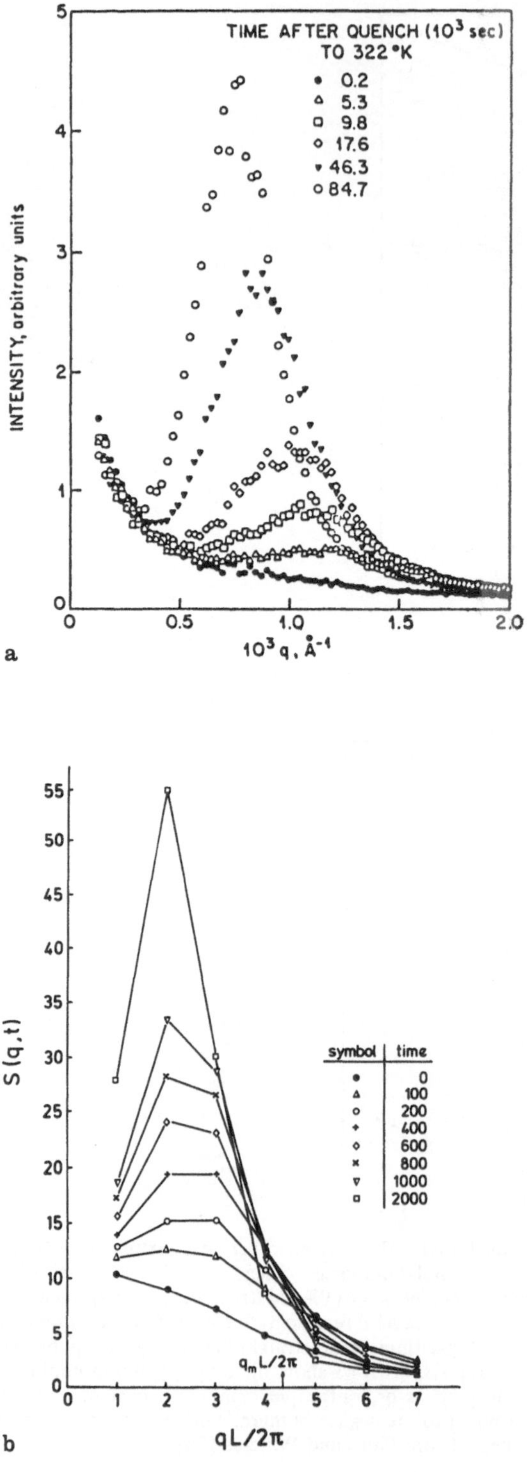

Fig. 7a. Plot of light scattering intensity, $\propto S(q, t)$ vs wavenumbers q for different times after the quench from $T_0 = 75\,°C$ to $T = 49\,°C$, for the same mixture as in Fig. 6a). From Bates and Wilzius [36]. **b** Structure factor S(q, t) vs wavenumber q for different times after a quench from infinite temperature (i.e. the initial state is an athermal blend) to $k_BT/\varepsilon = 1.0$, for $N = 32$, $\phi_v = 0.6$, $\varepsilon_{AB} = \varepsilon$, $\varepsilon_{AA} = \varepsilon_{BB} = 0$, and averages over 40 quenches in a simple cubic lattice of size $40 \times 40 \times 40$ (with periodic boundary conditions) are taken. *Arrow* shows the prediction for q_m of the linearized theory of spinodal decomposition. Time t after the quench is measured in attempted moves per monomer. From Sariban and Binder [155]

– until one reaches the long time coarsening behavior in the "final stage" described by [167] $x = 1$. While this result emphasizes a rather similar behavior of polymers and small molecules [168, 169], it must be noted that evidence exists [12, 158, 170–172] for nonuniversality of the curve $Q_m(\tau) \equiv q_m(t)/q_m(0)$ vs $\tau = D_{eff} q_m^2(0)\, t$: curves for different chain lengths N of the constituents do not superimpose on a master curve but split into a family of curves – "N-branching" [158, 170], see also the theoretical discussion by Onuki [171] – ; a similar effect occurs if a small amount of solvent [12] ("c-branching") or block copolymer corresponding to the constituent homopolymers is added ([172], "B-branching"). However, most spectacular is the effect that in quenching experiments of off-critical polymer mixtures a spontaneous pinning of further domain growth occurs, see Fig. 8b [173]. This pinning behavior is interpreted [173, 153] as a breakup of a percolating structure of the minority phase into well isolated droplets. Such a dynamic percolation phenomenon has indeed been observed in computer simulations of phase separation in Ising models of mixtures as well [174, 175] but since it has little effect on coarsening behavior there, the situation is different for strongly segregated polymers because there the "evaporation" of a chain from a minority droplet of the A-rich phase into the B-rich majority background is strongly suppressed [176]. Such processes would be necessary for a diffusion-controlled coarsening, where large A-rich domains grow on the expense of smaller ones [120, 152, 161, 177].

In the cases where the percolating networks of both A-rich and B-rich domains persist throughout the last stages, the data comply with the scaling hypothesis of Binder and Stauffer [178, 179]

$$S(q, t) = [q_m(t)]^{-3} F\{q/q_m(t)\} . \tag{89}$$

Figure 9 shows a comparison of the scaling function F(x) for several polymer mixtures [36, 180] and small molecules [181]. The data show clear evidence for Porod's law [182, 183], $S(q \gg q_m(t)) \propto q^{-4}$, which describes the scattering from sharp interfaces between the co-existing A-rich and B-rich regions. While the behavior $F(x \ll 1) \propto x^4$ also has been predicted theoretically [184–186], a theoretical calculation of the scaling function F(x) which would be fully satisfactory for all x is still a problem [187–190].

We conclude this section with a discussion of possible relations between interdiffusion in polymer blends and the self-diffusion coefficients of the polymer chains, which have been proposed to interpret the Onsager coefficient $\Lambda(0)$ in Eqs. (78), (79) or (85), respectively. According to the "slow mode theory" [9, 78, 191–192] the slowly diffusing component controls interdiffusion in a blend,

$$[\Lambda(0)]^{-1} = (N_A D_A^* \phi_0)^{-1} + [N_B D_B^* (1 - \phi_0)]^{-1} , \tag{90}$$

D_A^*, D_B^* being the respective self-diffusion coefficients of A-chains and B-chains in the blend. On the contrary, the "fast mode theory" [193–195] suggests that the fast diffusing component controls interdiffusion,

$$\Lambda(0) = \phi_0(1 - \phi_0)\, [N_A D_A^* (1 - \phi_0) + N_B D_B^* \phi_0] . \tag{91}$$

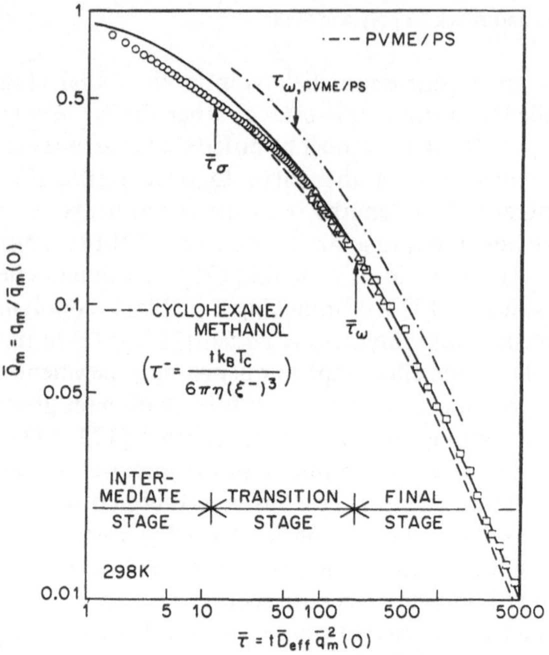

—·— PVME / PS
$\tau_{\omega, PVME/PS}$
$\bar{\tau}_\sigma$
$\bar{\tau}_\omega$
– – – CYCLOHEXANE / METHANOL
$\left(\tau^- = \frac{t k_B T_c}{6\pi\eta(\xi^-)^3}\right)$
INTER-MEDIATE STAGE
TRANSITION STAGE
FINAL STAGE
298K
$\bar{Q}_m = q_m / \bar{q}_m(0)$
1
0.5
0.1
0.05
0.01
1
5
10
50
100
500
5000
$\bar{\tau} = t\bar{D}_{eff}\,\bar{q}_m^{-2}(0)$

a

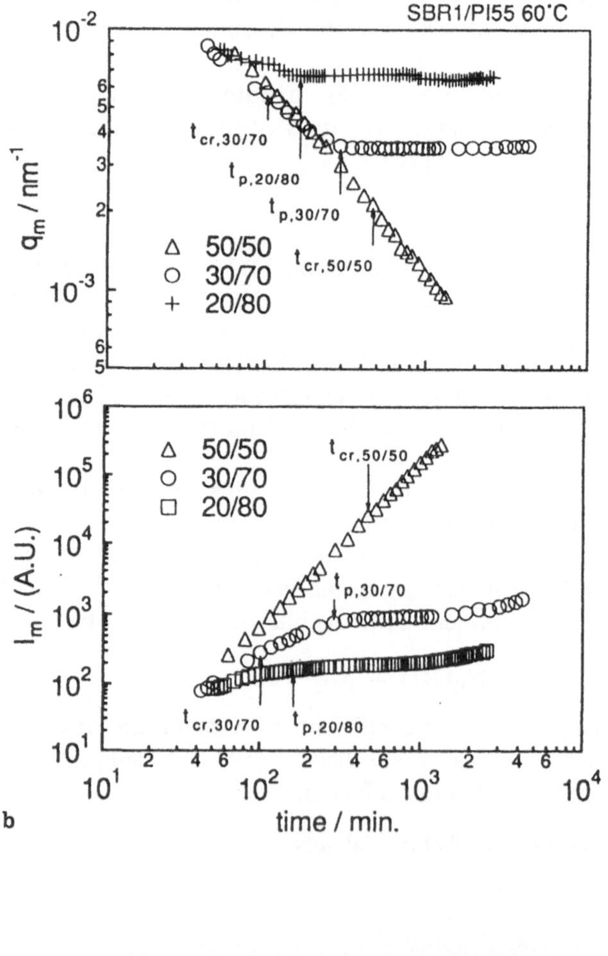

SBR1/PI55 60°C
q_m / nm^{-1}
10^{-2}
10^{-3}
$t_{cr,30/70}$
$t_{p,20/80}$
$t_{p,30/70}$
$t_{cr,50/50}$
△ 50/50
○ 30/70
+ 20/80
I_m / (A.U.)
10^6
10^5
10^4
10^3
10^2
10^1
△ 50/50
○ 30/70
□ 20/80
$t_{cr,50/50}$
$t_{p,30/70}$
$t_{cr,30/70}$
$t_{p,20/80}$
time / min.

b

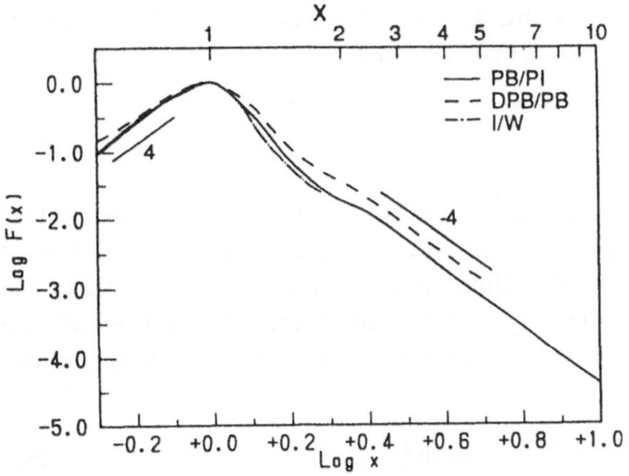

Fig. 9. A comparison of the late stage scaling function of near critical polybutadiene/polyisoprene (PB/PI) mixtures at $T = 55\,^\circ C$ with that of deuterated polybutadiene/polybutadiene mixture {[36], DPB/PB} and of an isobutyric acid/water mixture {[181] I/W}. Note the asymptotic exponents $F(x \ll 1) \propto x^4$ and $F(x \gg 1) \propto x^{-4}$. From Takenaka and Hashimoto [180]

While Eq. (90) can be derived from a vacancy model of interdiffusion [9, 196–197] if off-diagonal Onsager coefficients are neglected [90, 197] as well as from a dynamical version [192, 198] of the RPA, after a long discussion in the literature [199–210] it has now become clear [204–209] that the dynamic coupling between composition fluctuations and stress in polymer blends needs to be considered for a proper description of interdiffusion. With reasonable assumptions one then can justify Eq. (91), see [208–210]. While the experimental evidence is not entirely clear (see [211] for a review), the majority of experiments also seem to favor Eq. (91).

2.5 Ginzburg Criteria for the Validity of Flory-Huggins-Theory

We now recall the well-known Ginzburg criterion [148] for the validity of mean-field theory of polymer mixtures [76, 77] and its extension to nucleation and spinodal decomposition [78, 79]. For simplicity, we first consider a symmetrical mixture only. For the free energy functional $\Delta \mathscr{F}$ in Eq. (47), it is crucial to describe the relevant degrees of freedom by a slowly varying concentration field $\phi(\vec{r})$. One can imagine to construct $\phi(\vec{r})$ from the original Flory-Huggins lattice model by coarse-graining the local lattice occupation variable ϕ_i ($\phi_i = 1$ if

Fig. 8a. Scaling behavior of $q_m(t)$ vs the rescaled time $\bar{\tau} = t\bar{D}_{eff}\, q_m^2(0) = |2\lambda(0)\,[\chi_s(\phi_0) - \chi]$, for the polymer mixture as shown in Fig. 6a and a quench to $T = 25\,^\circ C$. Different symbols refer to different sample geometries. The *solid curve* is a fit to a formula obtained by Furukawa [168] corresponding data for polyvinylmethylether (PVME)-polystyrene (PS), *dash-dotted* [158] and cyclohexane/methanol, *dashed curve* [169], are included. From Bates and Wiltzius [36]. **b** Coarsening behavior of mixtures of SBR (a random copolymer of styrene and polybutadiene) and polyisoprene (PI) at various compositions, at $T = 60\,^\circ C$. **a** shows $q_m(t)$ and **b**. corresponding intensity $I_m(t)$, in arbitrary units, while *arrows* indicate the times where pinning (t_p) or crossover (t_{cr}) from intermediate to late stages occurs. From Hashimoto et al. [173]

lattice site i is taken by an A atom and $\phi_i = 0$ if it is taken by a B-atom) over a cell of volume L^d (in d dimensions) centered at $\vec{r}$:

$$\phi(\vec{r}) = (a/L)^d \sum_{i \in cell} \phi_i \,. \tag{92}$$

Near the critical point, the relevant scale for the concentration variation is set by the order parameter m_{coex}, Eqs. (14), (15). The condition for the validity of mean-field theory (for $T < T_c$) now is that the mean square fluctuation of the order parameter in a volume region of linear dimension L over which we average, Eq. (92), must be small in comparison with the order parameter square itself,

$$\langle [\delta\phi(\vec{r})]^2 \rangle_{T,L} = \langle [\phi(\vec{r}) - \phi_0]^2 \rangle_{T,L} \ll m^2_{coex} \,. \tag{93}$$

Using Eq. (92) one finds $\langle [\delta\phi(\vec{r})]^2 \rangle_{T,L} = (a/L)^{2d} \sum_{i,j \in cell} (\langle \phi_i \phi_j \rangle_T - \phi_0^2)$. Now the correlation function of concentration fluctuations $\langle \phi_i \phi_j \rangle_T - \phi_0^2$ {which is the spatial Fourier transform of the collective structure factor $S_{coll}(q)$ considered in Eq. (55)} in mean field theory has simple Ornstein-Zernike form [73] – Eq. (38), i.e. with $\vec{x} = \vec{r}_i - \vec{r}_j$ we can write

$$\langle \phi_i \phi_j \rangle_T - \phi_0^2 \propto (x/a)^{-(d-2)} \exp(-x/\xi) \,, \tag{94}$$

which yields for $L < \xi$

$$\langle [\delta\phi(\vec{r})]^2 \rangle_{T,L} \propto (L/a)^{\,2-d} \,, \tag{95}$$

where prefactors of order unity in Eqs. (94), (95) are omitted. Now the largest permissible choice for L is to choose $L = \xi$, since ξ is the length scale for concentration fluctuations near T_c. Thus the Ginzburg criterion is, noting $m^2_{coex} \propto \varepsilon$ {Eq. (15), $\varepsilon = (1 - T/T_c)$, for $\phi_0 = \phi_{crit}$}, for $d = 3$

$$\langle [\delta\phi(\vec{r})]^2 \rangle_{T,\xi} \propto \frac{a}{\xi} \propto \frac{a}{R_g \varepsilon^{-1/2}} \ll \varepsilon \,, \tag{96a}$$

or

$$N^{-1/2} \ll \varepsilon^{1/2}, \quad 1 \ll N\varepsilon \,, \tag{96b}$$

or (writing it as a comparison of lengths)

$$\xi \ll R_g \sqrt{N} \,. \tag{96c}$$

For large N hence mean field theory is very good, apart from a very narrow region close to the critical point. An intuitive interpretation to this finding can be given as follows [2]. Each polymer coil is a random walk of radius $a\sqrt{N}$ and hence takes a volume $a^d \sqrt{N^d}$. But the density of this polymer chain in this region taken by the polymer is rather small, namely of the order of $N/(a^d \sqrt{N^d}) = a^{-d}/\sqrt{N^{d-2}}$. Since the segment density is just a^{-d}, in the melt each cell of linear dimension a contains one segment, one concludes that each chain must interact with polymer segments of $\sqrt{N^{d-2}}$ *other* chains. Thus for $d = 3$ each chain interacts with $\sqrt{N} \gg 1$ "neighbors". In a sense the polymer mixture behaves as a mixture of small molecules but with a long range of interaction.

This argument can be made precise [79] by a mapping of the free energy functionals of both problems onto each other. Note that the partition function (even beyond mean-field!) is given by a functional integral involving the free energy functional $\Delta\mathscr{F}$,

$$Z \equiv \int \mathscr{D}\phi(\vec{r}) \exp[-\Delta\mathscr{F}[\phi(\vec{r})]/k_B T] , \tag{97}$$

where for the small molecule mixture an expression analogous to Eq. (47) holds,

$$\frac{1}{k_B T}\Delta\mathscr{F}_{\text{small molecules}} = \int d^d\vec{r}\{f[\phi(\vec{r})] + \frac{r_0^2}{6}[\nabla\phi(\vec{r})]^2\} , \tag{98}$$

r_0 being the interaction range, and the free energy density is given in mean field theory as a sum of entropy of mixing and enthalpy terms,

$$f(\phi) = \phi \ln \phi + (1-\phi)\ln(1-\phi) + 2\frac{T_c}{T}\phi(1-\phi) . \tag{99}$$

Similarly, for the polymer mixture near $\phi = 1/2$, Eq. (47) yields

$$\frac{1}{k_B T}\Delta\mathscr{F}_{\text{polymers}} = \int d^d\vec{r}\{f_{FH}[\phi(\vec{r})] + \frac{a^2}{9}[\nabla\phi(\vec{r})]^2\} , \tag{100}$$

with

$$f_{FH}(\phi) = \frac{1}{N}[\phi \ln \phi + (1-\phi)\ln(1-\phi) + N\chi\phi(1-\phi)] . \tag{101}$$

Thus we simply obtain

$$\Delta\mathscr{F}_{\text{polymers}} \equiv \frac{1}{N}\Delta\mathscr{F}_{\text{small molecules}} \tag{102}$$

if we identify

$$r_0^2 \equiv \tfrac{2}{3}Na^2, \; \chi = 2\left(\frac{T_c}{T}\right)N . \tag{103}$$

These arguments are simply extended to spinodal decomposition and nucleation [78, 79]. E.g., for the linearized theory of spinodal decomposition to hold we must require (Fig. 10) that the mean square amplitude of the growing concentration waves is small in comparison to the distance from the spinodal curve,

$$\langle[\delta\phi(\vec{r}, t)]^2\rangle_{T, L=\lambda_c} \ll [\phi - \phi_{sp}(\chi)]^2 , \tag{104}$$

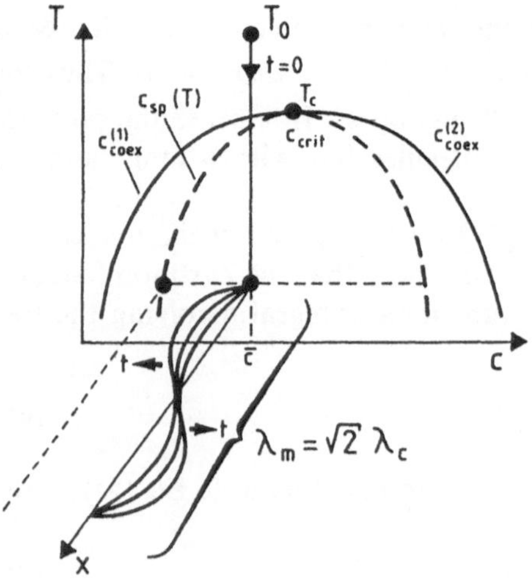

Fig. 10. Schematic plot of a quenching experiment at a concentration $\bar{c}$ in the unstable region, explaining the validity of the linearized theory of spinodal decomposition: spinodal decomposition is described by a growth of concentration waves, $\delta c(x, t) \propto \cos(q_m x) \exp[R(q_m)t]$ as long as $|\delta c(x, t)| \ll c - c_{sp}(T)$

where we note that the appropriate choice for the linear dimension L of the coarse graining cell is in fact the critical wavelength $\lambda_c = 2\pi/q_c$ of spinodal decomposition. Eq. (104) then can be approximated as, assuming that at the times of interest the maximum growing wavelength $\lambda_m = 2\pi/q_m$ dominates,

$$\langle[\delta\phi(\vec{r}, 0)]^2\rangle_{T_0, L=\lambda_c} \exp[2R(q_m)t] \ll [\phi - \phi_{sp}(\chi)]^2 . \tag{105}$$

Working this out in a similar way to that in Eqs. (93)–(96) yields [78, 79]

$$N^{(d-2)/2}\left[1 - \frac{\chi_{crit}}{\chi}\right]^{(4-d)/2}\left[\frac{\phi}{\phi_{sp}(\chi)} - 1\right]^{(6-d)/2} \gg \exp[2R(q_m)t]. \tag{106}$$

It can be seen that this condition (as well as Eq. (93), of course) can only be satisfied for polymers in $d = 3$ but not in two-dimensional geometry. Also the time range over which the linearized theory of spinodal decomposition holds increases only with the logarithm of the chain length,

$$R(q_m)t < \ln N . \tag{107}$$

In addition, the temperatures T_0, T involved in the quench must lie in the mean-field critical region, which for general d is given by the condition

$$N^{(d-2)/2}\left|1 - \frac{\chi_{crit}}{\chi}\right|^{(4-d)/2} \gg 1 . \tag{108}$$

Equation (108) also illustrates the well-known fact that in $d = 4$, mean field theory would be valid up to the critical point, at least if logarithmic corrections are disregarded [74]. But what is most interesting in the present context, is that Eq. (106) also describes how closely one can approach the spinodal curve without having excessively strong nonlinear effects. In d = 3 and not too close to the critical points, we thus must have

$$\phi/\phi_{sp}(\chi) - 1 \gg N^{-1/3} . \tag{109}$$

It thus follows that only for $N \to \infty$ do we expect a sharp spinodal curve to be meaningful. Note that the spinodal curve really plays the role of a line of critical points in the context of mean field theory: e.g. the critical wavelength λ_c diverges there, cf. Eq. (86) which also can be written as

$$\lambda_c = (R_g/2\pi)\,[2\phi_0(1-\phi_0)]^{-1/2}\,[\chi/\chi_s(\phi) - 1]^{-1/2} . \tag{110}$$

Similar Ginzburg criteria can also be worked out in the metastable region of the phase diagram (Figs. 1, 2): The correlation length in the metastable state diverges when the spinodal curve is approached, see Eq. (39). For a symmetric mixture one obtains

$$\xi = R_{gyr}\sqrt{2}[1 - \chi/\chi_s(\phi)]^{-1/2} = R_{gyr}\left(1 - \frac{\chi_{crit}}{\chi}\right)^{-1/2}\left(1 - \frac{\phi}{\phi_{sp}}\right)^{-1/2} \tag{111}$$

where $\phi_{sp}(\chi) = \frac{1}{2}[1 \pm (1 - \chi_{crit}/\chi)^{1/2}]$ was used. One can see that both characteristic lengths λ_c, ξ {Eqs. (110), (111) } diverge in exactly the same way as one approaches $\phi = \phi_{sp}$ from either side: the line $\phi = \phi_{sp}(\chi)$ {remember that $\phi_{sp}(\chi)$ is the inverse function of $\chi_{sp}(\phi)$} plays the role of a line of critical points in the phase diagram. Applying again the Ginzburg criterion, similar to Eq. (104),

$$\langle [\delta\phi(\vec{r})]^2 \rangle_{T,L=\xi} \ll [\phi_{sp}(\chi) - \phi]^2 , \tag{112}$$

yields a condition analogous to Eq. (106)

$$N^{(d-2)/2}\left[1 - \frac{\chi_{crit}}{\chi}\right]^{(4-d)/2}\left[1 - \frac{\phi}{\phi_{sp}(\chi)}\right]^{(6-d)/2} \gg 1 , \tag{113}$$

which for d = 3 in the mean-field critical regime requires that

$$1 - \phi/\phi_{sp}(\chi) \gg N^{-1/3} , \tag{114}$$

which is the counterpart of Eq. (109). While for $N \to \infty$ the singularities associated with the spinodal curve are well-defined, for finite N they are smeared out over a regime of width proportional to $N^{-1/3}$ on either side.

Of course, the increase of ξ in the metastable state as ϕ approaches $\phi_{sp}(\chi)$ can be observed only for short enough times: the lifetime of metastable states is limited, as is well known, metastable states decay via nucleation and growth. It turns out, however, that for polymer mixtures with high molecular weight, homogeneous nucleation is to a large extent suppressed, since the free energy barriers ΔF^* to form a droplet of critical radius R^* are very large [78]. In order to see this, we consider first the regime close to the coexistence curve, and apply classical nucleation theory [122, 123, 152] to estimate this barrier $\Delta F(R)$ which has to be overcome in a nucleation event. Assuming a spherical shape of the droplet, the excess free energy of a droplet of radius R is

$$\Delta F(R) = -(4\pi R^3/3)\,\delta\mu[\phi^{(2)}_{coex} - \phi^{(1)}_{coex}]/a^3 + 4\pi R^2 \Delta \mathscr{F}_{int}\,. \tag{115}$$

Here $\delta\mu$ is the chemical potential difference between the metastable state and the state at the coexistence curve, while $\Delta\mathscr{F}_{int}$ is the interfacial tension (considered in Eqs. (66), (67)). In a metastable state, the bulk (volume) term of $\Delta F(R)$ is negative but for small R the unfavorable surface term $4\pi R^2\Delta\mathscr{F}_{int}$ dominates and $\Delta F(R)$ increases with R up to the critical size R^*, where $\partial[\Delta F(R)\}/\partial R|_{R=R^*} = 0$: the value of $\Delta F(R^*) = \Delta F^*$ thus is the free energy barrier that needs to be overcome by droplet formation. Extremizing Eq. (115) with respect to R yields

$$R^* = 2a^3\Delta\mathscr{F}_{int}/\{\delta\mu[\phi^{(2)}_{coex} - \phi^{(1)}_{coex}]\}\,, \tag{116}$$

$$\Delta F^* = \frac{16\pi}{3}\,\frac{(\Delta\mathscr{F}_{int})^3\, a^6\,\{S^{coex}_{coll}(q=0)/k_BT\}^2}{(\delta\phi)^2\,[\phi^{(2)}_{coex} - \phi^{(1)}_{coex}]^2}\,, \tag{117}$$

Here we have assumed that the metastable state (at χ, ϕ) is so close to $\phi^{(1)}_{coex}$ that we can linearize $\delta\phi \equiv \phi - \phi^{(1)}_{coex} = \{S^{coex}_{coll}(q=0)/k_BT\}\,\delta\mu$, to eliminate $\delta\mu$ in favor of $\delta\phi$. Using Eqs. (15), (43) and (66) one finds [78]

$$\frac{\Delta F^*}{k_BT} = \frac{27\pi}{4}\left[N\left(1 - \frac{\chi_{crit}}{\chi}\right)\right]^{1/2}\left\{\delta\phi\Big/\left[\phi^{(2)}_{coex} - \phi^{(1)}_{coex}\right]\right\}^{-2}. \tag{118}$$

It is seen that the *scale* for the free energy barrier $\Delta F^*/k_BT$ is set by the parameter of the Ginzburg criterion, $\left[N\left(1 - \frac{\chi_{crit}}{\chi}\right)\right]^{1/2}$, which is much larger than unity in the mean field critical regime, described by Eqs. (96), (108). This fact holds not only close to the coexistence curve (Eqs. (115)–(118) are only valid for $\delta\phi/[\phi^{(2)}_{coex} - \phi^{(1)}_{coex}] \ll 1$), but throughout the metastable regime, where the free energy barrier is described by a more general function $h(\zeta)$ of the "supersaturation" $\zeta \equiv \delta\phi/[\phi^{(2)}_{coex} - \phi^{(1)}_{coex}]$,

$$\Delta F^* = \left[N\left(1 - \frac{\chi_{crit}}{\chi}\right)\right]^{1/2} h(\zeta)\,. \tag{119}$$

While we have $h(\zeta \to 0) \propto \zeta^{-2}$ the barrier vanishes at the spinodal in mean field theory [212, 213]

$$h(\zeta) \propto [\phi_{sp}(\chi) - \phi]/[\phi^{(2)}_{coex} - \phi^{(1)}_{coex}]^{3/2}\,, \tag{120}$$

in $d = 3$ dimensions. Figure 11 summarizes qualitatively the behavior of the nucleation barrier in the mean field regime of polymer mixtures (upper part) and compares it to the behavior of small molecule mixtures (lower part) {or the non-mean-field regime of polymer mixtures, respectively}. Here we draw attention to the regime of "spinodal nucleation", i.e. the phase separation starts by the

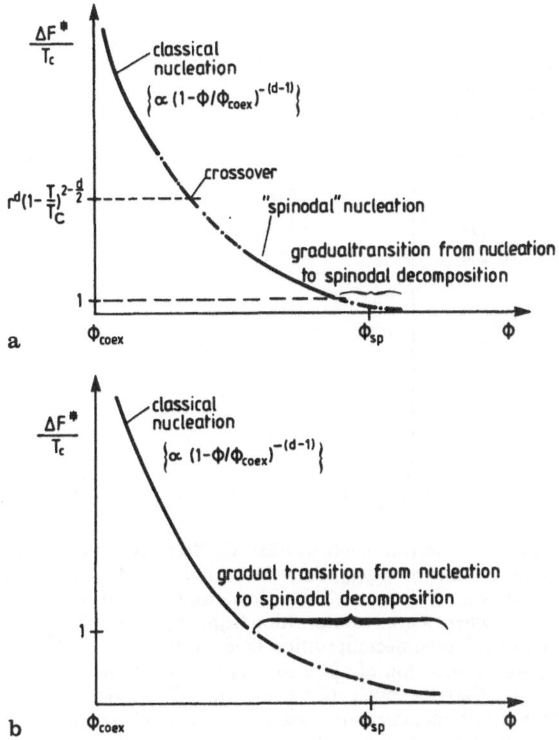

Fig. 11. Schematic plots of the free energy barriers for a d-dimensional system with a large range r of interaction – or a polymer mixture, respectively: $r^d(1-T/T_c)^{2-d/2}$ then has to be replaced by $a^d N^{d/2-1}(1-T/T_c)^{2-d/2}$. **A**) refers to the mean-field critical region, where $r^d(1-T/T_c)^{2-d/2} \gg 1$, **B**) to the non-mean-field region. In this figure, $k_B \equiv 1$, so the gradual transition from nucleation to spinodal decomposition occurs for $\Delta F^*/T_c \approx 1$. At $\Delta F^*/T_c \approx r^d (1-T/T_c)^{2-d/2]}1$ a crossover occurs from classical nucleation (i.e., compact spherical droplets) to spinodal nucleation (i.e., diffuse ramified droplets). From Binder [79]

formation of diffuse ramified clusters {rather than the compact spherical drops considered in Eqs. (115)–(118)}; these ramified clusters first have to compactify before they can grow. While such a behavior has been seen in computer simulations [214], polymer mixtures in metastable states close to spinodal curves would offer unique possibilities for detecting this behavior in real materials. Note that in mixtures of small molecule systems (as well as in the non-mean-field regime of polymer mixtures) the free energy barrier $\Delta F^*/k_B T$ becomes of order unity long before the spinodal curve is reached (Fig. 11, lower part): then homogeneous nucleation is very easy, droplets form everywhere in the system due to thermal fluctuations and the metastable state decays fast. Such a situation where many droplets grow simultaneously everywhere in the system qualitatively is not so different from the situation where a wavepacket of (interacting!) concentration waves grows – thus one can understand that a gradual transition from nucleation to spinodal decomposition occurs without any singular behavior at a spinodal line. Thus in the phase diagram of a polymer mixture one can distinguish several regions (Fig. 12). We emphasize that although Fig. 12 looks rather complicated it is somewhat simplified, since in reality the smeared-out spinodal curve is somewhat shifted from the location predicted by mean field theory toward the coexistence curve [174]. Therefore it is nontrivial to find experimentally the regions where "spinodal nucleation" and where the gradual

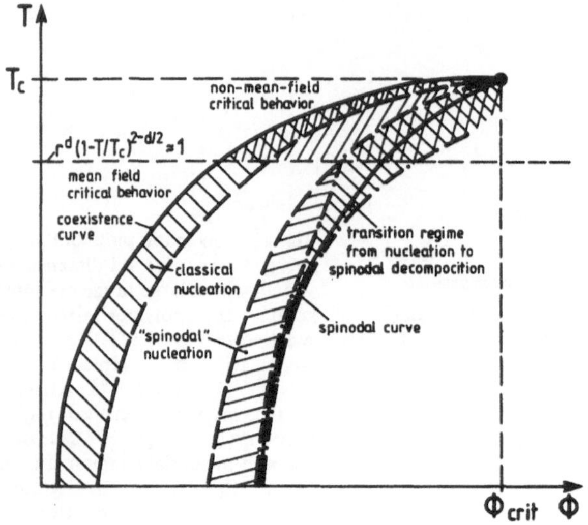

Fig. 12. Various regions in the temperature-composition plane near T_c. Since the situation for a symmetric mixture is symmetric around the line $\phi = \phi_{crit}$, only the regime $\phi < \phi_{crit}$, is shown. Coexistence curve and spinodal curve are shown as *full curves*. The regime inside the two *dash-dotted lines* around the spinodal curve is the regime where a gradual transition from nucleation to spinodal decomposition occurs. Linearized theory of spinodal decomposition holds to the right of the *right dash-dotted line*, while "spinodal nucleation" (formation of ramified droplets) occurs in the *shaded region* to the left of the *left dash-dotted line*. Classical nucleation (formation of spherical droplets) occurs between the *left dashed line* and the coexistence curve (in the *shaded region*). These regimes are all only observable in the mean-field critical region of the mixture. For $r_0^d(1 - T/T_c)^{2-d/2} \approx 1$ (for a d-dimensional system with a large but finite range of interaction r_0) or for $\sqrt{N}(1 - T_c/T) \approx 1$ (for a three-dimensional polymer mixture) the crossover to non-mean-field critical behavior occurs. In this regime (*shaded*) fluctuations are strong and nonlinear effects cause a crossover from classical nucleation to nonlinear spinodal decomposition which is spread out over a broad concentration region in the phase diagram, and the spinodal line has not much physical significance there. From Binder [79]

transition from nucleation to spinodal decomposition occurs. Detailed experimental studies of nucleation and growth in polymer mixtures have just begun [70]. These studies used relatively short chains (a polyisoprene-poly(ethylenepropylene) mixture with $N_{PI} = 29$, $N_{PEP} = 73$, cf. Fig. 1b) and found nonlinear spinodal decomposition behavior to persist into the regime which according to the mean field spinodal should be already metastable, consistent with the above remarks. So far, only nucleation of heterogeneous type seems to have been observed [70]. Since for high molecular weight polymers the barriers for homogeneous nucleations are generally expected to be rather high, as discussed above, one must expect that heterogeneous nucleation is the dominating mechanism for the decay of metastable states in polymer mixtures.

For a quantitative discussion of real polymer mixtures, it clearly is important to consider prefactors in evaluating the Ginzburg criterion, Eqs. (93)–(96) or (108), respectively. Also the asymmetry of the two types of chains needs to be taken into account. This problem was considered by Bates et al. [69] who suggest that the crossover occurs at a distance from T_c given by

$$\varepsilon_{cross} = Cv_m^2 \frac{[N_A^{-1}\phi_A^{-3} + N_B^{-1}\phi_B^{-3}]^2}{[N_A^{-1}\phi_A^{-1}+N_B^{-1}\phi_B^{-1}][\langle R_{gyr}^2\rangle_A N_A^{-1}\phi_A^{-1}+\langle R_{gyr}^2\rangle_B N_B^{-1}\phi_b^{-1}]^3} \tag{121}$$

where v_m is the segment volume and C a universal constant. Identifying ε_{cross} with the point, where a plot of $S_{coll}^{-1}(q \to 0)$ vs inverse temperature deviates from linearity, Bates et al. [69] estimate $C = 0.29 \pm 0.08$ from their data on PI-PEP mixtures, while their analysis of data due to Schwahn et al. [215] yields $C \approx 0.46$. In a recent paper [216] Janßen et al. estimate $C = 0.016$, however, for the deuterated polystyrene-polyvinylmethylether system, while Meier et al. [217] find $C = 0.32$ for PEMS/PDMS mixtures. Thus the problem of the crossover from mean field theory to the nontrivial Ising-like critical behavior [73–75] is not fully understood. We shall briefly return to this question in Sect. 4.3; here we now only summarize the critical behavior that one predicts for $|\varepsilon| \ll \varepsilon_{cross}$. The order parameter (we return to the case of symmetric mixtures) is predicted to behave as [107]

$$m = \hat{B}(N)\,\varepsilon^{\beta}, \quad \beta \approx 0.325, \quad \hat{B}(N) \propto N^{\beta - 1/2}\,. \tag{122}$$

The singular variation of the critical amplitude $\hat{B}(N)$ with chain length N is simple seen from the fact that the variable εN which controls the Ginzburg criterion, Eq. (96b), appears as a crossover scaling variable in the order parameter, assuming a smooth crossover between the order parameter in the mean field regime, Eq. (15), and in the Ising regime, Eq. (122):

$$m = \varepsilon^{1/2}\tilde{f}_m(\varepsilon N), \quad \tilde{f}_m(\varepsilon N \gg 1) = \sqrt{3}\,. \tag{123}$$

In order that Eq. (123) reduces to Eq. (122) for small εN, we must have $\tilde{f}_m(\xi) \propto \xi^{\beta - 1/2}$, which yields the N-dependence of $\hat{B}(N)$ noted in Eq. (122). Similarly, for the collective structure factor one writes [107]

$$S_{coll}(0) = N|\varepsilon|^{-1}\tilde{f}_s(|\varepsilon|N) \tag{124}$$

and assuming that $\tilde{f}_s(\zeta \ll 1) \propto \zeta^{1-\gamma}$ one finds [107]

$$S_{coll}(0) \propto \hat{\Gamma}(N)\,|\varepsilon|^{-\gamma}, \quad \text{with } \hat{\Gamma}(N) \propto N^{2-\gamma}, \ \gamma \approx 1.24\,. \tag{125}$$

Finally the correlation length ξ is written as

$$\xi = N^{1/2}|\varepsilon|^{-1/2}\tilde{f}_\xi(|\varepsilon|N) \tag{126}$$

and assuming $\tilde{f}_\xi(\zeta) \propto \zeta^{1/2-\nu}$ one finds

$$\xi = \hat{\xi}(N)\,|\varepsilon|^{-\nu}, \quad \hat{\xi}(N) \propto N^{1-\nu} \tag{127}$$

We also note that for the collective scattering function $S_{coll}(0)$ and correlation length it makes sense to consider both amplitudes $\hat{\Gamma}(N)$, $\hat{\xi}(N)$ referring to

$\phi = \phi_{crit}$ and $T > T_c$, as well as amplitudes $\hat{\Gamma}'(N)$, $\hat{\xi}'(N)$ referring to $T < T_c$, taking the limit $T \rightarrow T_c$ at a path along the coexistence curve. It is then implied that the critical amplitude ratios $\hat{\Gamma}(N)/\hat{\Gamma}'(N)$, $\xi(N)/\hat{\xi}'(N)$ are universal [75] and also as the critical exponents β, γ, ν etc. take the values of the Ising model – e.g. $\hat{\Gamma}(N)/\hat{\Gamma}'(N) \approx 5$, while in the mean field critical regime we have $\hat{\Gamma}(N)/\hat{\Gamma}'(N) = 2$ [75]. Apart from Monte Carlo simulations [92, 101, 107, 218], the chain length dependence of critical amplitudes and critical amplitude ratios has not found much attention yet.

3 Brief Discussion of More Sophisticated Theories

In this section, we mention very briefly some recent theoretical developments, which go far beyond the simple Flory-Huggins theory. As was emphasized above, the Flory-Huggins theory suffers from two basic defects: (i) Using a lattice model where polymers are represented as self-avoiding walks is a crude approximation, which neglects the disparity in size and shape of subunits of the two types of chain in a polymer blend, as well as packing constraints, specific interactions etc. (ii) Even within the realm of a lattice model, the statistical mechanics (involving approximations beyond the mean field approximation) is far too crude.

Consequently, more modern theories seek to avoid those shortcomings. A systematic and very impressive approach is the "lattice cluster theory" (LCT) of Freed and coworkers [51, 80–83, 219–222]. This theory is a systematic double expansion in z^{-1} (z being the coordination number of the lattice) and ε_{ij}/k_BT, ε_{ij} being the pairwise interaction energy. In the simplest case, the Flory-Huggins theory is recovered when one truncates these expansions at the lowest nontrivial order. A very interesting point is that this theory has been worked out for various types of monomer shapes covering several lattice sites: in this way one can, at least to some extent, eliminate the above problem (i), in spite of the use of a lattice model, and thereby describe the influence of monomer structure on polymer fluid properties. Dudowicz and Freed [80] discuss the experimentally observed volume fraction dependence of the χ-parameter of PS/PVME blends (extracted from the data using Flory-Huggins theory, of course) in the light of their approach, and find [221] that the concentration dependence of the correlation length ξ compares favorable to the data of Han et al. [223].

Another exciting development gets rid of the limitation (i) from the start by basing the approach on theoretical developments in the theory of fluids [44–49, 52, 224–225]. These theories all are based on the concept of the "direct correlation function" [226] related to the direct interaction between monomers; the total correlation function is expressed approximately in terms of the intramolecular structure factor and this direct correlation function, which needs a suitable decoupling approximation, which can be motivated by suitable factorizations of the (exact) Born-Green-Yvon (BGY) hierarchy [227, 228] of correlation functions. Approximations such as the Kirkwood superposition principle [229] are used [225] or the mean spherical approximation [44–49]. It turns out that the "reference interaction site model" (RISM) [230] extended to homopolymer melts (PRISM) [231–233] gives a very good account of intermolecular correlations, using the Gaussian character of intramolecular configurations as an input, as the comparison with extensive melt simulations shows [234]. For homopolymer melts, an Ornstein-Zernike like equation in Fourier space results [233]

$$h(q) = S(q)\, c(q)\, [S(q) + \rho_m h(q)] \,, \tag{128}$$

where h(q) is the Fourier transform of $h(r) = g(r) - 1$, g(r) being the intermolecular site-site pair correlation function, S(q) is the single chain structure factor, c(q) the intermolecular direct correlation function, and ρ_m the monomer density. For a polymer mixture, there are then three independent intermolecular site-site pair correlation functions: $g_{AA}(r)$, $g_{AB}(r)$ and $g_{BB}(r)$. In Fourier space, one obtains a matrix equation in full analogy with Eq. (128), namely [47]

$$\underline{H}(q) = \underline{\Omega}(q)\, \underline{C}(q)\, [\underline{\Omega}(q) + \underline{H}(q)] \,, \tag{129}$$

where the 2 × 2 matrices, $\underline{H}(q)$, $\underline{\Omega}(q)$ and $\underline{C}(q)$ are the analogues of h(q), S(q) and c(q). In particular, in real space the matrix element $H_{mm'}(r)$ {with m, m' = AB} can be written as $H_{mm'}(r) = \rho_m \rho_{m'}(r)\, h_{mm'}(r), \Omega_{mm'}(r) = \rho_m \delta_{mm'} W_m(r)$, with $S_m(q)$ being the Fourier transform of $W_m(r)$. If one enforces an incompressibility constraint, one can cast the theory in a form similar to the RPA, where [46]

$$S_{coll}^{-1}(q) = (v_A/v_B)^{-1/2}[\phi_A S_A(q)]^{-1} + (v_A/v_B)^{1/2}[\phi_B S_B(q)]^{-1} - 2\chi_{eff}(q) \tag{130}$$

where v_A, v_B are the segmental volumes and the segmental volume fractions ϕ_A, ϕ_B are then defined as

$$\phi_m = \rho_m v_m / (\rho_A v_A + \rho_B v_B), \quad m = A, B \,, \tag{131}$$

and the analog of the Flory-Huggins parameter now is expressed in terms of direct correlation functions,

$$\chi_{eff}(q=0) = \frac{\rho}{2}[\phi_A (v_A/v_B)^{-1/2} + \phi_B (v_A/v_B)^{1/2}]^{-1} \times \left[\frac{v_B}{v_A} C_{AA}(q=0) + \left(\frac{v_B}{v_A}\right)^{-1} C_{BB}(q=0) - 2C_{AB}(q=0)\right]. \tag{132}$$

Equations (130)–(132) are closure-independent results. For explicit results, one needs to evaluate these equations introducing a closure approximation. The mean spherical approximation (MSA) [226] used by Schweizer and Curro

[44–49] for a symmetric mixture ($N_A = N_B = N$, $\sigma_A = \sigma_B = \sigma$, hard core diameters $d_A = d_B = d$) can be cast in the simple form

$$\Delta h(r) = h_{AA}(r) - h_{AB}(r) = 0, \quad r < d, \tag{133a}$$

$$\Delta c(r) = c_{AA}(r) - c_{AB}(r) = -v_{AB}(r)/k_B T, \quad r > d, \tag{133b}$$

$v_{AB}(r)$ being the intermolecular potential. Working out Eqs. (129)–(132) with this closure, one finds with some algebra that the effective χ parameter becomes strongly renormalized in comparison to the "mean field" or "bare" chi parameter $\chi_0 = \rho \int v_{AB}(r)\, d\vec{r}/k_B T$,

$$\frac{\chi_{eff}(T)}{\chi_0(T)} \cong A/\{1 + (N/N_0)^{1/2}\}, \quad N \to \infty, \quad d = 3, \tag{134}$$

where A, N_0 are constants. Since the critical condition still is $\chi_{eff}(T_c)N = 2$, one concludes that $T_c \propto N^{1/2}$ for $N \to \infty$ instead of the Flory-Huggins prediction, $T_c \propto N$. This result has also been generalized to general dimensionality [49] as $T_c \propto N^{(d-2)/2}$. Unfortunately, the comparison of Eq. (134) with corresponding Monte-Carlo simulations [92, 99] shows that Eq. (134) is not correct, see Fig. 13. Experiments by Gehlsen et al. [235] on mixtures of protonated and deuterated poly(ethylene-propylene) confirmed the linear relation of T_c with N and thus corroborated the computer simulations of Deutsch and Binder [92, 99]. Fortunately, a reinvestigation of PRISM theory with a new molecular closure [224] yielded a linear relation again, $T_c \propto N$, with a prefactor in this relation distinctly reduced in comparison to the bare mean field (Flory-Huggins) result. In view of this sensitivity of the results on details of the closure approximations Yethiraj and Schweizer [224] conclude "that one must fundamentally re-evaluate the closure approximations used in integral equation theories of *both* small molecule and polymeric fluids when attractive forces (tail potentials) are present". Clearly, due to the additional control parameter (chain length N) that can be varied in polymer mixtures without changing the nature of intermolecular forces one can very stringently test theoretical concepts, much better than for small molecule systems. The fact that recent simulations [92, 99], experiment [235] and theory [224] have clarified the puzzling results of Refs. [44–49] is very gratifying, since the PRISM approach in principle is very powerful – e.g. the local chain structure (variable persistence length etc.) can be readily included [236, 237].

While in dense melts, correlation effects of the chain configuration may thus renormalize the Flory-Huggins χ-parameter only by a factor of order unity relative to its "bare" mean-field value – this renormalization factor can be of the order two or three, as Monte Carlo simulations show [92, 99, 101, 107] – the situation is different if one considers binary polymer mixtures A,B in a solvent good for both polymers [238–241]: in semidilute solutions, excluded volume interactions are screened only over distances larger than the screening length $\xi_s \propto \phi^{-v_e/(3v_e-1)}$, ϕ being the volume fraction taken by the polymers, and $v_e \approx 0.59$ the exponent describing coil dimensions in good solvents, $R \sim N^{v_e}$ [2].

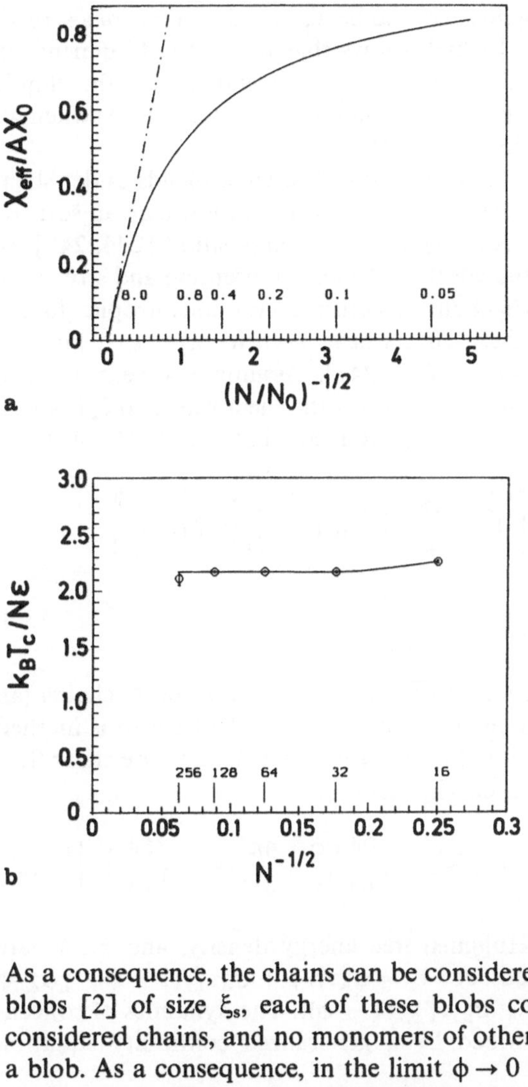

Fig. 13a. Predicted variation of $\chi_{eff}/(A\chi_0)$ vs $(N_0/N)^{1/2}$, Eq. (134). *Dot–dash line* shows the asymptotic behavior $\chi_{eff}/(A\chi_0) \propto (N/N)^{1/2}$, while the *marks* indicate the scale for (N/N_o) itself. **b** Variation of the observed normalized critical temperature $k_BT_c/(\varepsilon N)$ with $N^{1/2}$, from a simulation of the bond fluctuation lattice model (see Sect. 4). *Marks* indicate the values of N chosen. Note that for large enough N the integral equation theory of Schweizer and Curro [44–49] implies that this plot could be mapped on Fig. 13a), by multiplying the coordinate scales with suitable constants, without any other adjustable parameters being available. From Deutsch and Binder [92]

As a consequence, the chains can be considered as being rescaled into chains of blobs [2] of size ξ_{ss}, each of these blobs containing only monomers of the considered chains, and no monomers of other chains being present inside such a blob. As a consequence, in the limit $\phi \to 0$ the mixture becomes compatible! A renormalization group analysis [238, 239] shows that the effective renormalized Flory-Huggins parameter vanishes as

$$\chi_{eff}(\phi) \propto \phi^{x/(3\nu_e - 1)}, \quad x \approx 0.22 . \tag{135}$$

The interpretation of Eq. (135) is simply that the number of unfavorable contacts vanishes in the "semidilute" limit, the chains becoming more and more compatible because locally they do not "see" each other although they are still strongly interpenetrating.

Qualitatively, a strong reduction of the effective chi-parameter in solutions has indeed been seen in experiments [242], although a quantitative explanation of these data is still lacking.

Correlation effects in dense blends can also be discussed in terms of a screening effect of the interactions [243] which shows that the coil radii depend on both volume fraction and the chi-parameter of the blend, unlike the simple version of the RPA discussed in Sect. 2.3. We shall return to such effects when we discuss the Monte Carlo simulations (Sect. 4).

Finally, we will briefly discuss the properties of polymer blends under shear flow. In small molecule mixtures, shear flow is known to produce an anisotropy of critical fluctuations and anisotropic spinodal decomposition [244, 245]. In polymer mixtures, the shear has the additional effect of orienting and stretching the coils, thus making the single-chain structure factor anisotropic. In the framework of the Rouse model these effects have been incorporated into the RPA description of polymer blends [246, 247]. Assuming a velocity field $\vec{v} = \dot{\gamma} y \vec{e}_x$, where x, y, z are cartesian coordinates, $\dot{\gamma}$ the shear rate, and $\vec{e}_x$ is a unit vector in $\vec{x}$ direction, the single chain structure factor becomes [246, 247]

$$S_A(\vec{q}) = N_A\left\{1 - \frac{1}{18}N_A\sigma_A^2\left[q^2 + \frac{2\pi^2}{15}\tau_1^A\dot{\gamma}q_xq_y + \frac{4\pi^4}{315}(\tau_1^A\dot{\gamma})^2q_x^2\right] + o(q^4)\right\}, \tag{136}$$

τ_1^A being the largest relaxation time of the Rouse spectrum for A chains (an analogous expression holds for B-chains, of course). Via the RPA, it is implied that the effective free energy functional for mixtures in a steady state shear flow must have an anisotropic gradient energy term

$$\frac{\Delta F}{k_BT} = \int d\vec{r}\left\{f(\phi) + \frac{a^2}{36\phi(1-\phi)}\left[(\nabla\phi)^2 + \frac{2\pi^2}{15}\dot{\gamma}\bar{\tau}_1\frac{\partial\phi}{\partial x}\frac{\partial\phi}{\partial y} + \frac{4\pi^4}{315}(\dot{\gamma}\bar{\bar{\tau}}_1)^2\left(\frac{\partial\phi}{\partial x}\right)^2\right]\right\} \tag{137}$$

$f(\phi)$ being the standard (Flory-Huggins) free energy density, and $\bar{\tau}_1$, $\bar{\bar{\tau}}_1$ are "average" Rouse times {defined as $\bar{\tau}_1 = \sigma_A^2\tau_1^A/\phi + \sigma_B^2\tau_1^B/(1-\phi)$, $[\sigma_A^2/\phi + \sigma_B^2/(1-\phi)](\bar{\bar{\tau}}_1)^2 = \sigma_A^2(\tau_1^A)^2/\phi + \sigma_B^2(\tau_1^B)^2/(1-\phi)$}. The dynamics of concentration fluctuations then is described by a generalized continuity equation containing a "convective" term,

$$\frac{\partial}{\partial t}\phi(\vec{r}, t) + \vec{\nabla}\cdot[\phi(\vec{r}, t)\, v(\vec{r}, t)] = -\vec{\nabla}\cdot\vec{j}(\vec{r}, t) + \eta(r, t)\,, \tag{138}$$

instead of Eq. (68). While the phenomenological approach of Refs. [246, 247] clearly was a first step only, the basis for a more fundamental theory was developed in Refs. 204–209, but so far only the shear-induced phase separation of polymer solutions has been discussed in detail. Numerous interesting experiments [e.g. Refs. 248–250] showing a shear-induced homogenization of otherwise phase-separated polymer blends under shear thus must await a detailed theoretical study. Experiments typically are in the regime $\tau_1\dot{\gamma} \ll 1$ where the corrections proposed in Eqs. (136), (137) are negligible.

4 Monte Carlo Tests of Flory-Huggins Theory and Comparison to Experiment

Computer simulations are often carried out to model materials in full atomistic detail [251–253]. This is difficult for polymers (Fig. 14) since even a single chain [95, 106] exhibits structure from a scale of a single chemical bond (≈ 1 Å) to the persistence length (≈ 10 Å) to the coil radius (≈ 100 Å). Even larger lengths arise from collective phenomena in polymer mixtures: the correlation length ξ near T_c is much larger than the coil radius, see Eqs. (39), (42), (44) and (60); similar large lengths $\lambda_c(t)$, $\lambda_m(t)$ appear in the spinodal decomposition process, see Sect. 2.4 and Eqs. (86), (110) or Figs. 7a, 8. Also in block copolymers near their order-disorder transition similarly large correlation lengths occur (see Sect. 5). Also the time scales of interest are very long (up to 10^3 sec, again see Fig. 7a), and therefore a chemically faithful "molecular modelling" of the phase behavior of polymer blends by Molecular Dynamics (MD) methods [252, 253] is far out of reach of computational feasibility. On the other hand, with Monte Carlo methods it is possible [95, 106] to simplify matters by introducing coarse-grained models (Fig. 15), such as the simple self-avoiding walk model already used by Flory and Huggins (Fig. 3), or the bond fluctuation model [84–99]. In these coarse grained models, the fastest dynamical degrees of freedom (bond length and bond angle vibrations, which have characteristic times of 10^{-13} sec) and the small-scale information on chemical structure (on the scale of a monomer) is eliminated: basic "building blocks" of these models are effective subunits (of a size comparable to the persistence length [1, 2], for instance), and the elementary time step is a conformational change of the group of monomers represented by such a subunit (one attempted Monte Carlo step may then correspond to a physical time of 10^{-11} sec, for instance). This coarse-grained

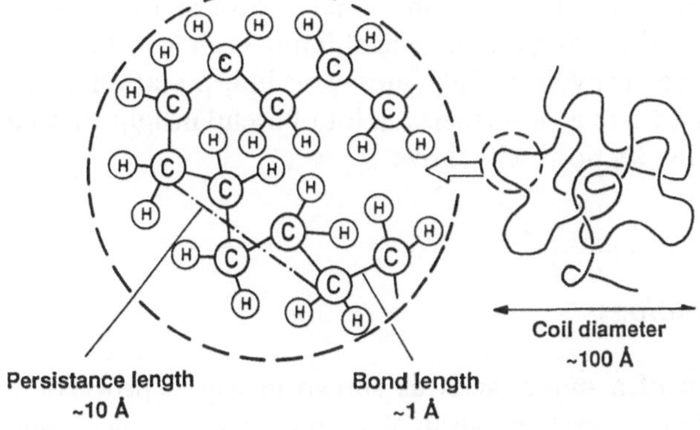

Fig. 14. Length scales characterizing the structure of a long polymer coil (example: polyethylene). From Binder [95]

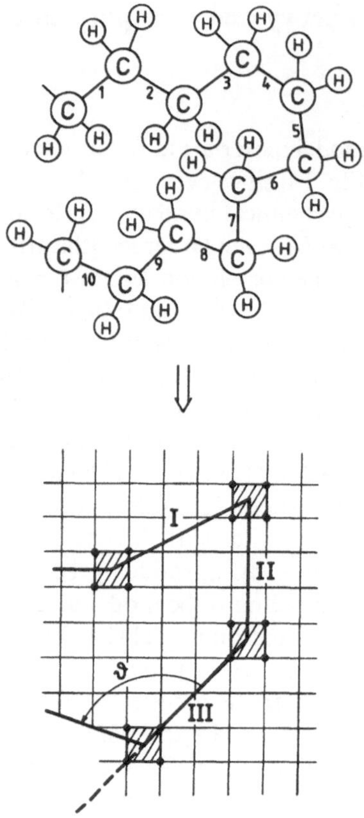

Fig. 15. Approximate mapping of a chemically realistic polymer (polyethylene in this example) to the bond fluctuation model on the (simple cubic) lattice. In this "coarse-graining" one integrates n successive chemical monomers (e.g. $n = 3$) into one effective monomer which blocks 8 adjacent sites on the simple cubic lattice (or 4 on the square lattice in $d = 2$ dimensions) from occupation by other monomers. The chemical bonds 1, 2, 3 then correspond to effective bond I, bonds 4, 5, 6 to effective bond II. Some information on the chemical structure can be kept indirectly by using suitable distributions $P_n(\vartheta)$ for the angle between subsequent effective bonds, but so far this has been done for homopolymer melts only [94–99]. In the simplest version of the bond fluctuation model [84–88] studied for blends in $d = 3$ dimensions [88, 91, 92, 99], bond lengths ℓ are allowed to fluctuate freely from $\ell = 2$ to $\ell = \sqrt{10}$, with $\ell = \sqrt{8}$ being excluded to maintain that chains do not cut through each other in the course of the random hops of the effective monomers. From Binder [95]

Monte Carlo modelling [95, 104, 106] is suitable for answering general questions such as the validity of Flory-Huggins theory or the Curro-Schweizer PRISM theory [44–50] etc.; it cannot address the molecular explanation of compatibility of specific polymers, of course, since information on the disparity in the size, shape, and interactions of monomers has simply been lost. Although these simulations cannot yet provide a fully atomistic interpretation of the Flory-Huggins χ-parameter, there is nevertheless a lot of useful insight that has been gained, as the following sections will show.

4.1 Simulation Methodology

A Monte Carlo simulation of a model such as shown in Fig. 3 proceeds in several steps: first one chooses system parameters N_A, N_B, ϕ_A, ϕ_B. Since every lattice site not taken by a monomer is vacant, the vacancy concentration ϕ_v then is $\phi_v = 1 - \phi_A - \phi_B$. The simulations that are described here in detail do require a nonzero vacancy concentration; however, we point out that different techniques for simulation of polymer mixtures with no vacancies have been successfully implemented [254]. Apart from studies of wetting phenomena [255], where one chooses a $L \times L \times D$ geometry with two hard walls of size $L \times L$ and otherwise periodic boundary conditions, one chooses an $L \times L \times L$ geometry with periodic boundary conditions in all lattice directions throughout. Of course, ϕ_A and ϕ_B have to be chosen such that the number of A-chains (n_A) and the number of B-chains (n_B) are both integer.

It is a nontrivial step to prepare a valid initial configuration of the athermal polymer melt (for which energy parameters ε_{AA}, ε_{BB}, ε_{AB} are still absent), because these chains must not cross and obey Gaussian statistics at distances large in comparison to the screening length [2]. For small enough N_A, N_B (and large enough L), one may start the system in a state where the chains initially are all in linear completely stretched-out state, and then relax the coil configurations by standard choices (Fig. 16, upper part) of local motions [104–106], such as combinations of kink-jumps, end rotations and crankshaft rotations for the model of Fig. 3 [101–103, 107, 155], or random hops of effective monomers by one lattice spacing in the case of the bond fluctuation model [84–89], Fig. 15. Since these motions are only accepted if they do not violate the self-avoidance constraints, and the initial regular configuration does not violate these constraints either, one works in the space of the "allowed" chain configurations. Watching the decay of the component of the end-to-end vector in the direction of the initial state, as disorder diffuses in from the ends of the chains into their interior as time passes, one can make sure that all "memory" of the initial nonrandom state is lost.

For longer chains (where the length of a fully stretched chain exceeds L) different techniques are needed: e.g., one can insert one chain after the other

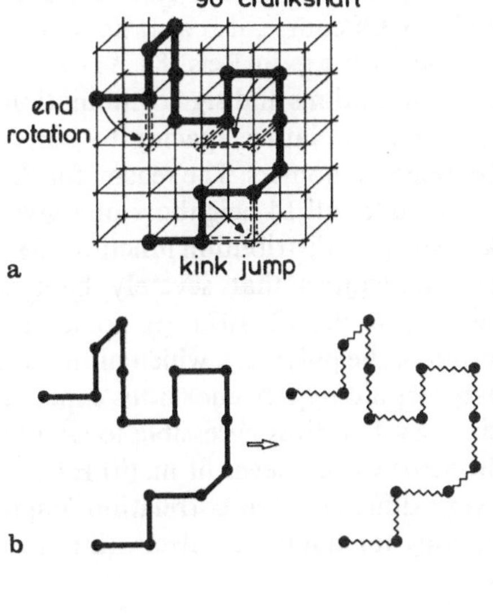

Fig. 16. Moves used to equilibrate coil configurations for the self-avoiding walk model of polymer chains on the simple cubic lattice (*upper part*): end rotations, kinkjump motions and crankshaft rotations [107]. From time to time these local moves alternate with a move (*lower part*) where one attempts to replace an A-chain by a B-chain in an identical coil configuration, or vice versa. In the transition probability of this move, the chemical potential difference $\Delta\mu$ as well as the energy change $\delta\mathcal{H}$ enter. From Binder [258]

using the Rosenbluth-Rosenbluth [256] inversely restricted sampling [107]. If such a chain insertion fails before the desired polymer volume fraction is reached, the attempt is repeated a few times, and if one is still unsuccessful, one has to start all over again with an initially empty simulation box. In practice, this technique works only for rather small N, and moreover the chain configurations do not have the right gaussian statistics, since one is applying a biased sampling procedure [104]. So the chain configurations also need to be relaxed (by the local motions mentioned above) for a long time.

The most efficient technique is to put into the system chains which initially are neither strictly self-avoiding nor mutually avoiding, but simple non-reversal-random walks (NRRW). This configuration is then relaxed, allowing only such moves that do not introduce any new overlap of monomers. After a long time, one ends up with chains which are both strictly self-avoiding and mutually avoiding, and comply with gaussian statistics. For the bond fluctuation model, it has been possible to produce at $\phi_v = 0.5$ equilibrated chain configurations for chain lengths up to $N = 512$ in lattices of size up to $L = 160$ [257]: this system, containing 256 000 effective monomers, holds the "world record" in polymer blend simulation.

Given an initial configuration of an athermal melt, one can introduce the energy parameters ε_{AA}, ε_{BB} and ε_{AB} occurring between the appropriate pairs of monomers, and take them into account in the acceptance rate of the Monte Carlo moves in the standard way [104–107]. In brief, for each attempted move one computes the energy change δE: if $\delta E < 0$, and the move does not violate any other constraints (self-avoidance, bond length constraints), it is accepted. If $\delta E > 0$, however, it is accepted only with a transition probability $w = \exp(-\delta E/k_B T)$. In practice, this is done by comparing w with a random number ζ, uniformly distributed between zero and unity: if $w > \zeta$ the move is executed while otherwise it is rejected and the old configuration is counted once more for the averaging.

Usually the range of the interactions ε_{AA}, ε_{BB} and ε_{AB} is restricted to nearest neighbors [101–103, 107, 155, 257], as in Fig. 3. Of course, it is also possible to work with a larger range of interaction, if desired: e.g. in Refs. 88, 91–93 the interaction was taken to be constant for all permutations and sign combinations of the vectors (2,0,0), (2,1,0) and (2,1,1) {putting the lattice spacing to unity}. Altogether, these are 54 distance vectors between monomers. The reason for this choice was that then the interaction range includes all 14 neighbors of a given monomer that contribute to the first peak of the pair distribution function [88].

Now, unfortunately, there are several difficulties that severely hamper Monte Carlo approaches to polymer blends [88, 91, 92, 107]: (i) To see the unmixing transition occur by diffusive motion of the polymers, which ultimately form A-rich and B-rich macroscopic domains would require enormous expenditure of computer time and huge system sizes. System sizes accessible to simulations in practice are so small that such domains would never fit in. (ii) Even in the one phase region, the simulation is very difficult – the correlation length being very large, one must watch out carefully for finite size effects. (iii) Since relaxation times near the critical point are very large ("critical slowing down" [144]) and acceptance rates for Monte Carlo moves in models of dense polymer systems are rather small [89, 107], one has to wait for an extremely long time until two configurations of the system are uncorrelated. The judgement of the statistical errors [104, 106] from "measurements" of quantities such as the order parameter $M = (n_A - n_B)/(n_A + n_B)$ is thus rather difficult.

The first of these difficulties can be avoided for *symmetrical* polymer mixtures ($N_A = N_B = N$) by working in the semigrandcanonical ensemble of the polymer mixture [107]: rather than keeping the volume fractions ϕ_A, ϕ_B and hence the numbers of chains n_A, n_B individually fixed, as one would do in experiment and in the "canonical ensemble" of statistical thermodynamics, we keep the chemical potential difference $\Delta\mu = \mu_A - \mu_B$ between the two types of monomers fixed as the given independent variable. While the total volume fraction $1 - \phi_v$ taken by monomers is held constant, the volume fractions ϕ_A, ϕ_B of each species fluctuate and are not known beforehand, but rather are an "output" of the simulation. Thus in addition to the moves necessary to equilibrate the coil configuration (Fig. 16, upper part), one allows for moves where an A-chain is taken out of the system and replaced by a B-chain or vice versa. Note that for the symmetrical polymer mixture the term representing the contributions of the chemical potentials μ_A, μ_B to the grand-canonical partition function Z

$$Z = \sum_c \exp[(\mu_A n_A N_A + \mu_B n_B N_B)/k_B T] \exp[-E(c)/k_B T], \tag{139}$$

where c is a configuration of the system and E(c) its internal energy, can be expressed in terms of $\Delta\mu$ as

$$\begin{aligned} &\exp[(\mu_A n_A N_A + \mu_B n_B N_B)/k_B T] \\ &= \exp[N(\mu_A + \mu_B) n/2k_B T)] \exp[N\Delta\mu M n/(2k_B T)] \end{aligned} \tag{140}$$

Since the number of chains n is kept constant and only the order parameter M, Eq. (2), can fluctuate, the first factor on the right hand side of Eq. (140) is constant and hence cancels out from the Monte Carlo averages. The transition probability for the "grandcanonical" moves where one goes from an old configuration (c) to a new configuration (c') via an "identity switch" of a chain ($A \rightleftharpoons B$), Fig. 16 (lower part), hence becomes,

$$W(c \to c') = \mathrm{Min}\left\{1, \exp\left[\frac{\Delta\mu(M' - M)Nn}{2k_B T}\right] \exp\left[-\frac{(E' - E)}{k_B T}\right]\right\}, \tag{141}$$

where E, M are energy and order parameter of the old configuration, E', M' refer to the new configuration. In principle, this semigrandcanonical procedure could be generalized to the asymmetric case ($N_A \neq N_B$), but the move where one would take out a short chain and try to insert a longer chain would almost always fail in a dense system due to excluded volume constraints.

Now an important observation is that the semi-grandcanonical partition function $Z(T, \Delta\mu)$ can be expressed in terms of a density of states $\Gamma(E, M)$ which does not depend on the parameters $(T, \Delta\mu)$ characterizing the considered thermodynamic state:

$$Z(T, \Delta\mu) = \int_{-1}^{+1} dM \int \exp[-E/k_B T] \exp[N\Delta\mu M n/2k_B T]\, \Gamma(E, M)\, dE\,. \qquad (142)$$

The Monte Carlo sampling now yields a number N of configurations that are distributed proportional to the distribution

$$P_{T,\Delta\mu}(E, M) = \frac{1}{Z(T, \Delta\mu)} \exp[-E/k_B T] \exp[N\Delta\mu M n/2k_B T]\, \Gamma(E, M)\,. \qquad (143)$$

In a simulation one can now record a "histogram" $H_{T,\Delta\mu}(E, M)$ just counting how often one observes the possible values of the observables (E, M). For a large number N of statistically independent observations this histogram precisely approximates $P_{T,\Delta\mu}(E, M)$, apart from statistical errors:

$$\mathcal{N}^{-1} H_{T,\Delta\mu}(E, M) \approx P_{T,\Delta\mu}(E, M)\,. \qquad (144)$$

Now an important observation (known as "reweighting technique" or "histogram extrapolation" [259–261, 91] is that $H_{T,\Delta\mu}(E, M)$ can be used to estimate $P_{T',\Delta\mu'}(E, M)$ at a whole range of neighboring values T', $\Delta\mu'$ around T, $\Delta\mu$:

$$P_{T',\Delta\mu'}(E, M) \approx \frac{Z(T, \Delta\mu)}{Z(T', \Delta\mu')} \exp\left[\frac{\Delta\mu' MNn}{2k_B T'} - \frac{\Delta\mu\, MNn}{2k_B T}\right] \exp\left[\frac{E}{k_B T} - \frac{E}{k_B T'}\right] \frac{H_{T,\Delta\mu}(E, M)}{\mathcal{N}}\,. \qquad (145)$$

Of course, this "single histogram-method" is practically useful only [91, 260] for such a range of parameters where the exponential functions in Eq. (145) do not emphasize values of E, M far out in the wings of the measured histogram H. For polymer mixtures near the critical point, the number of chains in the simulated systems are never extremely large, and then the distributions $P_{T',\Delta\mu'}(E, M)$ are all rather broad and this reweighting technique works over a reasonable broad range of temperatures and fields. As an example, Fig. 17 shows reduced order parameter distributions $P_T(M) \equiv \int P_{T,\Delta\mu=0}(E, M)\, dE$ which were obtained from such extrapolations using a single state point $(T, \Delta\mu = 0)$. The extension ("multi-histogram-analysis") to combine information from several state points $(T, \Delta\mu)$ is also very useful [91, 261] but will not be discussed further here.

The gradual change of the order-parameter distribution function with temperature in the critical region from a single peak structure above the critical temperature T_c to a double-peak structure below T_c already indicates that the finite size of the simulation box severely distorts the singular behavior that is expected at the phase transition in the thermodynamic limit, $L \to \infty$. As is

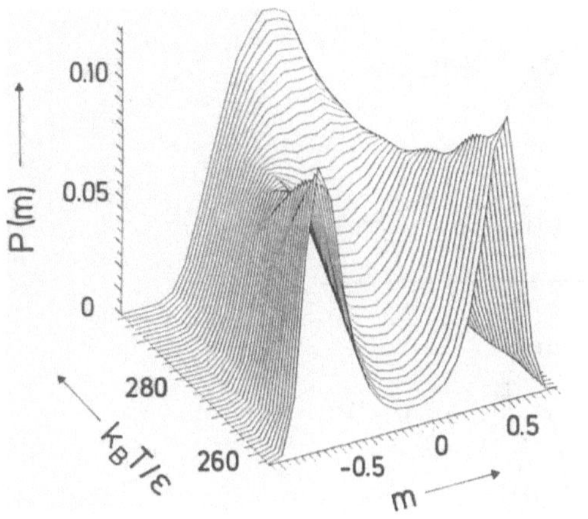

Fig. 17. Distribution function $P_T(M)$ of the order parameter M plotted over a range of temperatures for N = 128, L = 80, $\phi_v = 0.5$, using the bond fluctuation model of symmetric polymer mixtures and using the extrapolation formula Eq. (145) and data from a single temperature run at $k_BT/\varepsilon = 266.4$.The number of (statistically independent) samples was N = 16800. From Deutsch and Binder [92]

well-known [105, 262–264], finite size effects are responsible for a rounding and shifting of phase transitions, and hence need to be carefully analyzed here since the number of chains contained in the simulation box often is of order 10^2 only [91, 92, 101–107]. Figure 18 summarizes the situation qualitatively. The singular behavior of the order parameter, Eq. (122), is rounded off as soon as the correlation length ξ {Eq. (126)} becomes comparable with L. Finite size scaling [262–264] then implies that $m = \langle|M|\rangle$ does not depend on the two lengths L, ξ in general form, but basically only on their ratio L/ξ, apart from a scale factor

$$m = L^{-\beta/\nu}\tilde{M}(L/\xi), \quad L \to \infty, \quad \xi \to \infty, \quad L/\xi \text{ finite} . \tag{146}$$

Note that the power law prefactor $L^{-\beta/\nu}$ just enables that a sensible thermodynamic limit can result: the scaling function $\tilde{M}(\zeta)$ for large ζ behaves as $\tilde{M} \propto \zeta^{\beta/\nu}$, and so for large ζ the L-dependence cancels out and one is left with $m \propto \xi^{-\beta/\nu} \propto (1 - T/T_c)^\beta$, as it should be. A similar scaling behavior applies for the collective function $S_{coll}(q = 0)$ and $S'_{coll}(q = 0)$

$$S_{coll}(q = 0) = nN(1 - \phi_V)\,[\langle M^2\rangle - \langle M\rangle^2] , \tag{147a}$$

$$S'_{coll}(q = 0) = nN(1 - \phi_V)\,[\langle M^2\rangle - \langle|M|\rangle^2] . \tag{147b}$$

Note that in a finite system for $\Delta\mu = 0$ we have $\langle M\rangle = 0$ since the probability distribution $P_L(M)$ is always symmetric around $M = 0$ (Figs. 17, 18) and although for $T < T_c$ we have two peaks, corresponding to an A-rich phase and a B-rich phase, a spontaneous symmetry breaking does not occur: the system can always move from states in the vicinity of one peak to the other peak through the region of the minimum between the peaks, by creating a droplet of the other phase that grows until it fills the system. In long enough Monte Carlo runs such transitions are indeed observable. On the other hand, the quantity $\langle|M|\rangle$ approaches for $\Delta\mu = 0$ the spontaneous order parameter (corresponding to the coexistence curve) in a smooth fashion as $L \to \infty$. As a consequence,

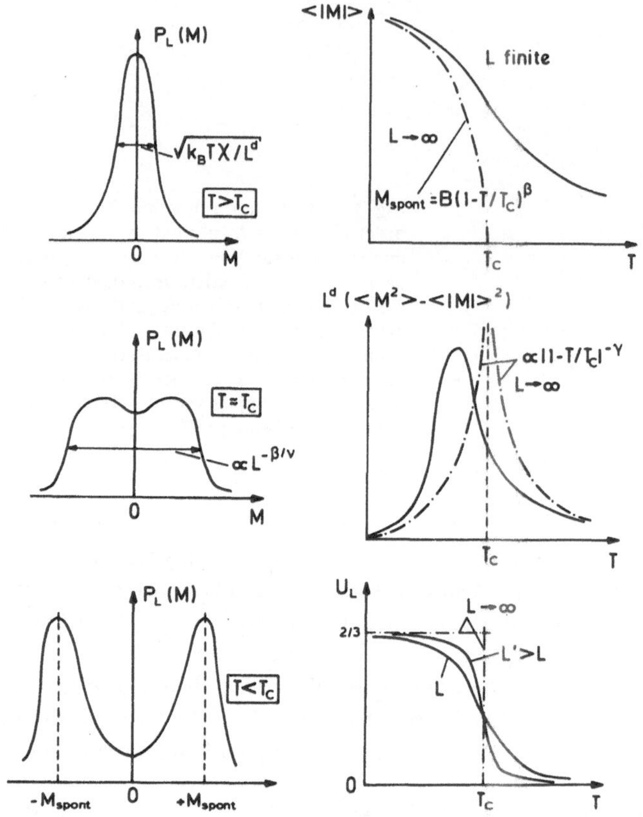

Fig. 18. Schematic evolution of the order parameter probability distribution $P_L(M)$ from $T < T_c$ (*bottom*) to $T = T_c$ (*middle*) to $T > T_c$ (*top*), and corresponding temperature dependence of the order parameter $\psi = \langle|M|\rangle$, collective structure factor at zero wave vector $k_B T\kappa' = L^d(\langle M^2\rangle - \langle|M|\rangle^2)$, d being the dimensionality of the box, and the reduced fourth-order cumulant $U_L = 1 - \langle M^4\rangle_L/(3\langle M^2\rangle_L^2)$. *Dash-dotted curves* indicate the singular variation which results in the thermodynamic limit, $L \to \infty$. From Binder [258]

$S'_{coll}(q = 0)$ describes the critical scattering along the coexistence curve, while $S_{coll}(q = 0)$ describes the critical scattering for $T > T_c$ and $\Delta\mu = 0$ (i.e., at critical composition $\phi = \phi_{crit}$). This fact that for the description of the response function, different expressions need to be used above and below the phase transition, in order to take into account the spontaneous symmetry breaking that appears in the thermodynamic limit between the coexisting phases, is well known [105] but has not been considered in some of the work on polymer mixtures [101–103, 107].

In order to relate Eq. (147) to expressions used for the collective scattering previously, we define the collective scattering for our model as follows

$$S_{coll}(q) \equiv \left\langle \left[\sum_j \exp(i\vec{q}\cdot\vec{R}_j)\,(\delta\phi_B^j - \delta\phi_A^j)\right]^2 \right\rangle \Big/ L^3 , \qquad (148)$$

where $\delta\phi_B^j - \delta\phi_A^j = \phi_B^j - \phi_A^j - \langle(\phi_B^j - \phi_A^j)\rangle$ (in the case of Eq. (147a)).

Using an Ising spin representation where $\phi_A^j = 1$ corresponds to $S_i = +1$, $\phi_B^j = 1$ corresponds to $S_i = -1$, and $\phi_v^j = 1$ corresponds to $S_i = 0$, one can transform [101–103] Eq. (148) to Eq. (147a).

Now the finite size scaling hypothesis [262–264] implies for the collective scattering functions

$$S_{coll}(0) = L^{\gamma/\nu}\tilde{S}(L/\xi),\ S'_{coll}(0) = L^{\gamma/\nu}\tilde{S}'(L/\xi),\ \Delta\mu = 0\ , \qquad (149)$$

where the scaling function $\tilde{S}(\zeta)$ and $\tilde{S}'(\zeta)$ behave for $T > T_c$ both as $\tilde{S}(\zeta) \propto \zeta^{-\gamma/\nu}$, $\tilde{S}'(\zeta) \propto \zeta^{-\gamma/\nu}$, to ensure a critical behavior $S_{coll}(0) \propto (T/T_c - 1)^{-\gamma}$, Eq. (125), in the thermodynamic limit. For $T < T_c$ also $\tilde{S}(\zeta)$ behaves as $\tilde{S}'(\zeta) \propto \zeta^{-\gamma/\nu}$, reflecting the critical behavior of the response function along the coexistence curve, while $\tilde{S}(\zeta) \propto \zeta^{2\beta/\nu}$, which gives a behavior $S_{coll}(0) \propto L^{(\gamma+2\beta)/\nu}\ \zeta^{-2\beta/\nu} \propto L^d(1 - T/T_c)^{2\beta}$ for $L \to \infty$, using the hyperscaling relation $d\nu = \gamma + 2\beta$ between the critical exponents. As is clear from Eq. (147a) using $\langle M \rangle \equiv 0$ for $\Delta\mu = 0$ and finite L, for $T < T_c$ we have $\langle M^2 \rangle \approx \langle |M| \rangle^2$ for large L and thus $S_{coll}(0)$ just tends to the product of order parameter square and system volume (note that $nN \propto L^d$, of course).

Another quantity useful for the location of phase transitions is the fourth order cumulant of the distribution $P_L(M)$ shown in Fig. 18,

$$U_L = 1 - \langle M^4 \rangle / [3\langle M^2 \rangle^2]\ , \qquad (150)$$

which has a finite size scaling behavior without any power law prefactor,

$$U_L = \tilde{U}(L/\xi)\ . \qquad (151)$$

The cumulant U_L is zero for a single symmetric gaussian distribution (which applies for all $T > T_c$ in the thermodynamic limit), while $U_L = 2/3$ for the double gaussian distribution in the limit $L \to \infty$. This step-function-like behavior of the cumulant makes it a good indicator for the transition: in fact, at $T = T_c$ where $\xi = \infty$, Eq. (151) predicts that all curves U_L versus T for different L should intersect in a common intersection point $\tilde{U}(0)$. Locating such a cumulant intersection one can find the critical temperature unbiased by any assumptions about the critical behavior. Also other dimensionless ratios such as $\langle M^2 \rangle / \langle |M| \rangle^2$ have a similar intersection property [91]. Figure 19 presents an example to show that these finite size scaling techniques work nicely in practice. On the other hand, from the "raw data" of the simulation such as plots of $\langle |M| \rangle$ vs. T for different sizes (Fig. 20) one hardly could estimate T_c with any reasonable precision. Fig. 21 shows that from the "data collapsing" suggested by Eqs. (146), (149) one can not only verify that finite size scaling indeed holds, but also estimate the asymptotic critical behavior of order parameter and response function including their prefactors ("critical amplitudes" [73, 75]).

While thus the estimation of coexistence curves and equilibrium scattering functions $S_{coll}(\vec{q})$ from Monte Carlo simulations is a well-controlled procedure, it is rather hard to estimate spinodal curves (Fig. 22). While is also possible to generate metastable states at different conditions and check whether they decay

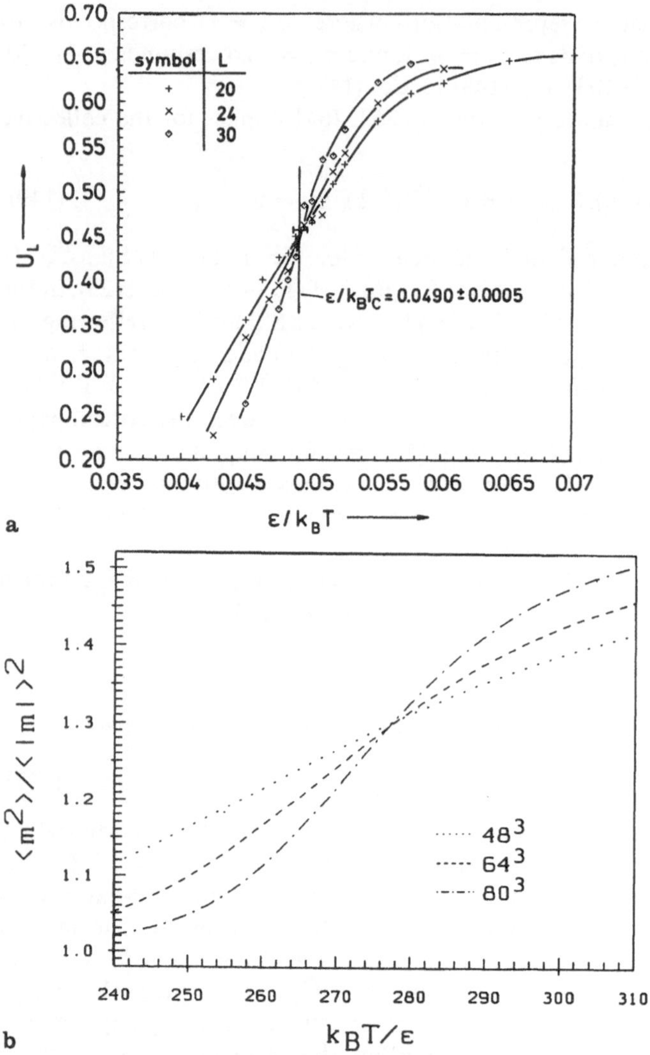

Fig. 19a. Cumulant U_L plotted vs inverse temperature ε/k_BT for the model of Fig. 3 with $\varepsilon_{AA} = \varepsilon_{BB} = -\varepsilon$, $\varepsilon_{AB} = 0$, $N = 64$, $\phi_v = 0.6$, and three different system sizes as indicated. *Vertical line* with error bar indicates the observed estimate for T_c and its uncertainty. *Curves* are guides to the eye only. From Sariban and Binder [101]. **b** $\langle M^2\rangle/\langle |M|\rangle^2$ plotted vs k_BT/ε for the bond fluctuation model with $\varepsilon_{AA} = \varepsilon_{BB} = -\varepsilon_{AB} = -\varepsilon/2$, $N = 128$, $\phi_v = 0.5$, and interaction range including distances up to $\sqrt{6}$ lattice spacings on the simple cubic lattice. Three lattice sizes are shown as indicated. Curves are due to multihistogram extrapolation from data taken at the temperatures shown in Fig. 20. From Deutsch and Binder [92]

fast or not [155], here we will discuss only the location of spinodals from plots $S_{coll}^{-1}(q)$ vs ε/k_BT, where spinodals are then defined from a linear extrapolation (Fig. 22). Since the curves that one extrapolates exhibit some nonlinearity, there is inevitably some ambiguity - which reflects the fact that for short chain length

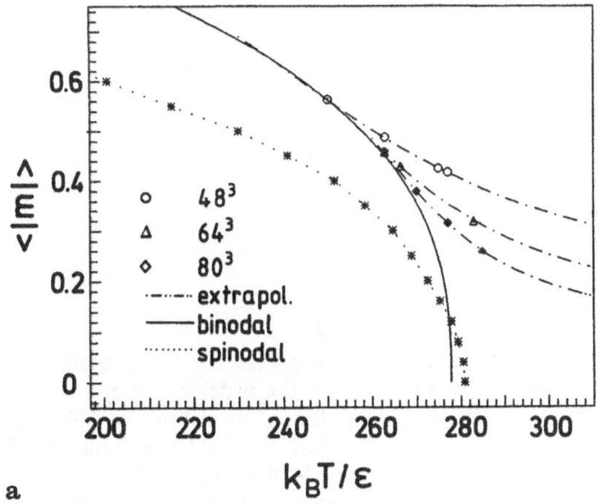

a

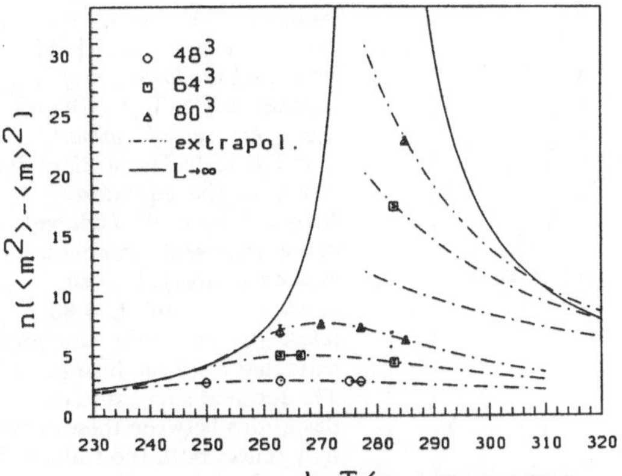

b

Fig. 20a. Order parameter $\langle|M|\rangle$ plotted vs k_BT/ε for the same case as shown in Fig. 19b. *Circles, triangles* and *diamonds* indicate data taken for the three lattice sizes L = 48, 64 and 80 at the shown temperatures. *Dash-dotted curves* are multihistogram extrapolations for these sizes over a wide temperature range. *Full curve* shows the predicted coexistence curve (binodal) resulting from the finite size scaling analysis (Fig. 21a below). *Dotted curve with stars* is the "mean-field"-spinodal, obtained from a linear extrapolation of $S_{coll}^{-1}(0)$ vs 1/T, as shown in Fig. 22. below. From Deutsch and Binder [92]. **b** Response functions $n(\langle M^2\rangle - \langle M\rangle^2)$ {curves shown for $k_BT/\varepsilon \geqq 280$ only} and $n(\langle M^2\rangle - \langle|M|\rangle^2)$ plotted vs k_BT/ε, for the same model as shown in Fig. 19b. *Circles, squares* and *triangles* indicate data actually taken directly for the three lattice sizes L = 48, 64 and 80 at the shown temperatures. *Dash-dotted* curves are the multihistogram extrapolations for these sizes over a wide temperature range. *Full curves* show the predicted response functions $T > T_c$ {corresponding to Eq. (147a)} and $T < T_c$ {corresponding to Eq. (147b)} in the thermodynamic limit, resulting from the finite size scaling analysis (Fig. 21b below). From Deutsch [91]

N the spinodal curve is not precisely defined (Sect. 2.5), while for long chain length one has to deal with finite size effects.

While the study of fully symmetric mixtures with Monte Carlo methods is relatively simple due to the fact that the coexistence curve occurs for $\Delta\mu = 0$ by symmetry against interchange of A and B, the problem is more difficult when this symmetry is destroyed. Deutsch [93, 266] has considered asymmetry in the

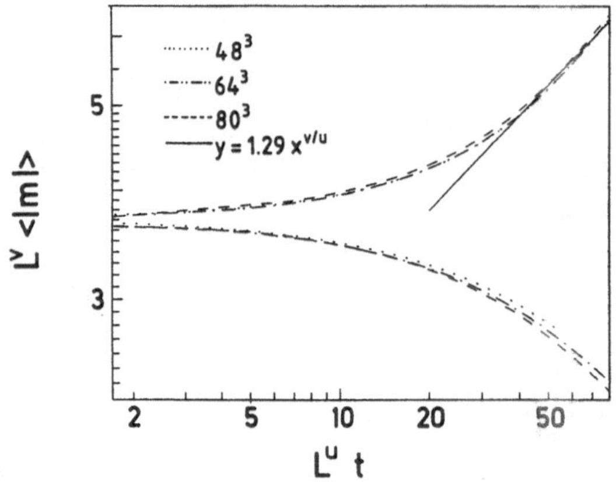

a

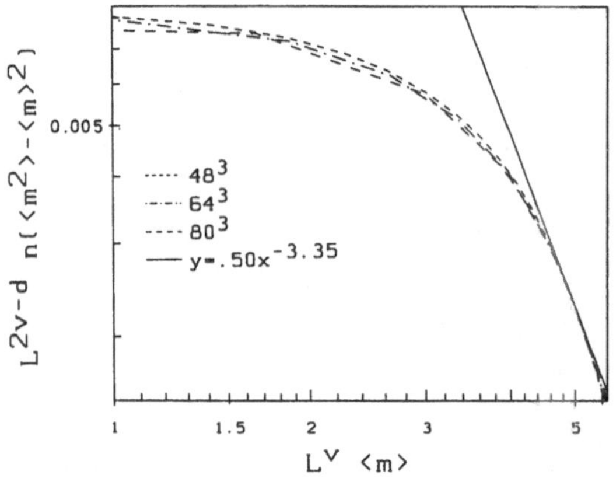

b

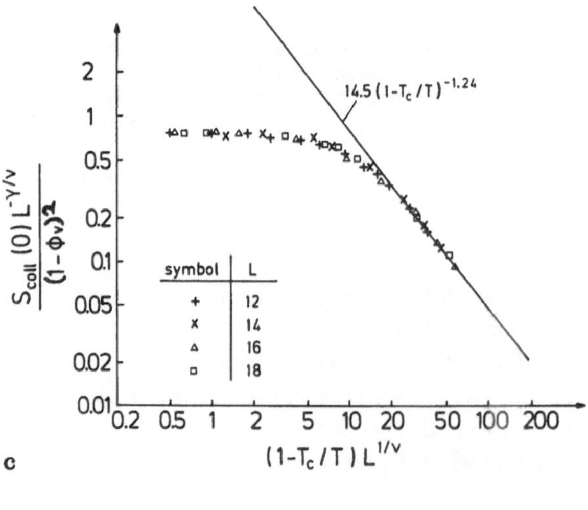

c

Fig. 21a. Finite size scaling plot of the order parameter $\langle|M|\rangle$ for the model of Fig. 19b, using $k_BT_c/\varepsilon = 277.7$, and measured effective exponents $u = 1/v^{eff} = 1.58$, $v = (\beta/\nu)_{eff} = 0.561$ – theoretical values are $1/\nu = 1.59$, $\beta/\nu = 0.51$ [74]. Note that the *lower set of curves* represents $T > T_c (t > 0)$ and the *upper set of curves* $T < T_c (t < 0)$. The *straight line* describes the equation $\tilde{M}(z) = 1.29(-z)^{\beta}$. Different curves represent three choices of $L = 48$ (*dotted*), $L = 64$ (*dash-dotted*) and $L = 80$ (*dashed*), and are the functions extracted from the histograms. The residual small systematic deviations between these curves may reflect both the statistical error of the histograms and corrections to finite size scaling. From Deutsch and Binder [92]. **b** Finite size scaling of the response function $L^{2v-d}n[\langle M^2\rangle - \langle M\rangle^2]$ vs $L^v\langle M\rangle$, for the model of Fig. 19b, using $k_BT_c/\varepsilon = 277.7$ and $v = 0.561$. Data for three lattice sizes are included as indicated. *Full curve* shows the power law resulting in the thermodynamic limit. From Deutsch [91]. **c** Finite size scaling plot of $S_{coll}(0)\ L^{-\gamma/\nu}(1-\phi_v)^{\ 2}$ vs $(1-T_c/T)\ L^{1/\nu}$, for the model of Fig. 3, with $\phi_v = 0.2$, $N = 32$ and using $k_BT_c/\varepsilon = 31.5$ as well as the theoretical Ising exponents. *Straight line* indicates the asymptotic power law for $L \to \infty$. From Sariban and Binder [107]

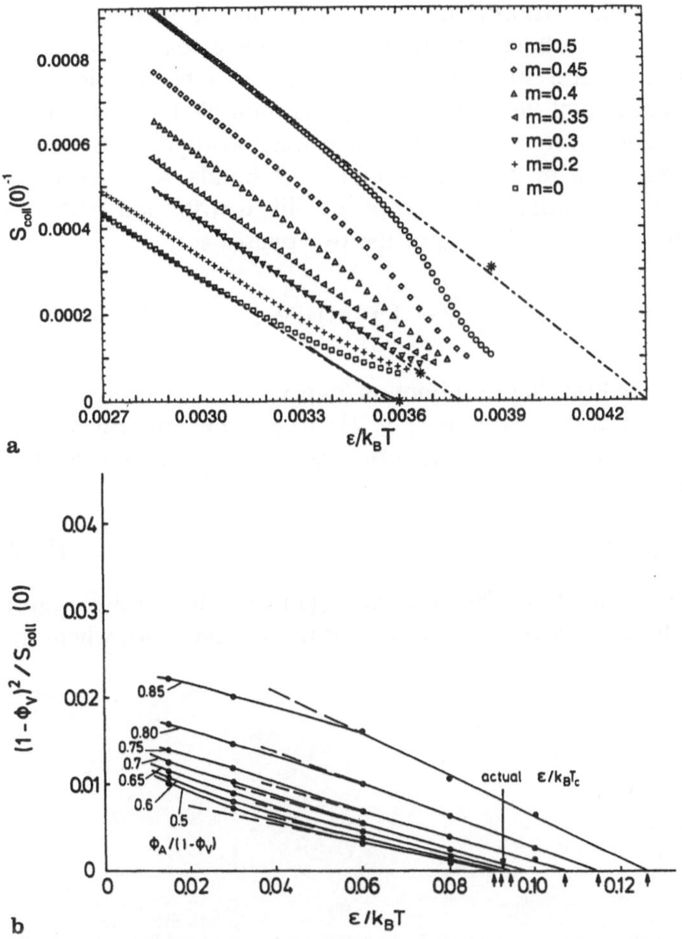

Fig. 22a. Plot of $S_{coll}^{-1}(q \rightarrow 0)$ vs ε/k_BT for the model of Fig. 19b, N = 128, L = 80 and several choices of $m = \langle |M| \rangle$ as indicated in the figure. For m = 0 the extrapolated curve for $L \rightarrow \infty$ is shown as a *full curve*, while the linear extrapolations are shown as *dash-dotted straight lines*. For m = 0.3 and m = 0.5 the linear extrapolations are also shown, the actual temperatures of the coexistence curve being shown by *stars* in all three cases. Note that in this immediate vicinity of T_c all the curvature seen in the data (which are generated by histogram extrapolation) is due to finite size effects. **b** Plot of $(1-\phi_v)^2/S_{coll}(q=0)$ vs ε/k_BT for the model of Fig. 3, N = 32, $\phi_v = 0.6$, and various choices of the volume fraction $\phi_A/(1-\phi_v)$ as indicated. Curves are a guide to the eye only. Since data over a very wide regime of temperatures are shown, curvature is due to an effective "renormalization" of the effective chi-parameter with temperature. Both the location of T_c and of the spinodal temperatures are shown with *arrows*. From Sariban and Binder [265]

pairwise interaction, defined in terms of an asymmetry parameter λ,

$$\varepsilon_{BB} = -\varepsilon_{AB} \equiv -\varepsilon\,, \qquad \varepsilon_{AA} = \lambda\varepsilon_{BB}\,. \tag{152}$$

The problem now is to locate for $\lambda > 1$ the line $\Delta\mu = \Delta\mu_{coex}(T)$ in the $(\Delta\mu, T)$ plane where the coexistence between the A-rich phase and the B-rich phase occurs. This first order transition line ends in a critical point – T_c, $\Delta\mu_c = \Delta\mu_{coex}(T)$ – which can only be located by a search in a two-dimensional parameter space! But also this problem can be handled very efficiently by a combination of finite size scaling and multiple histogram techniques – one now only has to record a four-dimensional histogram containing the numbers n_{AA}, n_{BB}, n_{AB} of A–A, B–B and A–B bonds, and the order parameter M. Obtaining the order parameter probability distribution P(M), one can define probabilities P_{A-rich}, P_{A-poor} of the two phases as

$$P_{A-rich} = \int_{M^*}^{1} P(M)\,dM\ , \quad P_{A-poor} = \int_{-1}^{M^*} P(M)\,dM\ , \tag{153}$$

M* being the value where P(M) has its minimum (in practice, one takes the value where $\langle M^2\rangle - \langle|M|\rangle^2$ has its maximum (Fig. 23), to avoid errors which come from strong statistical fluctuations of P(M) in the region of this minimum). Then a convenient quantity is the ratio R,

$$R \equiv \min\{P_{A-rich}/P_{A-poor},\ P_{A-poor}/P_{A-rich}\}\ , \tag{154}$$

which behaves for $L \to \infty$ as $R \to 1$ for $\Delta\mu = \Delta\mu_{coex}(T)$ while $R \to 0$ else. Figure 23a shows that one does find a very sharp ridge. But one cannot see where the

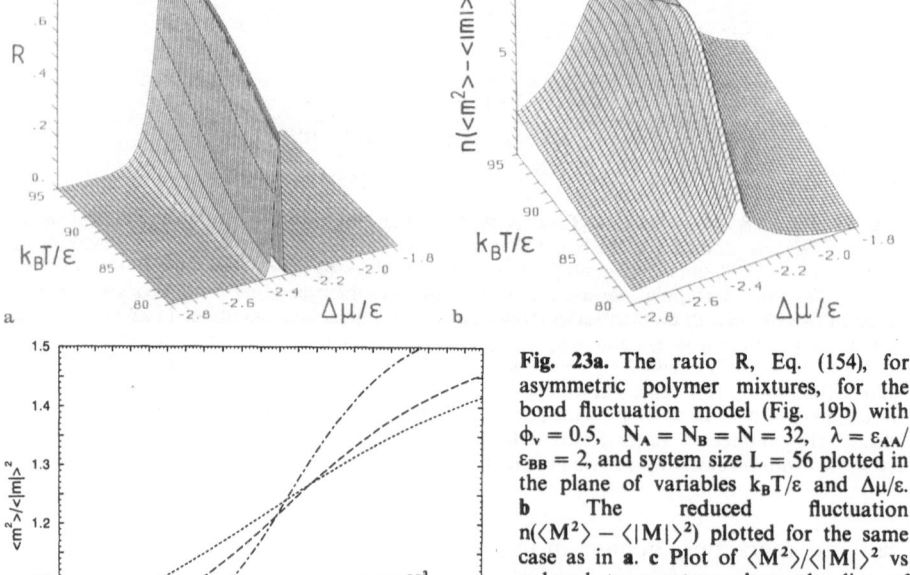

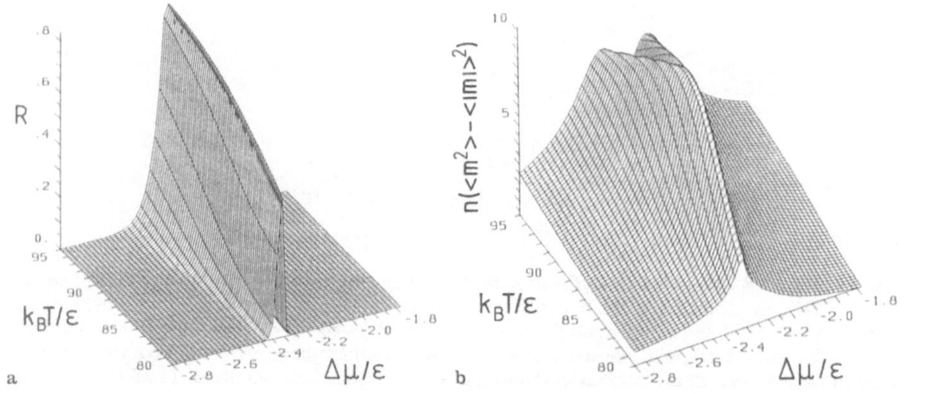

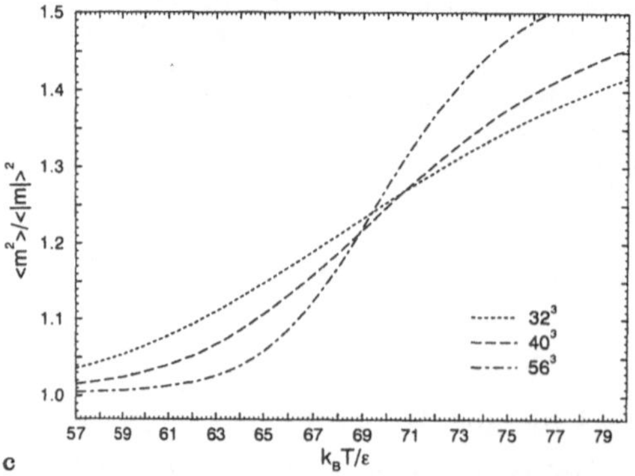

Fig. 23a. The ratio R, Eq. (154), for asymmetric polymer mixtures, for the bond fluctuation model (Fig. 19b) with $\phi_v = 0.5$, $N_A = N_B = N = 32$, $\lambda = \varepsilon_{AA}/\varepsilon_{BB} = 2$, and system size $L = 56$ plotted in the plane of variables k_BT/ε and $\Delta\mu/\varepsilon$. **b** The reduced fluctuation $n(\langle M^2\rangle - \langle|M|\rangle^2)$ plotted for the same case as in **a**. **c** Plot of $\langle M^2\rangle/\langle|M|\rangle^2$ vs reduced temperature along the line of maximal R in the $(\Delta\mu, T)$ plane, for $\lambda = 2.0$ and three sizes as indicated. From Deutsch and Binder [93]

critical point T_c occurs along this ridge. This is because above T_c there is of course also a $\Delta\mu$ for which the probabilities defined in Eq. (153) have equal magnitude, even though P(M) then has only a single peak. In this sense, one can view the line of maximal R in an asymmetric mixture as the analog of the temperature-axis of a symmetric mixture below and above T_c. However, the end point of the first-order phase transition can now be conveniently estimated, looking for the intersection of cumulants $U_L(T, \Delta\mu)$ for different sizes L, following a path $\Delta\mu = \Delta\mu_{coex}(T)$ in the $(T, \Delta\mu)$-plane along the ridge (Fig. 23c). We

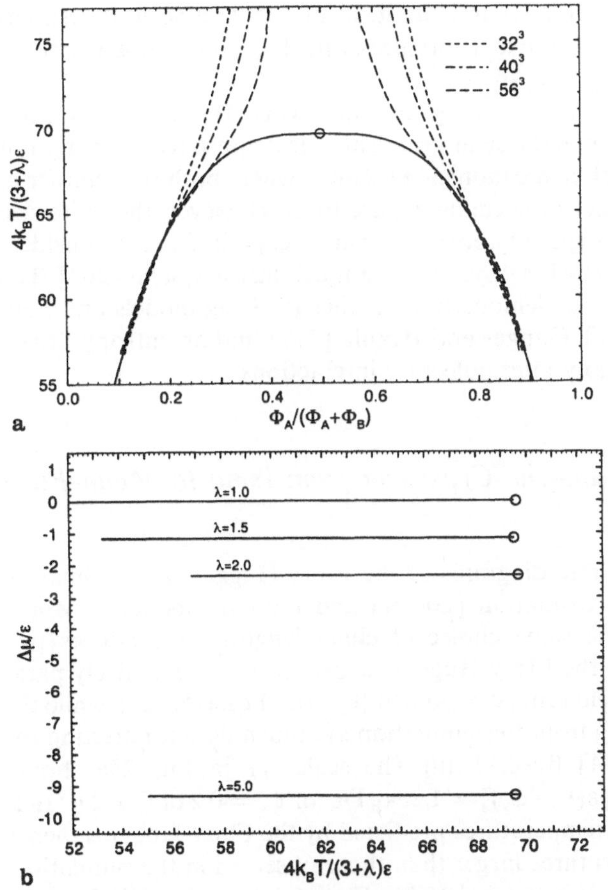

Fig. 24a. Phase diagram of the asymmetric polymer mixture ($\lambda = 2.0$, $N_A = N_B = N = 32$, $\phi_v = 0.5$) in the plane of variables reduced temperature and relative concentration $\phi_A/(\phi_A + \phi_B)$ of component A. The *dashed lines* are the histogram extrapolations for three simulated system sizes, the *full line* denotes the binodal, and the *circle* denotes the critical point. From Deutsch and Binder [93]. **b** Phase diagram of asymmetric polymer mixtures for $N_A = N_B = N = 32$, $\phi_v = 0.5$ in the $(T, \Delta\mu)$ plane. Three choices of the asymmetry parameter λ are shown as indicated. The first order transitions are shown as a *full line*, the critical points as *circles*. Temperature is normalized such that in the Flory-Huggins-approximation the critical temperature would occur for the same abscissa value. From Deutsch [266]

conclude this section noting that the concentrations of the coexisting phases are estimated as follows

$$\phi_A^r/(\phi_A + \phi_B) = \tfrac{1}{2}(1 + M_r), \ M_r = \int_{M^*}^{+1} MP_L(M)\,dM, (A - \text{rich}), \quad (155a)$$

$$\phi_A^p/(\phi_A + \phi_B) = \tfrac{1}{2}(1 + M_p), M_p = \int_{-1}^{M^*} MP_L(M)\,dM, (A - \text{poor}). \quad (155b)$$

Of course, again these data show a dramatic size effect (Fig. 24a) and thus a finite size scaling analysis of the type shown in Fig. 21a is necessary to extract the coexistence curve in the thermodynamic limit. In this way an accurate estimation of the phase diagram of polymer mixtures with $N_A = N_B$ but asymmetric interactions is possible (Fig. 24a, b).

For strongly asymmetric mixtures (e.g., mixtures where the A-chains are stiff while the B-chains are flexible) the semi-grandcanonical approach is clearly not feasible, and one must work in a canonical ensemble where both the number of A-chains n_A and the number of B-chains n_B are fixed. However, the finite size scaling ideas for $P_L(M)$ as exposed above still can be exploited if one considers the order parameter M in $L \times L$ *subsystems* of a much larger system [267]. The usefulness of this concept was demonstrated earlier for Ising models and Lennard-Jones fluids [268–271]. Gauger and Pakula [267] find an entropy-driven phase separation without any intermolecular interactions.

4.2 *Critical Behavior and the Crossover from Ising to Mean-Field Behavior*

Figure 25 compares the phase diagrams of the Flory-Huggins approximation [1], the Guggenheim approximation [20, 21] and the corresponding Monte Carlo results [107] for the same choice of chain lengths. One notices two pronounced differences: (i) the Flory-Huggins coexistence curve is nicely parabolic reflecting the mean field critical exponent $\beta = 1/2$, Eqs. (15, 122), while the coexistence curves resulting from the simulation are much flatter, reflecting the Ising critical exponent [74] $\beta \approx 0.32$. (ii) The scale z/χ in Fig. 25a should correspond to a scale $k_BT/[\varepsilon(1 - \phi_v)] = 1.25k_BT/\varepsilon$ for $\phi_v = 0.2$ of Fig. 25c: thus it is clear that the ordering temperatures predicted by the Flory-Huggins theory are about a factor of two to three larger than those observed in the simulation.

This problem is discussed in more detail in Fig. 26, where the ratio between the Flory-Huggins prediction, $T_c^{FH} = z\varepsilon(1 - \phi_v)N/(2k_B)$, Eqs. (4, 11), and the critical temperature observed in the simulation is plotted. It is seen that even for $\phi_v \to 0$ the Flory-Huggins-result for T_c seems to be too large by about a factor of two {this conclusion is corroborated by a direct calculation of polymer blends for $\phi_v = 0$ with the "bond-breaking" method [254] }. While for large ϕ_v (e.g. $\phi_v = 0.8$) the results strongly depend on the energy parameter ε_{AA}, ε_{AB}, ε_{BB} individually, for smaller ϕ_v ($\phi_v \leqq 0.5$) the single energy parameter approximation,

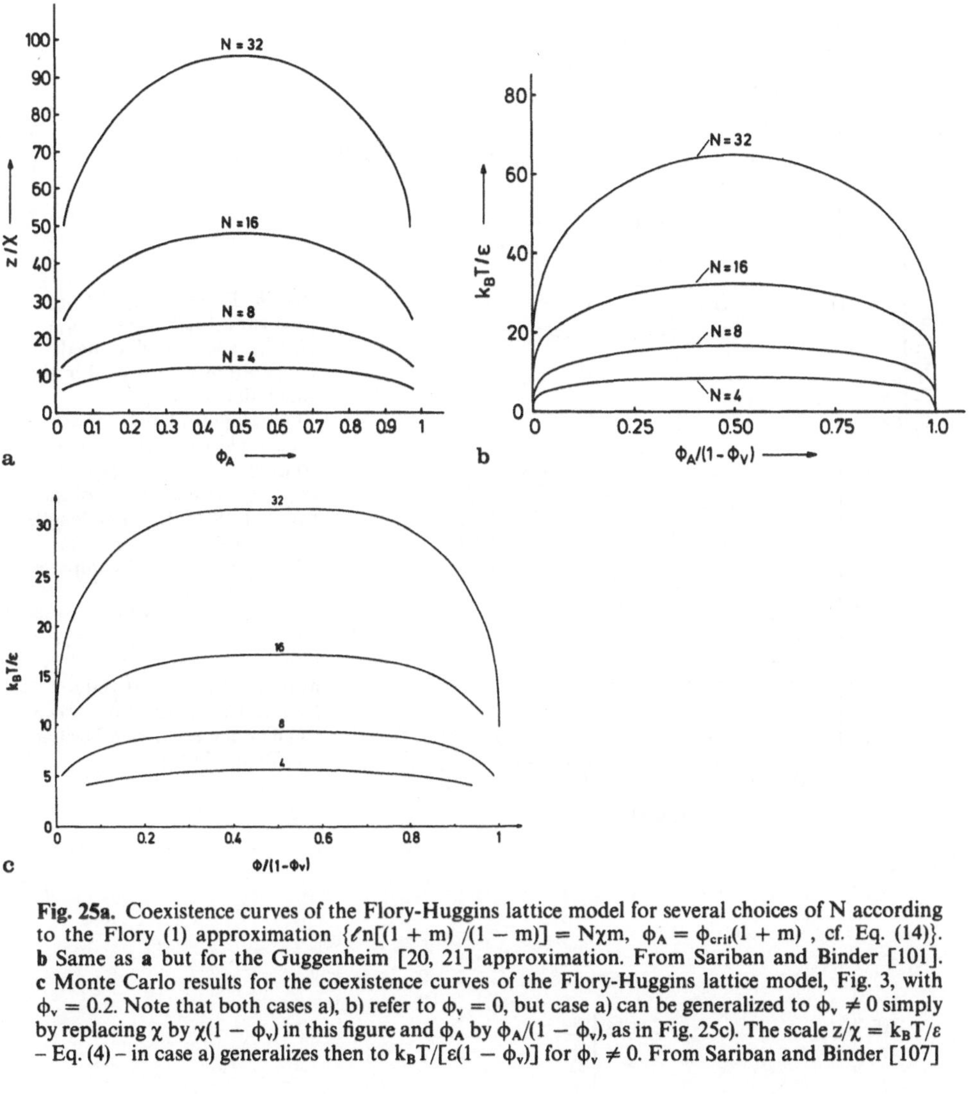

Fig. 25a. Coexistence curves of the Flory-Huggins lattice model for several choices of N according to the Flory (1) approximation $\{\ell n[(1+m)/(1-m)] = N\chi m,\ \phi_A = \phi_{crit}(1+m)$, cf. Eq. (14)$\}$. **b** Same as **a** but for the Guggenheim [20, 21] approximation. From Sariban and Binder [101]. **c** Monte Carlo results for the coexistence curves of the Flory-Huggins lattice model, Fig. 3, with $\phi_v = 0.2$. Note that both cases a), b) refer to $\phi_v = 0$, but case a) can be generalized to $\phi_v \neq 0$ simply by replacing χ by $\chi(1-\phi_v)$ in this figure and ϕ_A by $\phi_A/(1-\phi_v)$, as in Fig. 25c). The scale $z/\chi = k_BT/\varepsilon$ – Eq. (4) – in case a) generalizes then to $k_BT/[\varepsilon(1-\phi_v)]$ for $\phi_v \neq 0$. From Sariban and Binder [107]

claiming that T_c depends on the combination $\varepsilon_{AB} - (\varepsilon_{AA} + \varepsilon_{BB})/2$ only, Eqs. (4, 11), clearly becomes very accurate. Figure 24 shows that this remains true even for strongly asymmetric cases — for $\varepsilon_{AB} = \varepsilon/2$, $\varepsilon_{AA} = -\lambda\varepsilon/2$, $\varepsilon_{BB} = -\varepsilon/2$, the combination $\varepsilon_{AB} - (\varepsilon_{AA} + \varepsilon_{BB})/2 = (\varepsilon/4)(3 + \lambda)$ is exactly the factor used to normalize the temperature scale in Fig. 24. On the other hand, the strong increase of T_c^{FH}/T_c for the purely repulsive case (crosses) and large ϕ_v is a qualitative evidence for the prediction that polymer blends in a common good solvent become compatible in the semidilute limit [238–241]. This "renormalization" of the chi-parameter due to reduction of contacts via blob formation is checked in Fig. 27, where the "renormalized" interaction parameter (compare

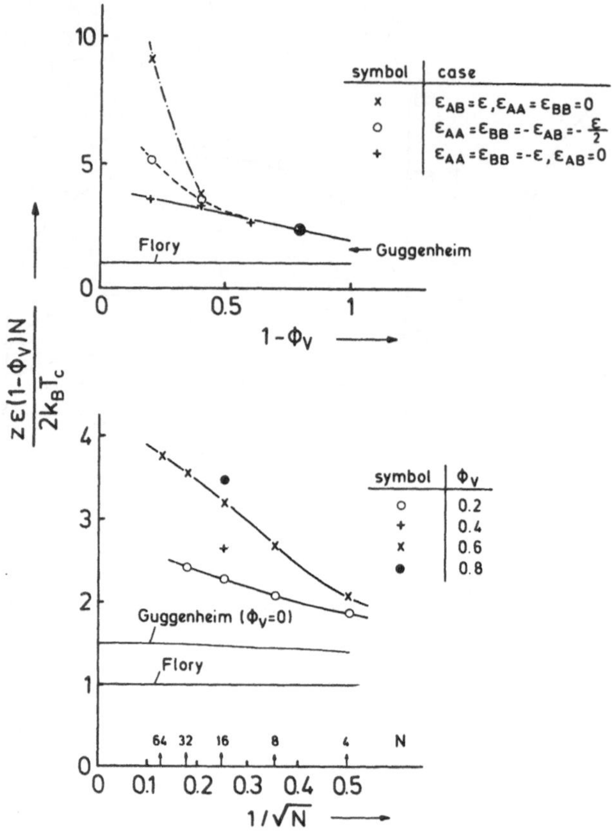

Fig. 26. Ratio of Flory-Huggins prediction $\{T_c^{FH} = z\varepsilon(1-\phi_v) N/2k_B\}$ and actual critical temperature T_c observed in the simulation of the Flory-Huggins lattice model (Fig. 3) plotted vs the monomer concentration $1-\phi_v$ (*upper part*) and vs the inverse of the square root of the chain length (*lower part*). The upper part refers to $N = 16$ and compares the different choices of $\varepsilon_{AA}, \varepsilon_{AB}, \varepsilon_{BB}$ that yield the same $\varepsilon = \varepsilon_{AB} - (\varepsilon_{AA} + \varepsilon_{BB})/2$. The Guggenheim [20, 21] result is available for $\phi_v = 0$ only (*arrow*). Curves are only drawn to guide the eye. From Sariban and Binder [101]

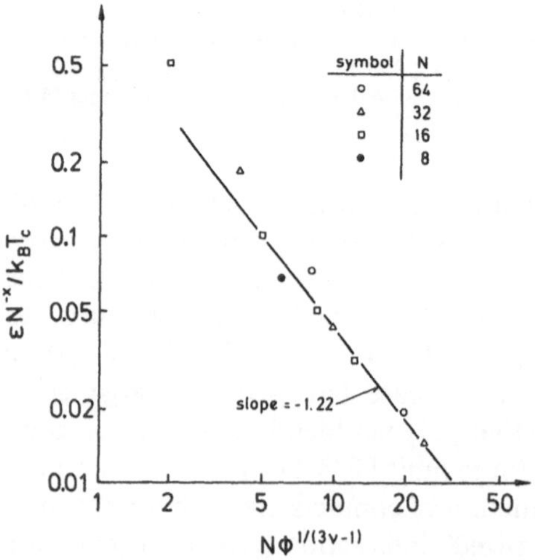

Fig. 27. Renormalized critical parameter $\varepsilon N^{-x}/k_BT_c$ (with $x \cong 0.22$ [238]) for the model of Fig. 3 plotted vs $N\phi^{1/(3\nu_e-1)}$ {this means a rescaling of N with the number of segments per "blob", $N_{blob} \propto \phi^{-1/(3\nu_e-1)}$, see de Gennes [2] }. *Straight line* shows the asymptotic exponent of the scaling function, $1 + x \approx 1.22$. From Sariban and Binder [272]

Eq. (135)) is plotted vs. $N/N_{blob}(\phi) \propto N\phi^{1/(3\nu_e-1)}$, $\nu_e \approx 0.59$ [2], using data for $N = 8$ to $N = 64$ for the model of Fig. 3 and $0.15 \leqq \phi = 1 - \phi_v \leqq 0.4$ [272].

The fact that T_c^{FH}/T_c increases with increasing N (Fig. 26, lower part) has been interpreted by Schweizer and Curro [47–49] as an indication for their PRISM prediction $N/T_c \propto N^{1/2} \rightarrow \infty$ as $N \rightarrow \infty$. As discussed in Sect. 3, see Fig. 13, this conclusion was in error, and rather a different explanation for the discrepancy between the prefactor in the $T_c \propto N$ – relation of Flory-Huggins theory and the simulations must be sought. Sariban and Binder [101, 103] suggested that this discrepancy is simply due to a strong overestimation of the number of B-neighbors that a given A-monomer can have. The total number of neighbors (i.e. monomers from chains different from the considered one) that monomers of a chain can have, is assumed to be $(n_c^{tot})_{FH} = zN(1 - \phi_V)$ in Flory-Huggins theory [1]. In Guggenheims theory [20, 21], account is taken of the fact that an inner monomer of a chain can have at most $z - 2$ and not z monomers of other chains. Correcting for the effect of chain ends, one thus predicts $(n_c^{tot})_G = [(z - 2)N + 2](1 - \phi_v)$. Even this reduced contact number is an overestimate, because it disregards that a chain has locally "crankshaft"-type configurations (Figs. 3, 16) where it makes "*self-contacts*", which do not contribute anything to the phase separation of different chains from each other, and hence must not be included in the neighbor count for the estimation of T_c. The advantage of computer simulations, of course, is that such considerations can be put to an immediate test [101, 103]. Figure 28a shows that there occurs indeed an appreciable number of such selfcontacts (intrachain contacts), and the total number of interchain contacts is correspondingly reduced: for the example shown, we would have $(n_c^{tot})_{FH} \approx 77$, $(n_c^{tot})_G \approx 54$, while the actual number $(n_c^{tot})_{simulation} = n_c\{AA(BB)\ \text{interchain}\} + n_c(AB) \approx 28(\varepsilon/k_BT = 0)$ to 30 (at the binodal curve). This reduction of n_c^{tot} from $(n_c^{tot})_{FH}$ to the value observed in the simulation is by a factor quite comparable to the ratio T_c^{FH}/T_c seen in Fig. 26.

A further interesting effect is that the number of contacts are slightly temperature dependent. In fact, the slight increase in the number of intrachain contacts with increasing ε/k_BT is due to a slight contraction of the chains (Fig. 28b). Figure 28b constitutes evidence that the simple form of the RPA exposed in Sect 2.2 is not strictly correct: there it was assumed that the chain gyration radii depend neither on volume fraction $\phi_A/(\phi_A + \phi_B)$ nor on the chi-parameter (or ε/k_BT, respectively). In fact, if one considers the case of very dilute mixtures, $\phi_A/(\phi_A + \phi_B) \ll 1$, and large values of the chi-parameter, $\chi \gg \chi_{crit}$, one finds even a gradual collapse of the minority chains [102].

We now discuss the critical point behavior of these Monte Carlo methods in more detail. Even in the finite size scaling analysis of the Monte Carlo data (Fig. 21), that short chains display Ising-like critical behavior was used (Fig. 21c). For longer chains, e.g. for $N = 128$, one finds "effective" critical exponents that deviate already somewhat from the Ising values, as expected due to the onset of crossover towards mean field critical behavior which should show up for very long chains (Sect. 2.5). However, using the mean field form of finite size scaling [262–264, 273] never produces a good "data collapsing", not even for the

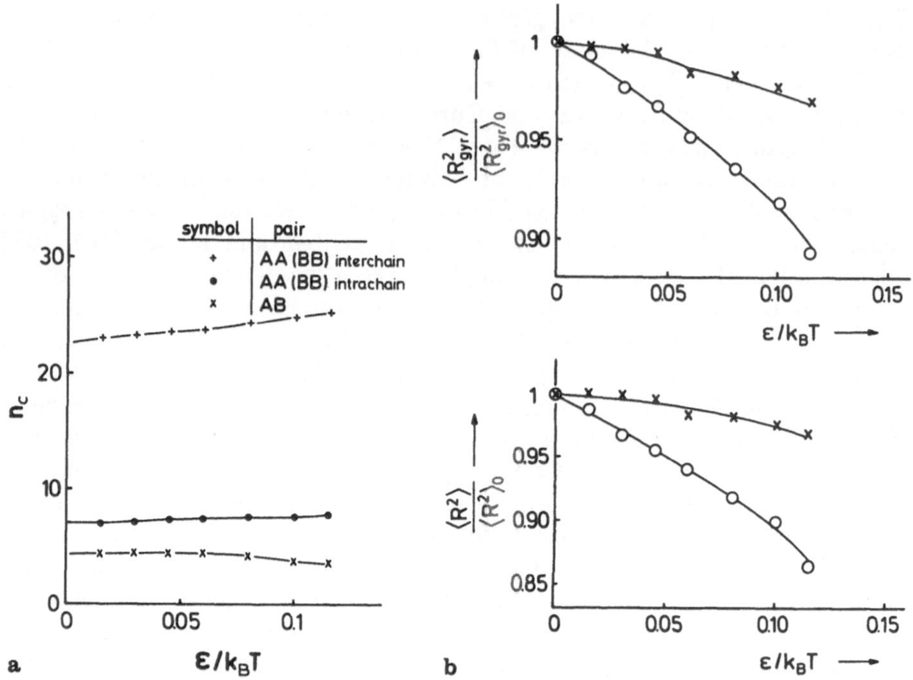

Fig. 28a. Number of contacts n_c for the model of Fig. 3 plotted vs inverse temperature for the case $\phi_v = 0.6$, $N = 32$, $\varepsilon_{AB} = 0$, $\varepsilon_{AA} = \varepsilon_{BB} = -\varepsilon$, $\phi_A/(1-\phi_v) = 0.9$, $\phi_B/(1-\phi_v) = 0.1$. From Sariban and Binder [103]. **b** Mean-square gyration radius (*upper part*) and mean-square end-to-end distance (*lower part*) plotted vs ε/k_BT), for the same model as a), *crosses* refer to the majority component (A), *circles* to the minority component (B). The largest value of ε/k_BT corresponds to a point at the binodal. All radii are normalized by their values at $\varepsilon/k_BT = 0$. From Sariban and Binder [101]

longest chains, $N = 256$, indicating that even these longest chains are still in the crossover regime [91, 92].

In order to investigate this crossover behavior, and to check predictions such as the relations Eqs. (122)–(125) for the critical amplitude prefactor, the Ising critical exponents were adjusted at their theoretical values [274] and $T_c(N)$ was fixed from the intersection methods discussed in Sect. 4.1 (see Fig. 19). While resulting critical amplitude estimates are fairly accurate for small N, Fig. 29, for large N huge error bars result, since the Ising exponents do not allow a good fit in the finite size scaling analysis due to crossover problems [91, 92]. While within these large error bars the data (Fig. 29b) are compatible with the theoretical prediction, for small N ($N \leq 32$) definitely a different behavior is observed: Fig. 29a shows that there power laws are found, as the straight lines on the log-log-plot imply, but the exponents are rather different from the predicted ones. It turns out, however, that the data of Fig. 29a can be explained if one takes the blob picture of polymer mixtures in a common solvent [238–241] into account. Broseta et al. [240] predicted for the critical amplitudes in the regime where N and N_{blob} are comparable a behavior that involves the exponent x (Eq.

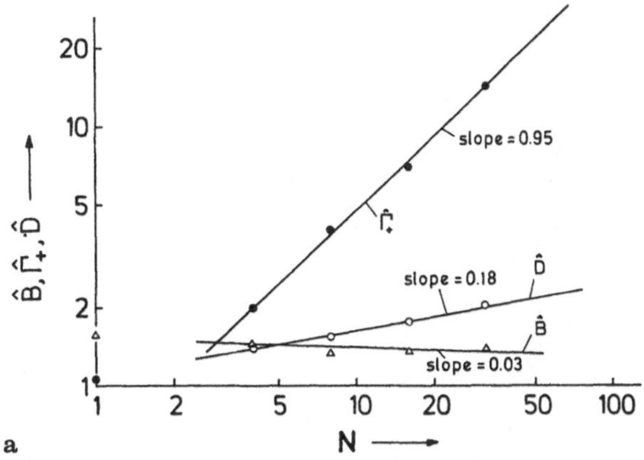

a

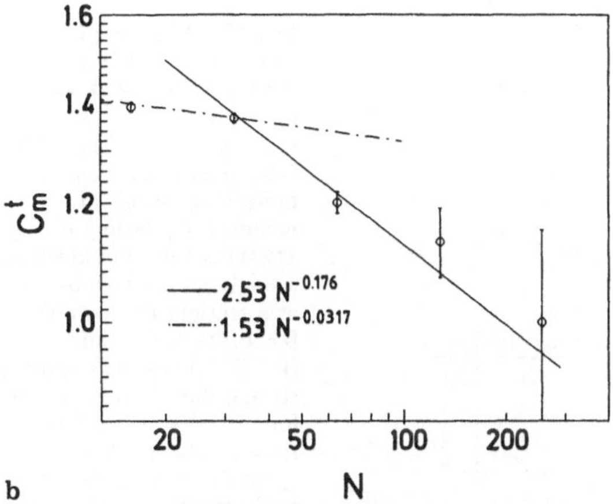

b

Fig. 29a. Log-log plot of critical amplitudes $\hat{B}$, $\hat{\Gamma}_+$ and $\hat{D}$ versus chain length N, for the model of Fig. 3 and $\phi_v = 0.2$. Here $\hat{\Gamma}_+$ means the amplitude of $S'_{coll}(q = 0)/(1 - \phi_v)^2$ for $T > T_c$ and $\hat{D}$ is the amplitude at the critical isotherm, $m = \hat{D}(\Delta\mu/k_BT)^{1/\delta}$, where $\delta = (\gamma + \beta)/\beta$. Points for N = 1 refer to the standard Ising model. From Sariban and Binder [107]. **b** Log-log plot of $\hat{B}$ (denoted as C_m^t in the figure) vs N. The *straight line* is the best fit to all data with $N \geqq 32$, using the theoretical exponent $\beta - 1/2 \approx -0.176$, Eq. (122). These data refer to the bond fluctuation model at $\phi_v = 0.5$ (raw data are shown in Fig. 20a). From Deutsch and Binder [92]

(135)) that describes the renormalization of the effective Flory-Huggins parameter:

$$\hat{B} \propto N^{x(\beta - 1/2)/(1+x)}, \quad \hat{\Gamma}_+ \propto N^{1 + x(1-\gamma)/(1+x)},$$
$$\hat{D} \propto N^{1/3 + [1/3 + (x/2)/(1+x)](3/\delta - 1)}. \tag{156}$$

The exponents in Eq. (156) are in striking agreement with the simulation, Fig. 29a. Also in the case of the bond fluctuation model for $N \leqq 32$ a behavior compatible with Eq. (156) is detected (dash-dotted straight line in Fig. 29b).

A different way of testing the crossover from Ising to mean field behavior is based on the use of Eq. (123), see Fig. 30. For small N distinct deviations from

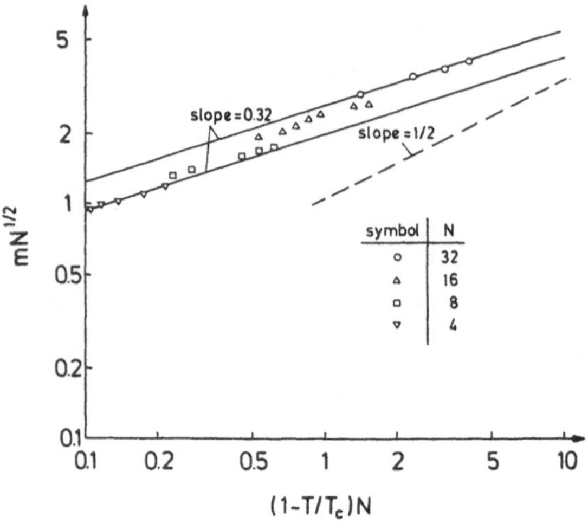

a

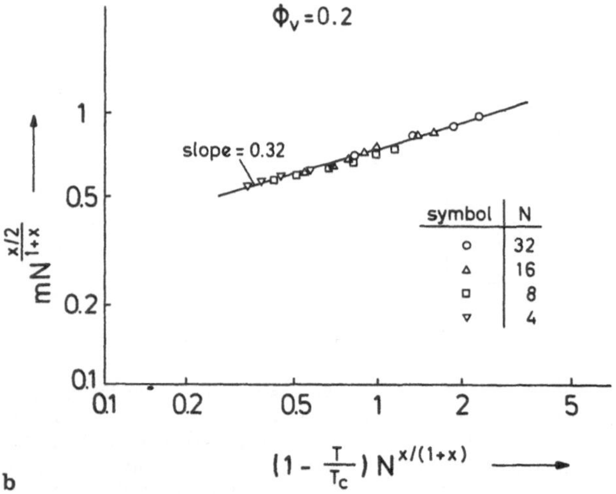

b

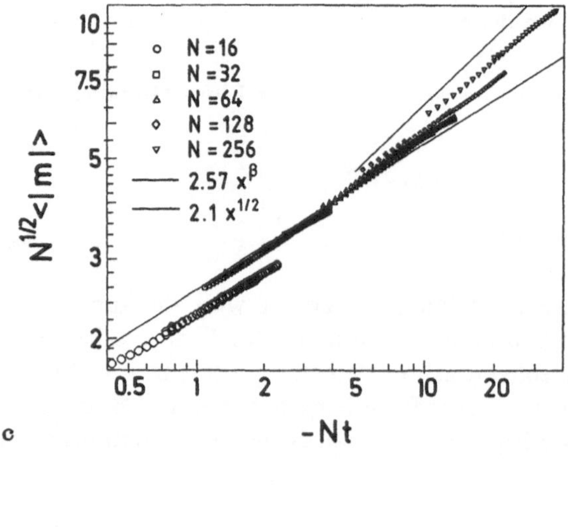

c

Fig. 30a. Log-log plot of $mN^{1/2}$ vs $(1 - T/T_c)$ N, for the model of Fig. 3, with $\phi_v = 0.2$ and $\varepsilon_{AA} - \varepsilon_{BB} = -\varepsilon$, $\varepsilon_{AB} = 0$. Only data not affected by finite size effects are included. According to Eq. (123) the data should fall on a single curve which behaves as a straight line with slope $\beta = 0.324$ for small $(1 - T/T_c)$ N and as another straight line with slope 1/2 for large $(1 - T/T_c)$ N, respectively. Different symbols show different chain lengths as indicated. From Sariban and Binder [107]. **b** Log-log plot of $mN^{x/[2(1+x)]}$ vs $(1 - T/T_c)$ $N^{x/(1+x)}$, using $x = 0.22$, for the same data as shown in Fig. 30a. From Sariban and Binder [272]. **c** Same as **a** but for the bond fluctuation model at $\phi_v = 0.5$. Different symbols show various chain lengths N, as indicated. Only such data were included for which the L-dependence was negligible. *Straight lines* indicate the Ising power law $\{\alpha(N|t|)^{\beta}$, valid for $N|t| \lesssim 1\}$ and the mean-field power law $\{\alpha(N|t|)^{1/2}$, valid for $N|t| \gg 1\}$. From Deutsch and Binder [92]

the predicted scaling form are observed (Fig. 30a, c), while for N in the range from $N \approx 32$ to $N = 128$ (Fig. 30c) the data are roughly consistent with the proposed collapse on a master curve, but still are far from the asymptotic mean field regime where $m \propto t^{1/2}$ holds (note that the deviation for $N = 256$ may be due to residual finite size effects and the inaccuracy in the location of T_c [92]). On the other hand, using a scaling valid for the semidilute regime taking into account the blob effects [272], one finds a reasonable data collapse for small N (Fig. 30b).

The best evidence for the crossover from Ising to mean field theory comes from an analysis of crossover phenomena in the context of finite size scaling [92, 275]. For a detailed analysis of this problem we refer to the literature, but rather mention here only that the critical value of the order parameter in a finite box at T_c is predicted to vary as [275]

$$m(T_c, L) \equiv \langle |M| \rangle_{T_c} = N^{-1/2} \tilde{f}_m(L/N) ,$$

$$\tilde{f}_m(\zeta) \propto \begin{cases} \zeta^{-3/4}, & \zeta \ll 1 \\ \zeta^{-\beta/\nu}, & \zeta \gg 1. \end{cases} \tag{157}$$

Figure 31 shows that the data for the bond fluctuation model and $N = 128$, 256 are already consistent with the mean field limit of this finite size scaling crossover law, where $m(T_c, L) \propto L^{-3/4}$, while for small N the standard finite size scaling behavior implied by Eq. (146), $m(T_c, L) \propto L^{-\beta/\nu}$ with $\beta/\nu \approx 0.515$ is observed. From the intersection of the two straight lines on this log-log-plot one can tentatively estimate that the crossover occurs for $L = L_{cross}$ with

$$L_{cross} \approx 0.8N . \tag{158}$$

Let us compare this estimate for a characteristic length with the correlation

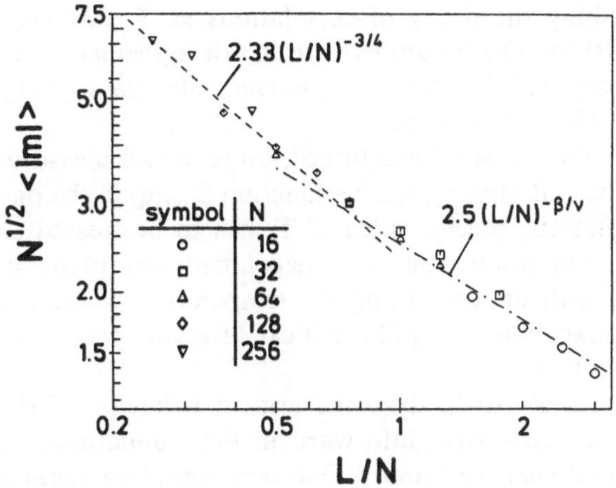

Fig. 31. Log-log plot of $mN^{1/2}$ at $T = T_c$ vs L/N, for the bond fluctuation model at $\phi_v = 0.5$. Different symbols show various chain lengths N, as indicated. The *dashed line* illustrates the power law expected in the mean-field regime of finite size scaling ($\langle |M| \rangle \propto L^{-3/4}$, for $L/N \ll 1$) and the *dash-dotted line* in the non-mean-field regime ($\langle |M| \rangle \propto L^{-\beta/\nu}$, with $\beta/\nu \approx 0.515$ for Ising exponents [274]. From Binder and Deutsch [275].

length ξ_{cross} – Eq. (42) for $N_A = N_B = N$ yields $\xi = \sqrt{2}R_{gyr}\varepsilon^{-1/2}$, Eq. (121) for $\phi_A = \phi_B = 1/2$ yields $\varepsilon_{cross} = Cv_m^2 N^2/R_{gyr}^6$ –

$$\xi_{cross} = \sqrt{2}R_{gyr}(\varepsilon_{cross})^{-1/2} = v_m^{-1}\sqrt{2/C}R_{gyr}^4/N$$
$$= v_m^{-1}\sqrt{2/C}(a/\sqrt{6})^4 N \,. \tag{159}$$

If one could identify $L_{cross} = \xi_{cross}$, Eq. (158) would yield an estimate of C, noting that for the bond fluctuation model at $\phi = 1/2$, with the lattice spacing as unit of length, one has $a \approx 2.63$, $v_m = 2$, and then $C \approx 1.37$. This is considerably larger than the experimental estimate of Bates et al. [69] who suggest $C \approx 0.29 \pm 0.08$. However, one must take into account that the crossover from mean field to Ising behavior really involves a rather broad regime, and so different quantities yield different constants. In fact, using the response S_{coll} at T_c rather than $m(T_c, L)$ one finds [257] $L_{cross} \approx 2N$ and then $C \approx 0.22$, in agreement with the experimental estimate.

4.3 Comparison to Experiment

Sariban et al. [101, 107, 276, 277] were the first to emphasize that the Ising critical behavior can be seen in polymer mixtures for not too long chains and verified it by their simulations. A consequence of Ising behavior that is easily verified by experiment is that the spinodal temperature (or mean field critical temperature T_c^{MF}, respectively) which is defined for $\phi_A = \phi_{Acrit}$ from a linear extrapolation of the inverse scattering intensity $S_{coll}^{-1}(q = 0)$ with temperature to the point where $S_{coll}^{-1}(q = 0) = 0$ must be offset from the actual critical temperature T_c (Fig. 32). This phenomenon has been seen in simulations [92, 101, 107] as well as in various experiments [69, 71, 215, 216, 278]. A detailed analysis of the non-mean field critical behavior has allowed the estimation of critical exponents $\gamma = 1.26 \pm 0.01$ [215–217, 69], $\nu = 0.59 \pm 0.01$ [215] or $\nu \approx 0.63$ [71], and also the exponent describing the decay of correlations at T_c has been estimated [215], $\eta \simeq 0.047 \pm 0.004$. These numbers are in fair agreement with the best numerical values proposed for the Ising universality class [274], $\gamma = 1.24$, $\nu = 0.63$ and $\eta = 0.039$, respectively.

There are still two caveats that must be mentioned, however: all measurements refer to an analysis of the collective scattering function $S_{coll}(q)$ in the one phase region of the blends, thus the precise value of T_c has to be treated as a fitting parameter and does not result from an independent measurement. Secondly, there are problems with understanding the temperature where the crossover from mean field behavior to Ising-like critical behavior occurs, as already discussed in the last section.

Clearly, it is also very desirable to analyze the critical behavior of the coexistence curve. While this is fairly straightforward in the simulations, as discussed above, it is difficult to do experimentally. One very appealing method

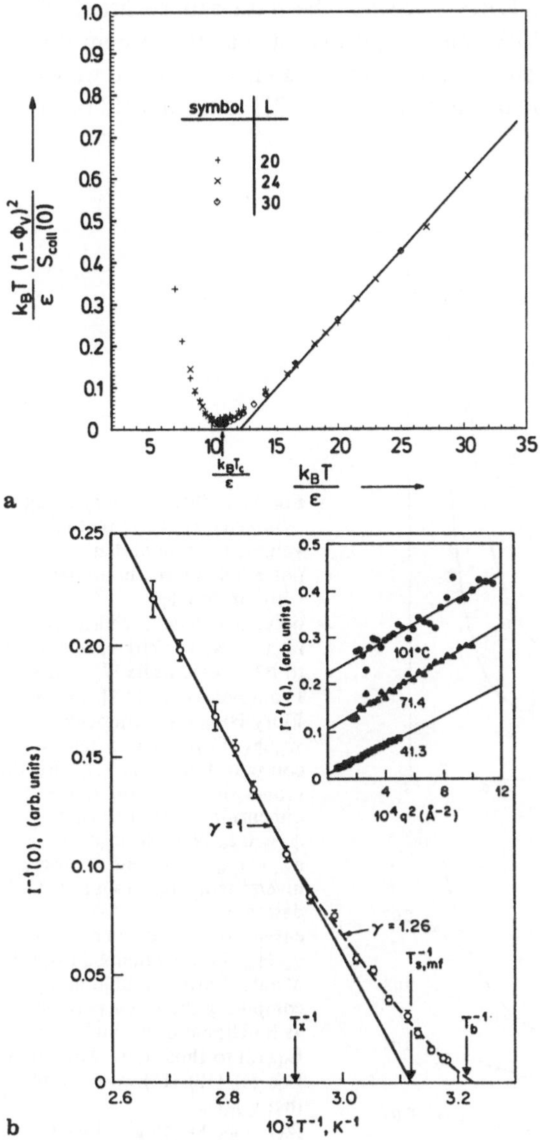

Fig. 32a. Inverse collective structure factor of the model shown in Fig. 3 (but in $d = 3$ spatial dimensions), for the case $\phi_v = 0.6$, $N = 16$, $\varepsilon_{AB} = 0$, $\varepsilon_{AA} = \varepsilon_{BB} = -\varepsilon$ at critical composition. Three lattice sizes are included as indicated in the figure. A straight line would imply a linear vanishing of $S_{coll}^{-1}(q = 0)$ at $T_c^{MF} > T_c$ (the actual critical temperature, as inferred from a finite size scaling analysis, is shown by an *arrow*). From Sariban and Binder [101]. **b** Inverse collective structure factor for PEP-PI mixtures plotted vs inverse temperature at critical composition (the phase diagram of this system is shown in Fig. 1b). From Bates et al. [69]

is to extract the compositions of coexisting phases from a nuclear reaction analysis of the interfacial concentration profile [279]. The accuracy of this method does not yet allow an estimation of the critical exponent β, however. The only analysis of the critical behavior of the coexistence curve has been done by Chu et al. [280] for Polystyrene/Poly-(2-Chlorostyrene) blends. Adding 22.6 weight percent of di-*n*-butylphthalate (DBP) which acts as plasticizer, they found from synchrotron scattering measurements combined with a centrifugal method that $\beta = 0.33$. Without DBP, however, the data can be fitted to the mean field exponent $\beta = 1/2$. While one might expect that the system diluted with the plasticizer may have a somewhat wider Ising critical region, we would expect to see crossover from Ising behavior (very near T_c) to mean field behavior

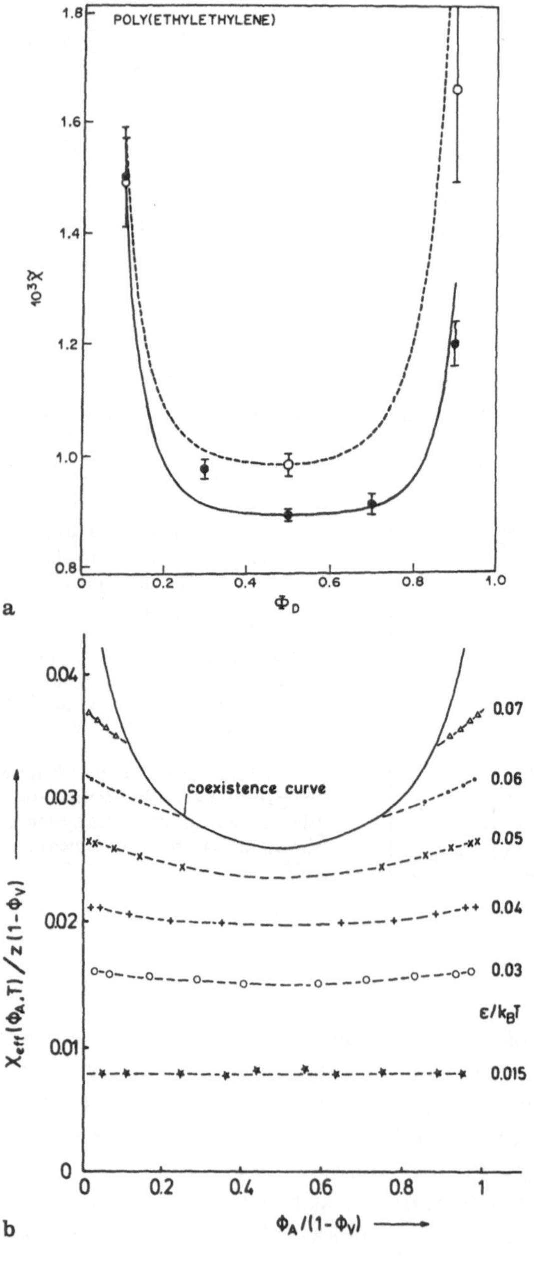

Fig. 33a. Effective Flory-Huggins parameter $\chi_{eff}(\phi, T)$ plotted vs volume fraction of deuterated polyethylethylene in a mixture with protonated polyethylethylene. *Filled symbols* refer to $N_w^p = 1710$, *open symbols* to $N_w^p = 818$, while $N_w^d = 1330$. From Bates et al. [68]. **b** Effective Flory-Huggins parameter $\chi_{eff}(\phi_A, T)$ plotted vs relative concentration $\phi_A/(1 - \phi_v)$ for the model of Fig. 3 (in $d = 3$ dimensions) and the case $\phi_v = 0.2$, $N = 16$, $\varepsilon_{AB} = 0$, $\varepsilon_{AA} = \varepsilon_{BB} = -\varepsilon$ and a variety of inverse temperatures ε/k_BT. Only data outside of the coexistence curve are well defined. Here $\chi_{eff}(\phi_A, T)$ was obtained from the Monte Carlo simulation by computing the order parameter as functional of $\Delta\mu$ and fitting Eq.(12) to those data. Note that Eqs. (4), (10), (12) would imply that $\chi_{eff}(\phi_A, T)/[z(1 - \phi_v)] = \varepsilon/k_BT$. From Sariban and Binder [101]. **c** Effective Flory-Huggins parameter χ_{eff} plotted vs the order parameter m for the bond fluctuation model with $\phi_v = 0.5$, $N = 128$, and three different lattice sizes. Here χ_{eff} is obtained fitting Eq. (37) to data as shown in Fig. 22. From Deutsch and Binder [92]

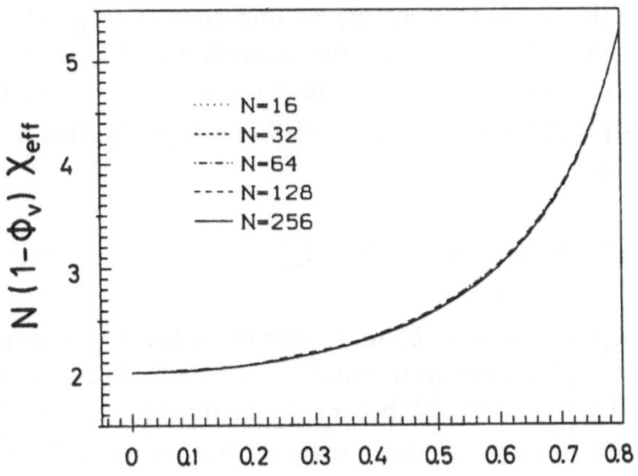

Fig. 33a–c. (Contd.)

(further away from T_c), and this is not seen, although a rather wide temperature range is covered. Thus more experiments on this point seem very desirable.

Another very interesting problem concerns the concentration dependence of the effective Flory-Huggins parameter $\chi_{eff}(\phi, T)$ that can be extracted from small angle scattering data by fitting Eqs. (37), (46) or (55) to them. It has been found that $\chi_{eff}(\phi, T)$ for symmetric mixtures (such as mixtures of protonated and deuterated polyethylene, Fig. 33a) is nearly independent of volume fraction ϕ for ϕ around ϕ_c but shows a sharp upturn both near $\phi \to 0$ and near $\phi \to 1$. A qualitatively similar behavior is seen in the simulations (Fig. 33b, c). It is rather clear, that neither for the experiments nor for the simulation an analysis of the data in terms of a model of a strictly incompressible binary mixture is adequate [63, 80, 221, 281, 282]. In this context, we would draw attention to a recent simulation of an off-lattice model of polymer blends in the isothermal-isobaric ensemble [283] relying on the incremental chemical potential method [284–286].

4.4 Interdiffusion Revisited

Motivated by experiments where the broadening of interfacial profiles is studied, when a layer of polymer A is brought on top of a layer of polymer B {see e.g. [287–293]}, Monte Carlo simulations have been performed in an $L \times L \times D$ geometry, with $D \gg L$ and two impenetrable $L \times L$ walls, while periodic boundary conditions are applied in the directions parallel to the walls only [88, 90]. First a homopolymer melt in this geometry (e.g. with $L = 20$, $D = 80$, using the bond fluctuation model) is equilibrated, and then a clock is set at time $t = 0$, and all chains whose center of gravity is at a coordinate in the left half of the box ($1 \leqq z \leqq 40$) are labelled as A chains while chains whose center of gravity is in the right half of the box ($41 \leqq z \leqq 80$) are labelled as B chains. One then simply follows the time evolution of the monomer profile of one species (Fig. 34a). Similar to experiment [290], one then uses the "interquartile width", i.e. the distance where $\phi_B(z, t)$ has increased from $(1 - \phi_v)/4$ to $3(1 - \phi_v)/4$, to estimate the interdiffusion constant. $W(t) = 2\beta\sqrt{D_{int}t}$ where $\beta \approx 0.96$ if the profile has the well-known error function form,

$$\phi_B(z, t) = (1 - \phi_v)\{1 - \tfrac{1}{2} \text{ erfc } [(z - 40)/2\sqrt{D_{int}t}]\}. \tag{160}$$

Figure 34b presents the time-dependence of such interquartile widths for several choices of an attractive energy ε_{AB} between monomers of different chains. Due to the curvature of these plots it is not so clear whether the asymptotic regime of late times where $W(t) \propto \sqrt{t}$ holds has actually been reached. Assuming that this

symbol	χ
◇	5.185
△	3.889
□	2.593
○	1.296
▷	0

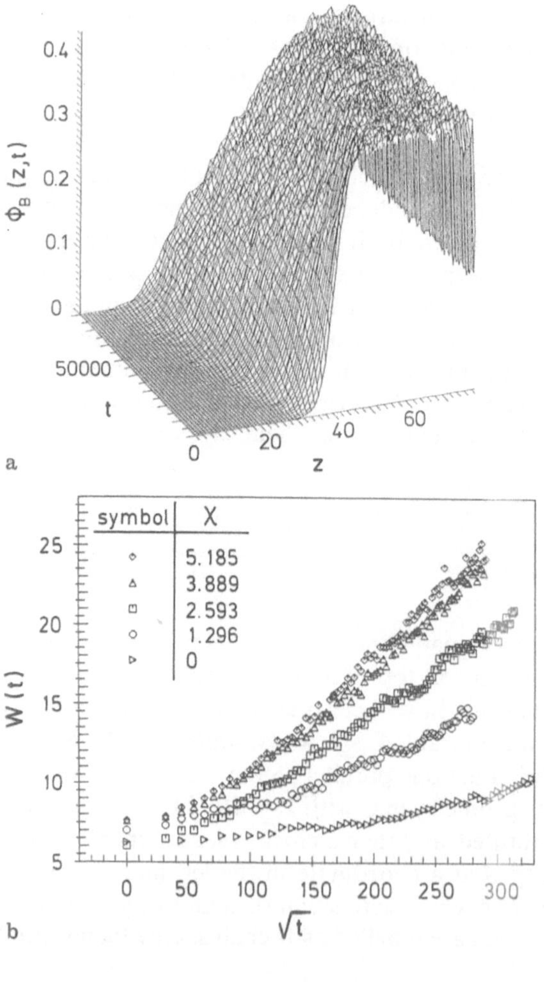

Fig. 34a. Time evolution of the concentration profile $\phi_B(z, t)$ for $\phi_v = 0.58$, $N = 20$, monomeric jump rates $\Gamma_A = \Gamma_B = 1$, and $\varepsilon_{AB}/k_BT = -(5/18)$, using the bond fluctuation model in a $20 \times 20 \times 80$ geometry. To gain statistics 48 samples are run in parallel and averaged together. From Deutsch and Binder [88]. **b** Interquartile width $W(t)$ for $N = 40$, $\phi_v = 0.58$, $\Gamma_A = \Gamma_B = 1$, and different choices of the χ parameter ($\chi = q\varepsilon_{AB}/k_BT$ with $q = 14$) as indicated in the figure. Here lattice linear dimensions $30 \times 30 \times 80$ were chosen, to avoid self-overlap of the chains due to periodic boundary conditions. From Deutsch and Binder [88]

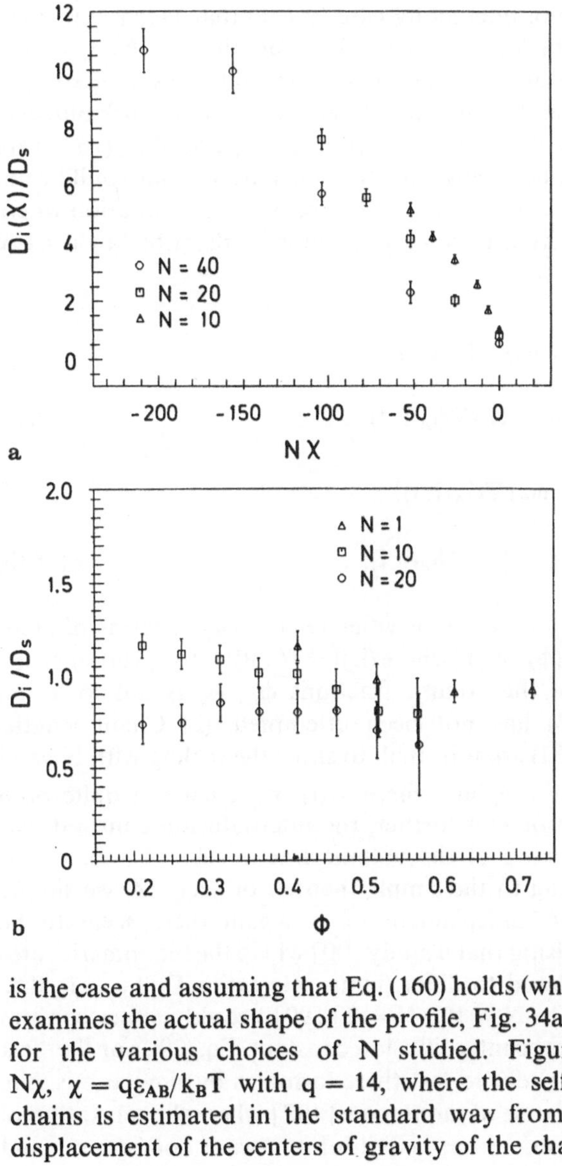

Fig. 35a. Interdiffusion constants $D_{int}(\chi)$ of the three-dimensional bond fluctuation model at $\phi_v = 0.58$ plotted vs $N\chi$, where $\chi = q\varepsilon_{AB}/k_BT$ and $q = 14$, for three choices of $N_A = N_B = N$. All data for $D_{int}(\chi)$ are scaled with the corresponding self diffusion constants D_s of the chains. **b** Same as a) but plotting data for $\chi = 0$ vs the volume fraction $\phi = 1 - \phi_v$. From Deutsch and Binder [88]

is the case and assuming that Eq. (160) holds (which seems reasonable when one examines the actual shape of the profile, Fig. 34a), one thus can estimate $D_{int}(\chi)$ for the various choices of N studied. Figure 35a) plots $D_{int}(\chi)$ /D_s vs $N\chi$, $\chi = q\varepsilon_{AB}/k_BT$ with $q = 14$, where the self-diffusion constant D_s of the chains is estimated in the standard way from a study of the mean square displacement of the centers of gravity of the chains,

$$D_s = \lim_{t\to 0}\langle[\vec{r}_{cg}(t) - \vec{r}_{cg}(0)]^2\rangle/(6t) \; . \qquad (161)$$

Only chains with centers of gravity far from the hard walls are included in Eq. (161) to avoid artefacts due to the constraining geometry.

According to the simple theory outlined in Sect. 2.4, cf. Eq. (79), one would expect

$$D_{int}(\chi)/D_{int}(\chi = 0) = 1 - 2\phi_A\phi_B N\chi \; , \qquad (162)$$

and since one expects in the noninteracting case ($\chi = 0$) that $D_{int}(\chi = 0) \propto D_s$, one expects that plotting data for $D_{int}(\chi)/D_s$ data for different N, χ should superimpose on a master curve and this master curve should be a simple straight line. Figure 35a shows, however, that neither of these predictions work out well. There are several reasons why there may be a problem: (i) Eqs. (79), (162) were derived in the linear regime where concentration deviations from equilibrium are small. In fact, Jilge et al. [90] pointed out that the nonlinear situation where an A-rich layer interdiffuses with a B-rich layer should be described by a set of two coupled equations

$$\frac{\partial \phi_A(\vec{r}, t)}{\partial t} = \nabla\{D_{AA}(\phi_A, \phi_B)\, \nabla\phi_A(\vec{r}, t)\} + \nabla\{D_{AB}(\phi_A, \phi_B)\, \nabla\phi_B(\vec{r}, t)\} , \tag{163a}$$

$$\frac{\partial \phi_B(\vec{r}, t)}{\partial t} = \nabla\{D_{BA}(\phi_A, \phi_B)\, \nabla\phi_A(\vec{r}, t)\} + \nabla\{D_{BB}(\phi_A, \phi_B)\, \nabla\phi_B(\vec{r}, t)\} , \tag{163b}$$

and they derived expressions for the diffusivities $D_{ij}(\phi_A, \phi_B)$ in terms of associated Onsager coefficients $\Lambda_{ij}(\phi_A, \phi_B)$ {where (i, j) = (A, B)}. However, since the dependence of $\Lambda_{ij}(\phi_A, \phi_B)$ on the volume fractions ϕ_A, ϕ_B is unknown, an explicit solution of Eq. (163) has not been attempted. (ii) Chain lengths $N = 10, 20$ included in Fig. 35a) are too small to show the scaling with N valid for $N \to \infty$. (iii) The asymptotic regime where $W(t) \propto \sqrt{t}$ has not quite been reached. To check for these problems further, the interdiffusion constant was studied also for $\chi = 0$ as function of the volume fraction $\phi = 1 - \phi_v$ of the monomers (Fig. 35b). According to the simple theories of Sect. 2.4, we simply expect $D_i/D_s = 1$ in this case, independent of ϕ. Again there seem to be deviations. Thus it is not surprising that a study [90] where the monomeric rates for attempted jumps were allowed to differ (jump rate ratios $\Gamma_A/\Gamma_B = 1$, 0.5, and 0.1 were studied for $N = 1$, $N = 10$, $N = 20$ [90]) could neither confirm the "slow mode" prediction for the interdiffusion constant, Eq. (90), nor the "fast mode" prediction, Eq. (91). In order to test these formulas by simulations, one really should proceed as in the case of monomers [197] where the relaxation of small concentration deviations from equilibrium was considered and D_i, D_s^A, D_s^B, Λ_{AA}, Λ_{BB}, Λ_{AB} were independently measured. This study revealed that off-diagonal Onsager coefficients Λ_{AB} must not be neglected when $\Gamma_A/\Gamma_B \ll 1$ [197]. We expect that a similar conclusion is true for polymers, too, and thus neither Eq. (90) nor Eq. (91) should be trusted. Of course, one also must keep in mind that all these Monte Carlo studies do not include any effects due to bulk flow and thus in principle are also unable to describe the dynamics of concentration fluctuations near the critical point correctly, where Eqs. (78, 79) fail and rather Eqs. (83), (84) apply, as was confirmed in several experiments [71, 217, 294]. Thus the Monte Carlo studies of interdiffusion at this point cannot be compared to experimental data directly. But some of the problems which hamper their interpretation – Eq. (163) needs to be used and hence it is not clear what the constant D_{int} extracted from W(t) really means; unknown concentration dependence of self diffusion constants $\Gamma_A/\Gamma_B \neq 1$; etc – also hamper corresponding experiments where interdiffusion of layers is studied.

4.5 Simulation of Spinodal Decomposition in Blends

Simulations of spinodal decomposition have used the model of Fig. 3, using [155, 295] in d = 3 dimensions local motions of bonds ("kink jumps", "crankshaft motions", "end rotations", see Fig. 16) that yield chain dynamics in accord with the Rouse model [139, 140] or using [296, 297] slithering-snake algorithms [298, 299] in d = 2 dimensions. The latter provides an unphysically fast single-chain diffusion, but it should yield a reasonable description of collective dynamics if time is suitably rescaled. Finally, a rather different approach is to carry out simulations on a rather coarse grained level., where one numerically integrates [300–302] Eqs. (72), (73) – or their real space counterparts. The latter approach has the distinct advantage that one easily can reach relatively late times and observe the scaling of the structure factor, Eq. (89). In fact, the simulations yield $q_m(t) \propto t^{-1/3}$ at late times, both for critical and for off-critical mixtures [300–302]. This type of power law behavior (with exponent x = 1/3 in the relation $q_m(t) \propto t^{-x}$) was first predicted by Lifshitz and Slyozov [177] for solid binary mixtures. Of course, as discussed in Sect. 2.4, Eqs. (72), (73) describe (nonlinear) diffusion phenomena only, the fluid character of the system (where pressure gradients can relax via fluid flow) is not included. Thus these equations clearly cannot reproduce the experimentally observed crossover towards the scaling behavior with x = 1 (Fig. 8a) predicted theoretically first by Siggia [167]. Since Eqs. (72, 73) incorporate effects specific for polymers only via the use of the Flory-Huggins free energy functional, Eq. (47), rather than the free energy functional of a small molecule mixture, Eq. (98), effects specific for polymers are included only rather indirectly (in particular, through the composition dependence of the gradient energy term). Thus the simulations of Refs. 300–302 really assume the validity of a generalized Flory-Huggins description from the start, and test the validity of this description at best indirectly. It thus is no surprise that also the "pinning phenomena" for off-critical quenches (Fig. 8b), which are attributed to the break-up of the interconnected (percolating) network of the segregated minority phase into isolated droplets has not been observed in these simulations.

Heermann and coworkers [296, 297] were the first to carry out simulations of spinodal decomposition in two space dimensions. In this case chains cannot penetrate into each other, so each chain can interact only with a few neighboring chains around it, and our discussion of the Ginzburg criterion (Sect. 2.5) implies that nonlinear phenomena are very important even during the early stages of the quench, and a stage where the structure factor increases exponentially fast with the time after the quench, as predicted by the Cahn-type theories [9–12, 78, 120, 129, 149] does not occur. The simulations [296, 297] are certainly compatible with these expectations, but a complete systematic study (where the chain length N is varied over a wide range and the associated equilibrium phase diagrams are determined as well) still remains to be done in $d = 2$ dimensions. Heermann et al. [296, 297] rather study the approach toward scaling behavior of the structure factor at later stages after the quench and find an approach towards Lifshitz-Slyozov type growth [177]. This is expected, since their lattice model – as do all Monte Carlo lattice models as well – lacks the hydrodynamical mechanism, and thus cannot reproduce the crossover towards faster mechanism of the type proposed by Siggia [167].

A particular advantage of the simulations in $d = 2$ is that one can straightforwardly analyze the morphology of the structure that evolves during the coarsening process (Fig. 36). This morphology looks strikingly similar to the

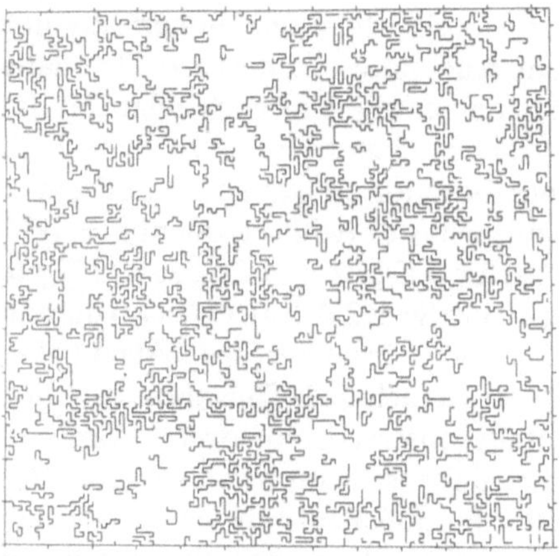
a

c

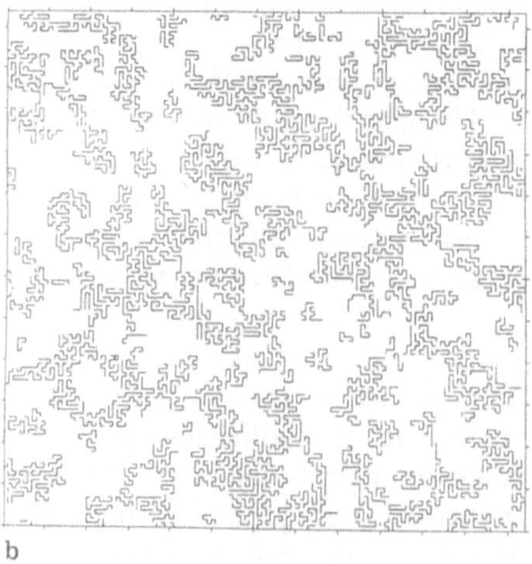
b

Fig. 36a–c. Configurations of a binary (AB) polymer blend in two space dimensions, containing 738 A-chains of chain length $N_A = 10$, 738 **B**-chains of the same chain length $N_B = 10$, and $\phi_v = 0.024$ vacant sites. A 123×123 square lattice with periodic boundary conditions is used, and a repulsive energy ε_{AB} between monomers of different kinds. The quench is carried out from $\varepsilon_{AB}/k_BT = 0$ as initial configuration ($t = 0$, part **a**) to $\varepsilon_{AB}/k_BT = 1.0$. Two times are shown after the quench: $t = 3750$ attempted Monte Carlo moves per vacancy (**b**) and $t = 44500$ such steps (**c**). Only one type of chain (e.g. the A-chain) is displayed, B-chain and vacancies are not shown. From Baumgärtner and Heermann [296]

morphology of small molecule mixtures, stressing again the universitality of the spinodal decomposition mechanism.

The simulations of Sariban and Binder [155, 295] have the advantage that they deal with a three-dimensional model of polymer mixtures, where the chains move with a Rouse-like dynamics (Figs. 3, 16), and hence, in principle, one can expect a physically realistic description of the very early stages of spinodal decomposition. Late stages would also suffer from the lack of hydrodynamic flow effects, as stressed above, but anyway are inaccessible in these simulations since the dynamics of the model is very slow. Also for their model phase diagrams are known from previous work [101, 107], as described in Sects. 4.2, 4.3, and two choices of the chain length were studied: $N = 8$, $N = 32$ [155]. However, these chain lengths are rather short and since the model contained a rather large fraction of vacancies ($\phi_v = 0.6$), it is no surprise that the mean-field critical regime where the linearized theory of spinodal decomposition would hold was not observed in this simulation either. As already shown in Fig. 7b, the resulting time evolution of the collective structure factor resembles corresponding experiments qualitatively (Fig. 7a), but a quantitative agreement should not be expected, of course. In order to be able to simulate such slow processes at all (note that the time scale of Fig. 7a is 1000 s, and the physical time corresponding to one Monte Carlo move per monomer in Fig. 16a may be as short as 10^{-11} s!), it clearly is crucial to work with much shorter chains and much more free volume than is physically realistic, and to consider much deeper quenches. As is clear from our discussion in Sect. 2.4, for shallow quenches (into the critical region) a critical slowing down of spinodal decomposition occurs as well. So the simulations described in Fig. 5b, 7b do not refer to a typical polymer blend of high molecular weight, but rather oligomers A, B in a common solvent, and in addition very deep quenches ($\chi/\chi_{crit} \geqq 2$) are considered. The choice of this limiting case clearly was crucial in order to make such a study feasible at all. As shown in Fig. 5, deep quenches for polymer mixtures are very interesting in their own right, since then phase separation starts at a characteristic wavelength $\lambda_m(t) = 2\pi/q_m(t)$ which is smaller than the coil size. In this case, the linear Cahn-type theory [78] predicts pronounced deviations from linearity in the "Cahn plot", $R(q)/q^2$ vs q^2 (Fig. 5a).

Another feature observed in these simulations of deep quenches with many vacancies was a segregation of vacancies, if one chooses a model with attractive interactions between monomers ($\varepsilon_{AA} = \varepsilon_{BB} = -\varepsilon$, $\varepsilon_{AB} = 0$). This phase separation between polymers and solvent shows up via the growth of the structure factor $S_\rho(q, t)$ measuring the polymer density fluctuations, Fig. 37,

$$S_\rho(\vec{q}, t) = \left\langle \left\{ \sum_J \exp(i\vec{q}.\vec{R}_J)[\phi_B^j(t) + \phi_A^j(t) - \langle \phi_B^j(t) + \phi_A^j(t) \rangle] \right\}^2 \right\rangle \Big/ L^3 . \quad (164)$$

Note that $\langle \phi_B^j(t) + \phi_A^j(t) \rangle = 1 - \phi_v$ is strictly constant here, of course. In the case where the model is purely repulsive $\varepsilon_{AA} = \varepsilon_{BB} = 0$, $\varepsilon_{AB} = \varepsilon$, on the other

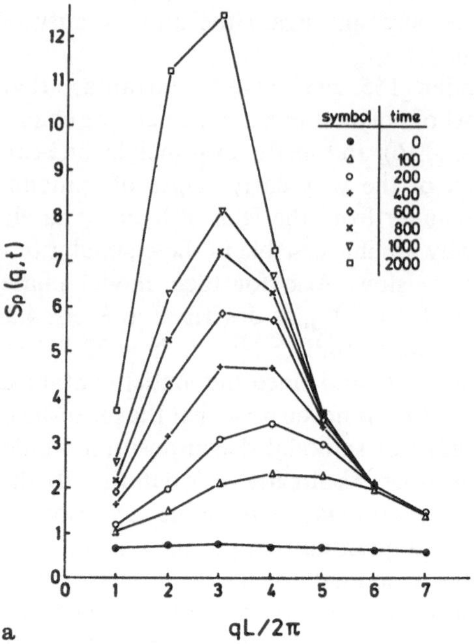

a

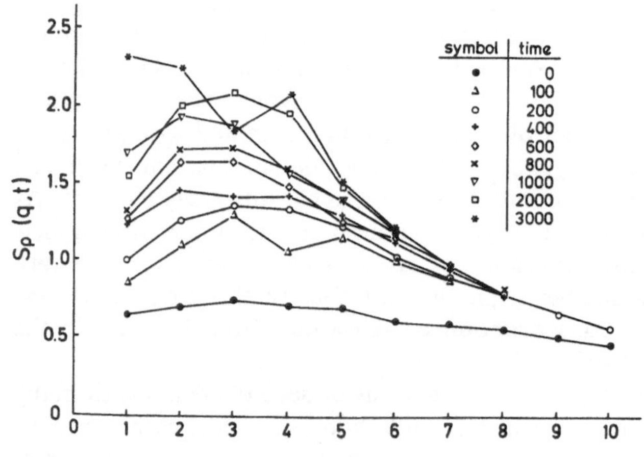

b

Fig. 37a, b. Collective structure function $S_\rho(q, t)$ describing the polymer density fluctuations of the model shown in Figs. 3, 16, for $\phi_v = 0.6$, $N = 32$, $\phi_A = \phi_B$, plotted vs wavenumber for various times after the quench. Case **a** corresponds to $\varepsilon/k_BT = 1.0$ and case **b** to $\varepsilon/k_BT = 0.3$, for $\varepsilon_{AB} = 0$, $\varepsilon_{AA} = \varepsilon_{BB} = -\varepsilon$. Note that a lattice size $L = 40$ with periodic boundary conditions is used. Since $S(\vec{q}, t)$ is defined only for the discrete values $q_v = (2\pi/L)\, v$, $v = 1, 2, 3$ (choosing $\vec{q}$ in lattice directions), the discrete points are connected by straight lines to guide the eye. From Sariban and Binder [155]

hand, the data rather suggest that the structure factor saturates at an Ornstein-Zernike behavior,

$$S_\rho(q, t \to \infty) \propto (1 + q^2\xi_\rho^2)^{-1} \ , \tag{165}$$

where ξ_ρ is a correlation length for density fluctuations that remains of the order of a few lattice spacings. Also for shallow quenches ($\varepsilon/k_BT = 0.3$) and attractive interactions a behavior as described in Eq. (165) seems to occur (Fig. 37b).

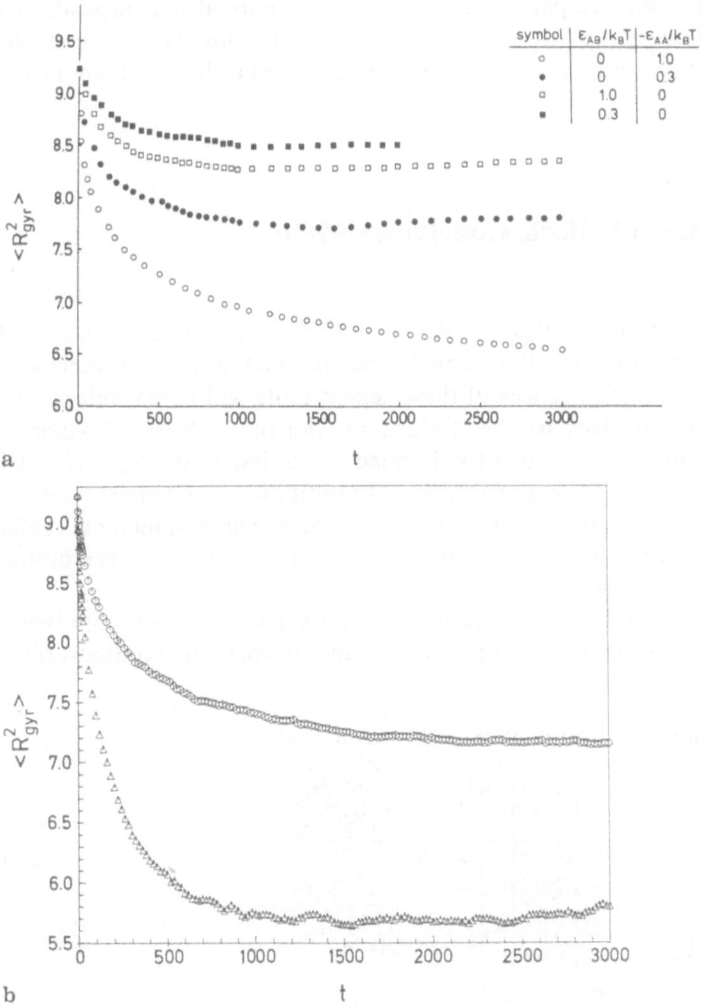

Fig. 38a, b. Mean square gyration radius plotted vs time after the quench for $N = 32$, $\phi_v = 0.6$, in quenching experiments of the model defined in Figs. 3, 16a). Case **a** shows various choices of energy parameters for $\phi_A = \phi_B$, case **b** shows a case of asymmetric concentrations for $\varepsilon/k_BT = 0.6$ and attractive interactions ($\varepsilon_{AA} = \varepsilon_{BB} = -\varepsilon$), namely $\phi_A = 0.32$ (*octagons* show the radii of this majority component) and $\phi_B = 0.08$ (*triangles* show the radii of this minority component). From Sariban and Binder [155]

In these deep quenches, one observes a variety of other interesting effects that have not been considered by theory so far: the tendency to avoid unfavorable contacts between chains as fast as possible leads to a chain contraction during very early stages after the quench (Fig. 38). This interpretation is corroborated by a direct study of the time-dependence of different types of nearest-neighbor contacts [155]. This reduction in coil size also leads to a decrease of the effective self-diffusion constant of the chains with time after the quench. We refer the reader to the original literature [155] for further details. It is quite clear, that with respect to simulations of spinodal decomposition of polymers only modest first steps could be taken, but this work already has yielded stimulating insight into various effects that need to be considered in the future.

5 Phase Behavior of Block Copolymer Melts

Copolymers are macromolecules consisting of different repeat units; here only two types (A,B) of monomers will be considered, and neither regular alternating (ABAB...) nor random sequences of these repeat units will be considered, but rather we restrict ourselves to the diblock architecture where a "block" of $N_A = fN$ A-monomers is covalently bonded to a block of $N_B = (1 - f)N$ B-monomers (Fig. 39a). Although triblock and multiblock copolymers have also been synthesized [39, 40], they are not considered here. The parameter f is called the "composition" of the diblock copolymer, and in the case $f = 1/2$ the diblock copolymer is called "symmetric".

As for polymer blends, the phenomenological theory of block copolymers attributes the thermodynamic properties to an interplay of configurational

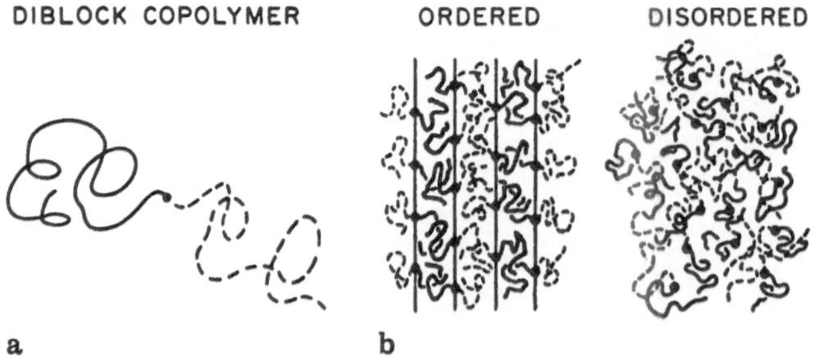

Fig. 39a. Chemical architecture of a diblock copolymer. A diblock copolymer consists of a polymerized sequence of A monomers (A-Block) covalently attached to a similar sequence of B-monomers. **b** The microphase separation transition occurs when a compositionally disordered melt of copolymers (*right side*) transforms to a spatially periodic compositionally inhomogeneous phase (*left side*) on lowering the temperature. For nearly symmetric copolymers (composition f near $f \approx 1/2$) the ordered phase has the lamellar structure shown. From Fredrickson and Binder [61]

entropy and enthalpic contributions, the latter being due to effective interactions between monomers, lumped into the Flory-Huggins χ parameter as in Sect. 2. While in binary blends a large enough (positive) χ parameter induces phase separation on a macroscopic scale, this is clearly prevented in block copolymer melts due to the covalent junction: only a local unmixing is possible, the system forms a concentration wave (Fig. 39b) or a periodic structure that can be viewed as a suitable superposition of such concentration waves (Fig. 40). Thus the number of energetically unfavorable pairwise interactions between monomers is reduced, and thus the melt reaches a state of lower free energy by "mesophase formation" – the fluid, disordered on the microscopic (atomistic) scale develops long range order on a "mesoscopic scale" (which is of order of the coil radius, e.g. a scale of 100 Å). The transition from the fully disordered melt (Fig. 39b, right part) to this locally unmixed state is hence called "mesophase transition" or "microphase separation transition" (MTS) or "order-disorder transition" (ODT). As will be discussed in Sect. 5.2, Leibler's [43] mean field theory predicts this transition to occur roughly for $\chi N \approx 10$ for $f = 1/2$.

Now two different limiting cases are distinguished: if N exceeds the critical value for this MST only slightly, the ordered state can be described essentially by a single sinewave (Fig. 41a). In this case one assumes the coil configurations to be essentially undisturbed gaussian coils. Since the gyration radius of both A-blocks and B-blocks then is of the order of $R_g \approx aN^{1/2}/6$, a being the length of a characteristic segment, we conclude that also the wavelength of the sinusoidal variation in Fig. 41a must be of this order $\chi^* = 2\pi/q^* \propto N^{1/2}$. On the other hand, if N exceeds by far the critical value for microphase separation, we have "strong segregation" (Fig. 41b): the A-rich and B-rich regions are separated by fairly sharp interfaces, to reduce the number of unfavorable A-B contacts as much as possible. In this limit the chain configurations are no longer ordinary random walks, but rather somewhat stretched in the direction of the modulation, $\lambda^* \propto R_g \propto N^{2/3}$ (Sect. 5.1). After a brief discussion of this strong segregation limit (SSL) in the next section, we shall review Leibler's theory [43] of the ODT in Sect. 5.2, and discuss various extensions (including fluctuation effects) in Sect.

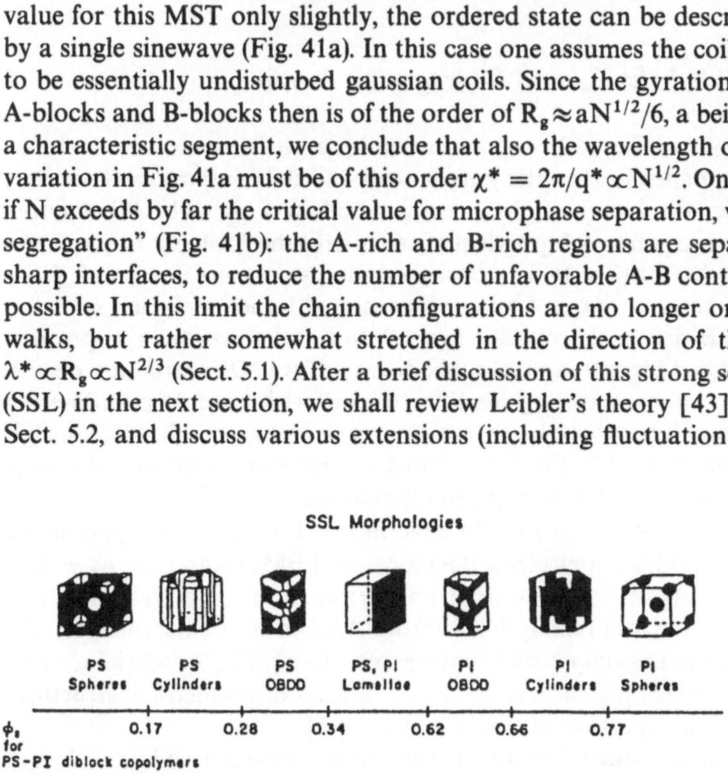

Fig. 40. Equilibrium morphologies of A-B diblock copolymers, and the composition region where they are predicted to exist in the strong segregation limit ($\chi N \to \infty$). The quoted values of polyisoprene (PI) composition refer to PI-PS (polystyrene) diblocks. From Bates and Fredrickson [39]

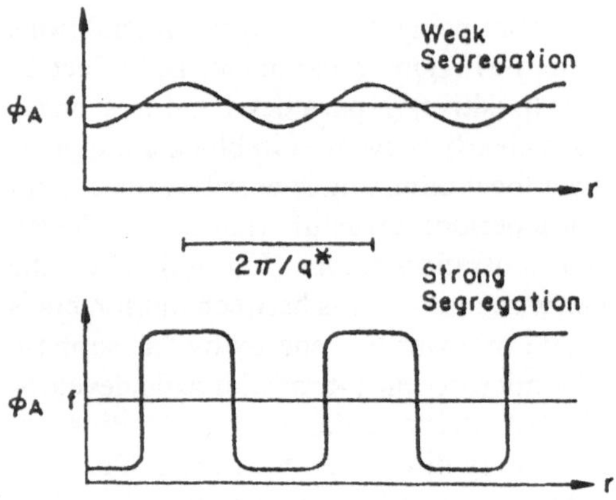

Fig. 41. Concentration profiles for the lamellar phase in the "weak segregation limit" and in the "strong segregation limit". From Bates and Fredrickson [39]

5.3. Section 5.3 is then devoted to a comparison with various Monte Carlo computer simulations to test these predictions, and also discuss the experimental evidence.

5.1 Overview and the Strong Segregation Regime

As already shown in Fig. 40, various mesophase structures occur in block copolymer melts: depending on the χ parameter and the composition f, we may have (apart from the lamellar phase emphasized in Figs. 39b, 41) a body-centered cubic arrangement of spherical micelles with the A-blocks in the center while the B-blocks form a B-rich background (this occurs for small f, i.e. $N_A \ll N_B$); for somewhat larger f a two-dimensional lattice of A-rich spherical cylinders on the B-rich background is formed. In a rather narrow composition interval the "ordered bicontinuous double diamond" (OBDD) structure occurs, consisting of two interpenetrating rigid "networks" with fourfold coordination like in the diamond lattice, one being A-rich and the other B-rich. The lamellar phase then occurs for $f \approx 1/2$. For $f > 1/2$ one has the same sequence of phases again, but now the roles of A and B are interchanged.

All the structures shown in Fig. 40 have in fact been seen in experimental work applying transmission electron microscopic (TEM) techniques as well as small angle scattering of X rays or neutrons. While most structures have been known for a long time and readily follow from theoretical predictions [41, 42], the OBDD structure has only been found recently (see [303] for a brief review). This structure is interesting since it belongs to a class of geometrical structures with constant mean curvature of the interface [303].

In the SSL, the structure (Figs. 40, 41b) always is characterized by two length scales: e.g. in the lamellar structure we have the period of the lamella D, and the width δ of the interface between the blocks. Of course, the period D experimentally is deduced from the position of the Bragg peaks (observation of higher order Bragg reflections elucidates the type of ordered phase, of course, and the

Bragg intensities yield information on the degree to which segregation locally is achieved). The information on the interface thickness δ can be extracted from observations at large scattering vectors $\vec{q}$, where the intensity is predicted [304] to vary as

$$I(q) \propto q^{-4} \exp(-q^2\delta^2/2\pi) . \tag{166}$$

The prefactor q^4 in Eq. (166) describes the scattering from sharp interfaces ("Porods law" $I(q) \propto q^{-4}$ [182]). In the strong segregation limit, the interfacial profile between unmixed phases (see also Sect. 2.3) is obtained as [132, 305–313]

$$\phi(z) = \tanh((z - z_0)/\delta) \tag{167}$$

where $\phi(z)$ is the volume fraction of A across an interface oriented perpendicularly to the z-direction and centered at $z = z_0$. Incorporation of the profile, Eq. (167), into the description of the scattering yields Eq. (166). The interfacial thickness δ in this limit is controlled by the χ parameter only [305–313], see also Sect. 2.3

$$\delta \approx \sqrt{1/6}\, a\chi^{-1/2} \tag{168}$$

where we have assumed the simple special case that both A and B blocks have the same segment length a. Using experimentally values of χ in Eq. (168) yields for the systems of interest typical values of $\delta \approx 20$ Å, and this is compatible with results extracted from scattering data via Eq. (166).

While δ hence does not depend on the chain length N, the period D depends on both N and χ. The theoretical analysis of this problem [41, 42, 56, 57, 59, 313–316] is rather complicated and requires somewhat restrictive assumptions, such as the "self-consistent field" – approach [56] for polymers. We shall not describe these theories here, but restrict ourselves to a scaling-type plausibility argument, using the fact that the interfacial tension between fully segregated phases is also depending on the χ parameter only but not on chain length [305–313]

$$f_{int} \propto \chi^{1/2} a^{-2} . \tag{169}$$

We now consider the free energy per chain for a lamellar structure of period D, splitting it into contributions due to the A-B interface and due to the interior of the domains. The interfacial contribution is given by the product of f_{int} and the interfacial area per chain, the latter being Na^3/D: Multiplying this area with the domain thickness D must give N times the volume a^3 of a statistical segment. In this argument, constants of order unity are neglected throughout, of course. Thus the total interfacial free energy is $f_{int}/k_BT \propto Na\chi^{1/2}/D$.

Now let us assume that the contribution to the free energy per chain in the domain has the same variation with the linear dimension D as in a free gaussian chain, namely a quadratic variation: $F_{domain}/k_BT \propto D^2/a^2N$. The selfconsistent field theory [56] shows that the chains actually are stretched *nonuniformly* in the domains, and also one finds a nonuniform distribution of chain ends, the latter are found more frequently near the domain centers rather than near the A-B interfaces, where the junction points are concentrated. However, all these effects only affect the (disregarded) prefactor, but not the exponents, of the Flory-type laws that we are going to derive. As always with Flory-type arguments, one has to seek a minimum of the total free energy per chain, varying D:

$$\frac{\partial}{\partial D}\left(\frac{F_{int} + F_{domain}}{k_B T}\right) \propto \frac{\partial}{\partial D}\left(\frac{Na\chi^{1/2}}{D} + \frac{D}{a^2 N}\right) = 0 . \tag{170}$$

This yields $-Na\chi^{1/2}/D^2 + 2D/a^2N = 0$, i.e.

$$D \propto a\chi^{1/6}N^{2/3} . \tag{171}$$

Thus the chains deviate strongly from simple gaussian statistics, since the gyration radius in the direction perpendicular to the interfaces must have the same order as D: the chains are considerably stretched.

A more complete theory must compute the prefactors of the above free energy contributions, of course, in order to allow a comparison between the free energies of the various ordered structures that compete with each other, and find the structure at given χ, N, f that yields the minimum free energy. In this way, the theory has yielded predictions for the compositions f at which first-order phase transitions from one structure to the other occur [39–42, 56, 57, 59, 316]. These predictions seem to be roughly compatible with the available experimental observations, although more work clearly is desirable. Just as the OBDD structure first was overlooked and only recently discovered [303] and included in the theoretical discussion [316], there may also exist the need to treat still other structures [64] than hitherto considered. A parameter that may prove important is asymmetry between the blocks (different persistence lengths, or sizes of the effective subunits, for instance). Clearly a different behavior is to be expected if one block is rather rigid and the other rather flexible. Such phenomena are still rather unknown and will not be discussed here.

5.2 Summary of Leibler's RPA Theory of the Microphase Separation Transition

In a seminal paper, Leibler [43] presented the first mean-field-like theory of the ODT transition and the phase diagram of block copolymer melts in the weak segregation limit. This work still is the basis for more elaborate theories [58–64] and for the discussion of recent experiments (e.g. [317–323]). As shown in Fig. 42, the quantitative details of the resulting predictions are still subject of current research, but nevertheless we try to sketch this theory here, since this derivation gives a good insight into the relevant physical aspects of this problem.

Again we assume both blocks have the same statistical segment length a, and that the melt is incompressible. The relevant variables then are local (relative)

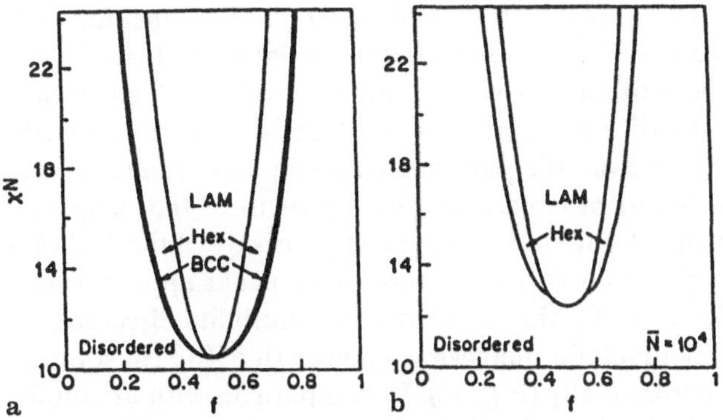

Fig. 42. Theoretical phase diagram for diblock copolymers in the weak segregation limit. The *left side* shows the mean field result of Leibler [43], the *right side* the theory of Fredrickson and Helfand [58] which includes fluctuation corrections, for an effective degree of polymerization $\bar{N} = 10^4$. LAM, Hex, BCC denote the various mesophases: lamellar, hexagonal (i.e., cylindrical morphology) and body-centered cubic (i.e., spherical micellar morphology). From Bates and Fredrickson (39)

densities $\rho_A(\vec{r})$, $\rho_B(\vec{r})$ of A, B monomers at position $\vec{r}$, measured in units of the constant melt density. Thus $\rho_A(\vec{r}) + \rho_B(\vec{r}) = 1$, and since in the disordered phase the system is spatially uniform, we also have

$$\langle \rho_A(\vec{r}) \rangle = f, \langle \rho_B(\vec{r}) \rangle = 1 - f . \tag{172}$$

The interaction energy density

$$u(\vec{r}) = k_B T \chi \rho_A(\vec{r})\, \rho_B(\vec{r}) . \tag{173}$$

favors for $\chi > 0$ an inhomogeneous state: in the A-rich domains we have $\langle \rho_B(\vec{r}) \rangle \ll 1 - f$, in the B-rich domains $\langle \rho_A(\vec{r}) \rangle \ll f$, and hence at all positions $\vec{r}$ (apart from interfaces) the product $\rho_A(\vec{r})\, \rho_B(\vec{r})$ in Eq. (173) is smaller than the factor $f(1-f)$, applicable for random mixing, Eq. (172), and hence the free energy gets reduced for large χ due to mesophase formation (for large enough χ the enthalpy gain, Eq. (173), overwhelms the loss of configurational entropy of mixing).

This consideration suggests to introduce an order parameter field $\psi(\vec{r})$ which describes the deviations of the local densities $\rho_A(\vec{r})$, $\rho_B(\vec{r})$ from their averages, Eq. (172): with $\delta\rho_A(\vec{r}) \equiv \rho_A(\vec{r}) - f = \psi(\vec{r})$ we have

$$\langle \psi(\vec{r}) \rangle = \langle (1-f)\, \rho_A(\vec{r}) - f\rho_B(\vec{r}) \rangle = \langle \rho_A(\vec{r}) - f \rangle = \langle \delta\rho_A(\vec{r}) \rangle . \tag{174}$$

It is of interest to consider the correlation function of the local order parameter,

$$\tilde{S}(\vec{r} - \vec{r}\,') = \frac{1}{k_B T} \langle \delta\rho_A(\vec{r})\, \delta\rho_B(\vec{r}) \rangle . \tag{175}$$

In the disordered phase $\langle \psi(\vec{r}) \rangle = \langle \delta\rho_A(\vec{r}) \rangle = 0$, but the correlation function

$\tilde{S}$ will get long ranged if we approach a second order phase transition. The Fourier transform $S(\vec{q})$ of $\tilde{S}(\vec{r} - \vec{r}')$ is proportional to the intensity $I(q)$ of X-ray or neutron scattering. The situation is somewhat reminiscent of an antiferromagnet, where the local order parameter is the local density of magnetic moments: the phase transition from the paramagnet to the long range ordered structure is announced in the disordered phase by a growth of the "staggered susceptibility" (which is proportional to $S(\vec{q})$) at such points $\vec{q}^*$ of the Brillouin zone, where in the ordered phase the superstructure Bragg peaks appear. Just as in the antiferromagnet for $T \to T_N$ the staggered susceptibility diverges, we expect in the block copolymer melt for increasing χ a growth of $S(\vec{q})$ at some q^*; indeed this is observed experimentally (Fig. 43). In comparison with an antiferromagnet, two qualitative distinctions should be emphasized: the antiferromagnet is a solid with a perfect crystal lattice, for which the order is characterized by discrete points in $\vec{q}$-space, corresponding to the symmetry of the reciprocal lattice (e.g. superstructure Bragg spots occur at special points at the boundary of the Brillouin zone, depending on the type of the antiferromagnetic structure). The block copolymer melt is a fluid phase, of course, hence the disordered phase is fully isotropic and there is no preferred direction in $\vec{q}$ space. Hence the instability of the system associated with the microphase separation transition, which shows up in the divergence of $S(q^*)$, does not occur at special points $\vec{q}^*$ in $\vec{q}$-space but on a whole *sphere*, defined by $|\vec{q}| = q^*$. Also the value of q^* is not fixed by any symmetry, unlike the antiferromagnet where the symmetry requires the peak to appear at the Brillouin zone boundary. Therefore we expect that the value of q^* where the spherically averaged structure factor $S(q)$ has its peak will depend on all relevant parameters (N, f, χ) in a nontrivial way. In addition, we usually expect a phase transition of first order, as will be justified below. The temperature defined by $[S(q^*)]^{-1} = 0$ then is not the temperature T_{MST} where the phase transition occurs, but rather has the meaning of a spinodal: in mean field theory, this is the limit of metastability for disordered block copolymer melts in the region of the ordered phase, similar to the spinodal curve of polymer mixtures (which is defined by $[S(q = 0)]^{-1} = 0$, cf Eqs. (5, 35)). At T_{MST} we then have $[S(q^*)]^{-1} > 0$, see Fig. 43b. In this case the character of the instability does not determine the nature of the order that appears discontinuously at T_{MST}: the order corresponding to $|\vec{q}| = q^*$ would always be lamellas with $\lambda^* = 2\pi/q^*$ being the period of the structures, while the orientation of the lamellar in space would be arbitrary. However, to characterize a first order phase transition one has to find the phase of minimum free energy, and this may be a different structure, see Fig. 42.

For treating such problems one wishes to develop a Landau-like theory, where the free energy is expanded in powers of the order parameter ψ [43, 123, 324]. It is useful to introduce the field $U(\vec{r})$ conjugate to the order parameter and carry out a Legendre transformation to relate the free energy $F(\psi)$ to a thermodynamic potential $F'(U)$,

$$F(\psi) = F'(U) - \int U(\vec{r})\, \psi(\vec{r})\, d^3\vec{r}, \quad F' = -k_B T \ln Z\{[U(\vec{r})]\}\,. \tag{176}$$

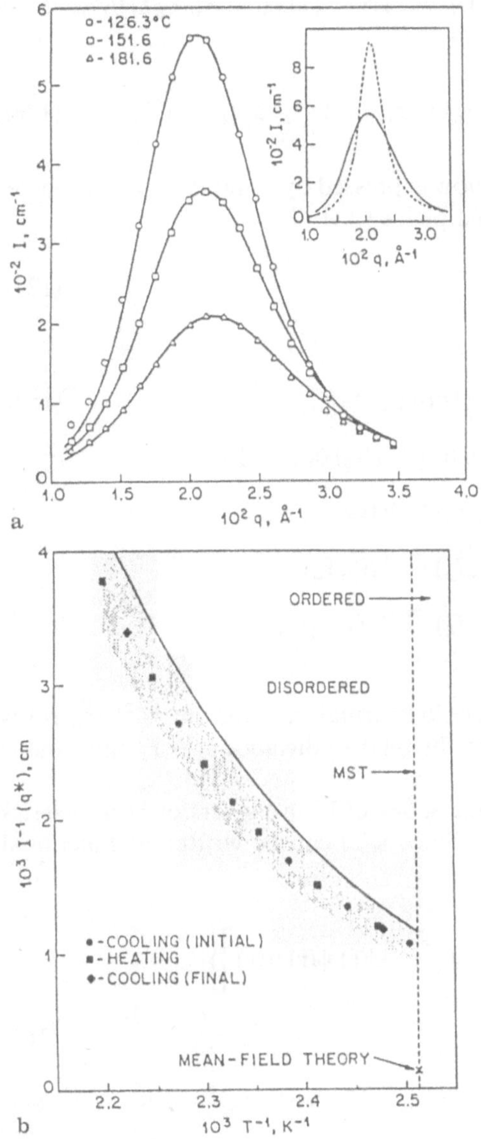

Fig. 43a. Neutron small angle scattering intensity I(q) plotted vs q for three temperatures T above T_{MST} (*main graph*), for a polyethylenepropylene(PEP) – polyethylethylene(PEE) diblock copolymer, with $f = 0.55$, molecular weight $M_W = 57.500$, polydispersity index $M_W/M_N = 1.05$. The microphase separation transition occurs for $T_{MST} = 125\,°C$. For further explanations cf: Text.**b** Inverse peak intensity $I^{-1}(q^*)$ plotted vs inverse temperature.The full curve is a one-parameter fit to the theory of Fredrickson and Helfand [58], while Leibler's [43] prediction for the intensity at the transition is marked as"mean field theory". From Bates et al. [317]

Here the partition function Z is written in terms of the Hamiltonian $\mathscr{H}_0$ (for $U \equiv 0$) as

$$Z = \operatorname{Tr} \exp\left[-\frac{1}{k_B T}\left(\mathscr{H}_0 + \int U(\vec{r})\, \psi(\vec{r})\, d^3\vec{r}\right)\right]$$

$$= Z_0 \left\langle \exp\left[-\frac{1}{k_B T}\left(\mathscr{H}_0 + \int U(\vec{r})\, \psi(\vec{r})\, d^3\vec{r}\right)\right]\right\rangle_0 , \qquad (177)$$

with $Z_0 = \mathrm{Tr}\exp(-\mathscr{H}_0/k_BT)$, $\langle(\dots)\rangle_0 = [\mathrm{Tr}(\dots)\exp(-\mathscr{H}_0/k_BT)]/Z_0$. Now the functional F′ is expanded in powers of U,

$$\frac{F'}{k_BT} = \frac{F_0}{k_BT} + \sum_{n=1}^{\infty} \frac{1}{n!} \frac{1}{(k_BT)^n} \int \tilde{G}(\vec{r}_1, \dots, \vec{r}_n)\, U(\vec{r}_1) \dots U(\vec{r}_n)\, d^3\vec{r}_1 \dots d^3\vec{r}_n \,. \tag{178}$$

The coefficients of this expansion are now expressed by suitable order parameter correlation functions of the disordered phase [43]:

$$\tilde{G}^{(1)}(\vec{r}_1) = \langle \delta\rho_A(\vec{r}_1)\rangle_0 = 0 \,, \tag{179}$$

$$\begin{aligned} \tilde{G}^{(2)}(\vec{r}_1, \vec{r}_2) &= \langle \delta\rho_A(\vec{r}_1)\, \delta\rho_A(\vec{r}_2)\rangle_0 \,, \\ \tilde{G}^{(3)}(\vec{r}_1, \vec{r}_2, \vec{r}_3) &= \langle \delta\rho_A(\vec{r}_1)\, \delta\rho_A(\vec{r}_2)\, \delta\rho_A(\vec{r}_3)\rangle_0 \,, \\ \tilde{G}^{(4)}(\vec{r}_1, \vec{r}_2, \vec{r}_3, \vec{r}_4) &= \langle \delta\rho_A(\vec{r}_1)\, \delta\rho_A(\vec{r}_2)\, \delta\rho_A(\vec{r}_3)\, \delta\rho_A(\vec{r}_4)\rangle_0 \\ &\quad - \tilde{G}^{(2)}(\vec{r}_1, \vec{r}_2)\, \tilde{G}^{(2)}(\vec{r}_3, \vec{r}_4) \\ &\quad - \tilde{G}^{(2)}(\vec{r}_1, \vec{r}_3)\, \tilde{G}^{(2)}(\vec{r}_2, \vec{r}_4) \\ &\quad - \tilde{G}^{(2)}(\vec{r}_1, \vec{r}_4)\, \tilde{G}^{(2)}(\vec{r}_2, \vec{r}_3) \,, \end{aligned} \tag{180}$$

etc.

Of course, in the disordered phase in thermal equilibrium $\tilde{G}^{(2)}(\vec{r}_1, \vec{r}_2)$ is translationally invariant, it depends on the relative distance $\vec{r}_2 - \vec{r}_1$ only, and is trivially related to $\tilde{S}(\vec{r} - \vec{r}')$, Eq. (175).

Finally one wishes to transform the series of F′ in powers of U to a series where F is expressed in powers of ψ. Since $\psi(\vec{r})$ can be written as functional derivative of F with respect to U,

$$\begin{aligned} \psi(\vec{r}) &= \tfrac{1}{Z}\mathrm{Tr}\left\{\delta\rho_A(\vec{r}) \exp\left[-\frac{1}{k_BT}\left(\mathscr{H}_0 + \int U(\vec{r})\, \psi(\vec{r})\, d^3\vec{r}\right)\right]\right\} \\ &= \delta F/\delta U(\vec{r}) \,, \end{aligned} \tag{181}$$

we obtain from Eqs. (178), (181)

$$\begin{aligned} \psi(\vec{r}) = \sum_{n=0}^{\infty} \frac{1}{n!} \frac{1}{(k_BT)^n} \tilde{G}^{(n+1)} \\ (\vec{r}, \vec{r}_1, \dots, \vec{r}_n)\, U(\vec{r}_1) \dots U(\vec{r}_n)\, d^3\vec{r}_1 \dots, d^3\vec{r}_n \,. \end{aligned} \tag{182}$$

For making use of the translational invariance of the correlations defined in Eqs. (179), (180), it is convenient to work with Fourier transforms: $O(\vec{q}) = \int O(\vec{r}) \exp(i\vec{q}\cdot\vec{r})\, d^3\vec{r})$ is the Fourier transform of a function $O(\vec{r})$. Thus (V is the system volume)

$$\psi(\vec{r}) = \sum_{n=1}^{\infty} \frac{1}{(n-1)!} \frac{1}{(k_B T)^{n-1}} V^{-n}$$

$$\sum_{\vec{q}_2, \ldots, \vec{q}_n} \delta(\vec{q}_1 + \vec{q}_2 + \ldots + \vec{q}_n)\, \tilde{G}^{(n)}(\vec{q}_1, \vec{q}_2, \ldots, \vec{q}_n)\, V(\vec{q}_1) \ldots, V(\vec{q}_n) \tag{183}$$

The δ-function in Eq. (183) thus expresses translational invariance. Integrating Eq. (181) with respect to U and using Eq. (183) finally yields the desired expansion

$$\frac{F}{k_B T} = \frac{F_0}{k_B T} + \sum_{n=2}^{\infty} \frac{1}{n!} \frac{1}{V^n} \sum_{\vec{q}_2, \ldots, \vec{q}_n} \Gamma_n(q_1, q_2, \ldots, q_n)\, \psi(\vec{q}_1) \ldots \psi(\vec{q}_n) \ . \tag{184}$$

Here the coefficients $\Gamma_n(\vec{q}_1, \vec{q}_2, \ldots, \vec{q}_n)$ are expressed in terms of the Fourier transforms of the $\tilde{G}^{(n)}$ as follows,

$$\Gamma_2(\vec{q}_1, \vec{q}_2) = [\tilde{G}^{(2)}(\vec{q}_1, \vec{q}_2)]^{-1} ,$$

$$\Gamma_3(\vec{q}_1, \vec{q}_2, \vec{q}_3) = -\tilde{G}^{(3)}(\vec{q}_1, \vec{q}_2, \vec{q}_3) / \prod_{i=1}^{3} \tilde{G}^{(2)}(\vec{q}_i, -\vec{q}_i), \ \text{etc.} \tag{185}$$

So far the development of the theory is very general – the fact that we deal with polymers has not yet been used. Now the important point is that the RPA [2, 114] allows an explicit calculation of the Γ_n's, using gaussian chain statistics. The central assumption that the interactions, Eq. (173), do not affect the coil configurations, will be critically examined in the light of computer simulation results in Sect. 5.4. We shall not give the details of this RPA calculation here, but just quote its most relevant results [43]:

$$\frac{1}{V}\Gamma_2(\vec{q}_1, \vec{q}_2) = \delta(\vec{q}_1 + \vec{q}_2)\, \tilde{S}^{-1}(\vec{q}_1) , \tag{186}$$

$\tilde{S}(\vec{q})$ being the Fourier transform of the correlation $\tilde{S}(\vec{r} - \vec{r}')$ defined in Eq. (175). It is explicitly found as

$$\tilde{S}(\vec{q}) = W(\vec{q}) / [S(\vec{q}) - 2\chi W(\vec{q})] , \tag{187}$$

where $W(\vec{q})$, $S(\vec{q})$ can be further expressed by partial structure factors $S_{11}(\vec{q})$, $S_{22}(\vec{q})$ of the A-blocks (or B-blocks), respectively, and the "mixed" structure function of a single chain,

$$S_{12}(\vec{q}) \left\{ S_{11}(\vec{q}) \equiv N^{-1} \sum_{i=1}^{fN} \sum_{j=1}^{fN} \langle \exp[-i\vec{q} \cdot (\vec{r}_i - \vec{r}_j)] \rangle \ \text{etc.} \right\}$$

$$S(\vec{q}) = S_{11}(\vec{q}) + 2S_{12}(\vec{q}) + S_{22}(\vec{q}), \quad W(\vec{q}) = S_{11}(\vec{q})\, S_{22}(\vec{q}) - S_{12}(\vec{q}) \tag{188}$$

$$S_{11}(\vec{q}) = N g_D(x, f), \quad x = N q^2 a^2/6 = N R_g^2, \quad S_{22}(\vec{q}) = N g_D(1 - f, x) , \tag{189}$$

with $g_D(x, f)$ denoting the well-known Debye function,

$$g_D(x, f) = 2[fx + \exp(-fx) - 1]/x^2 . \tag{190}$$

The "mixed" structure function is

$$S_{12}(\vec{q}) = \frac{1}{N}\sum_{i=1}^{fN}\sum_{j=fN+1}^{N} \langle \exp[i\vec{q})\cdot(\vec{r}_i - \vec{r}_j)] \rangle$$

$$= N[g_D(1, x) - g_D(f, x) - g_D(1-f, x)]/2 . \tag{191}$$

Also the higher order terms such as $\Gamma_3(\vec{q}_1, \vec{q}_2, \vec{q}_3)$ can be worked out in detail, thanks to the Gaussian statistics all higher probabilities can be reduced to two particle probabilities. We shall not write them down here, but rather discuss the general results that follow from this Landau expansion.

In the disordered phase terms higher than second order can be neglected, and hence

$$F/k_BT = F_0/k_BT + 1/(2V)\sum_{\vec{q}} \tilde{S}^{-1}(\vec{q})\, \psi(\vec{q})\, \psi(-\vec{q}) . \tag{192}$$

This result demonstrates that $\tilde{S}(\vec{q})$, indeed plays the role of a collective structure factor for the order parameter – for a ferromagnet we would have $F = F_0 + \frac{1}{2}\chi^{-1}M^2$, χ being the susceptibility and M the magnetization; for an antiferromagnet, $F = F_0 + \frac{1}{2}\chi^{-1}(\vec{q}_0)|M(\vec{q}_0)|^2$, $\chi(\vec{q}_0)$ being the staggered susceptibility, $M(\vec{q}_0)$ is the staggered magnetization; etc. Of course, Eq. (192) still contains a sum over all Fourier components, only some of them correspond to the order parameter: these are exactly those modes $\psi(\vec{q}^*)$ for which the coefficient $\chi^{-1}(q^*)$ changes its sign first, when χ increases. At the point χ_s where $\chi^{-1}(q^*) = 0$ $\{\psi(\vec{q}^*)\}$ then exhibit critical fluctuations, while all other modes $\psi(\vec{q})$ with $|\vec{q}| \neq q^*$ have still ordinary Gaussian fluctuations, and hence are irrelevant uncritical variables in the framework of a Landau theory: they can be simply integrated out, just as other degrees of freedom that were left out from the start. Of course, this discussion needs to be reexamined when we consider fluctuation corrections (Sect. 5.3). One can also show that only the coefficient of the quadratic term $|\psi(\vec{q})|^2$ can change its sign when χ is increased, while higher order terms are independent of χ: this is true because the local enthalpy, Eq. (173), is simply quadratic in the local order parameter. All terms of higher order than quadratic hence are purely entropic in origin: since we assume undisturbed Gaussian chain conformations, which do not depend on χ, these higher order terms must be independent from χ.

To find the order parameter we thus look for the minimum q^* of $\tilde{S}^{-1}(\vec{q})$,

$$0 = \frac{d}{dq}\tilde{S}^{-1}(q) = \frac{d}{dq}\left(\frac{S(q)}{W(q)}\right), \tag{193}$$

using the fact that $\tilde{S}(\vec{q})$, $S(\vec{q})$, $W(\vec{q})$, do not depend on the direction of $\vec{q}$, and

Eq. (187). Hence it is implied that the solution (q^*) of Eq. (193) does not depend on χ! E.g., for symmetrical block copolymers one finds

$$q^* R_g \cong 1.945 \,, \tag{194}$$

and the critical condition where $\tilde{S}(q^*)$ diverges is given by

$$\chi_s N = 10.495 \,. \tag{195}$$

Experimental data (Fig. 43a) indeed are fit by the structure factor of Leibler's theory [43] nicely - the curves shown in this figure from a convolution of Eq. (187) with the resolution function, as shown in the insert (the dashed curve is Eq. (187), the full curve is result of the convolution, for T = 126.3 °C). However, at each temperature both $R_g(T)$ and $\chi(T)$ are used as adjustable parameters – therefore a nearly perfect fit is possible although the peaks at the different temperatures do not occur at precisely the same q^*. Thus the agreement shown in Eq. (187) should not be taken as a proof for the accuracy of Leibler's theory - such a *proof would require an independent measurement of* $R_g(T)$ *and* $\chi(T)$.

A more distinctive test of Leibler's theory has been possible via Monte Carlo simulations [325, 326], as will be discussed in detail in Sect. 5.4, though these simulations are restricted to very short chains only (modelled by self-avoiding walks with N = 16 to N = 60 steps on the simple cubic lattice). We here only anticipate one example (Fig. 44) that also the simulation results can be adjusted perfectly to Leibler's theory if one treats R_g (in Eq. (189) as an effective parameter

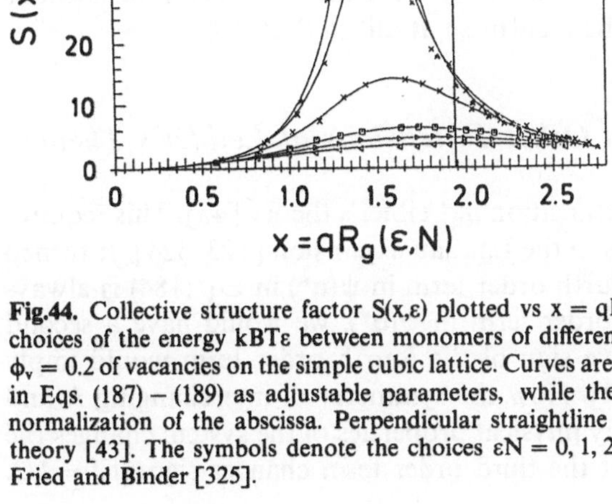

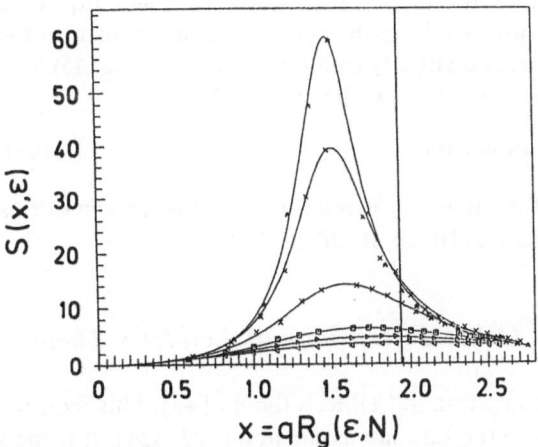

Fig.44. Collective structure factor S(x,ε) plotted vs x = qRg(ε,N) for f = 1/2, N = 20 and various choices of the energy kBTε between monomers of different kinds, allowing for a volume fraction ϕ_v = 0.2 of vacancies on the simple cubic lattice. Curves are a fit to Eq. (187), treating both χ and Rg in Eqs. (187) – (189) as adjustable parameters, while the actual gyration radius is used for the normalization of the abscissa. Perpendicular straightline shows the value x* = 1.945of Leibler's theory [43]. The symbols denote the choices εN = 0, 1, 2, 3, 4 and6 (from bottom to top). From Fried and Binder [325].

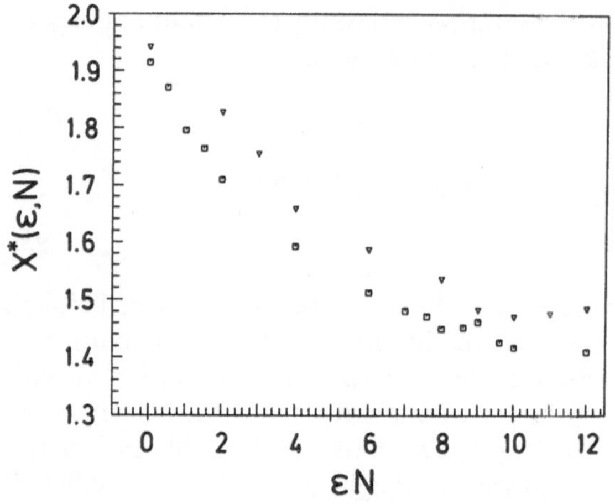

Fig. 45. Normalized position of the structure factor maximum, $x^*(\varepsilon,N) = q^*Rg(\varepsilon,N)$, plotted vs εN for $N - 20$ (squares) and $N = 40$ (triangles), for the model of block copolymers with only repulsive interactions between nearest neighbor A, B-pairs on the simple cubic lattice with $\phi_v = 0.2$. Note that the order-disorder transition is estimated to occur for $\varepsilon N \approx 7$–8 in this Monte Carlo simulation (due to finite size effects and equilibration problems for $\varepsilon N \gtrsim 6$ no more accurate estimation was possible). From Fried and Binder [325].

that can be adjusted by the fit. However, the actual gyration radius $R_g(\varepsilon, N)$ can be determined in the simulation readily and agrees with the effective R_g from the fit only if the interaction parameter ε is very small. On the scale of $x = qR_g(\varepsilon, N)$ the maximum of the structure function occurs at the value $x^* = 1.945$, Eq. (194), only for $\varepsilon \to 0$ (i.e. $\chi \to 0$), while approaching the order-disorder transition the maximum position is shifted towards distinctly smaller values of x (Fig. 45). It is seen that near the order disorder transition we have (Fig. 45)

$$x^*(\varepsilon, N) = q^*R_g(\varepsilon, N) \approx 1.45 \pm 0.05 . \tag{196}$$

indicating that for $\chi N \gg 1$ the RPA no longer is reliable. We shall return to this result (which really should not be a surprise at all) in Sect. 5.4.

5.3 Fluctuation Effects and Other Extensions of Leibler's Theory

We first discuss the order of the transition in Leibler's theory [43]. This requires a study of the higher order terms in the Landau expansion [123, 324]. It turned out that the coefficient of the fourth order term in $\psi(q^*)$ in Eq. (184) is always positive: if there were no third order term in $\psi(q^*)$, we would have a second order transition, while a negative sign of the fourth order term would imply a first order transition [123, 324]. Now the symmetry of interchanging A and B at $f = 1/2$ without changing any physical properties of the system changes the sign of $\psi(q^*)$, which implies that the third order term changes sign at $f = 1/2$.

Hence the Leibler theory [43] indeed predicts a second order transition for $f = 1/2$, while for $f \neq 1/2$ where the third order term is present, a first order transition is predicted, $S^{-1}(q^*) = 0$ then only yields the limit of metastability of the disordered phase ("spinodal curve"). Thus using the higher order terms in Eq. (184) to actually compute the free energies of various "candidates" for the ordered structure, one finds which phase has the lowest free energy, and in this way the phase diagram shown in Fig. 42 (left part) has resulted [43].

However, for all second order phase transitions it is well-known that one must pay attention to fluctuations of the order parameter – taking them into account often invalidates Landau-like theories [74]. This also happens here (Fig. 42, right part): it turns out that no longer any second order transition occurs at all, rather we encounter a "fluctuation-induced first order transition" [58, 327].

Here we wish to work out only the most essential features and physical content of these fluctuation effects. For this purpose we restrict attention to $f = 1/2$ and consider the analog of Eq. (184) in real space, treating only $\psi(\vec{q})$ with $|\vec{q}|$ near q^*. Omitting F_0, the effective "Hamiltonian" of the system reads

$$\frac{1}{k_BT}\mathscr{H}\{\psi(\vec{x})\} = \int d^3\vec{x}\left\{\frac{1}{2}\psi[\tau_0 + e_0(\nabla^2 + q^{*2})]\psi + \frac{u_0}{4!}\psi^4\right\}. \tag{197}$$

Here $u_0 \propto \rho_c N$ is a constant, ρ_c being the chain density, $\tau_0 = 2\rho_c[\chi_s N - \chi N]$, and $e_0 = 3c^2\rho_c R_g^2/(2q^{*2})$, c being a constant of order unity that can be calculated ($c \cong 1.1019$ for $f = 1/2$ [58]). Introducing an effective chain length $\bar{N} = 6^3(R_g{}^3\rho_c)^2$, we rescale Eq. (197) as follows

$$\vec{r} = \vec{x}/(\sqrt{6}R_g), \quad \phi(\vec{r}) = c\bar{N}^{1/4}\psi(\vec{x}/\sqrt{6}R_g), \tag{198}$$

to obtain

$$\frac{1}{k_BT}\mathscr{H}\{\phi(\vec{r})\} = \int d^3\vec{r}\left\{\frac{1}{2}\phi[\tau + e(\nabla^2 + q_0^2)^2]\phi + \frac{u}{4!}\phi^4\right\}, \tag{199}$$

where $\tau = 2\ [\chi_s N - \chi N]/c^2$, $e = 1/(24q^{*2}R_g^2)$, $q_0^2 = 6(q^{*2}R_g^2)$, $u = \lambda(\bar{N})^{-1/2}$ ($\lambda = 106.18$ for $f = 1/2$ [58]). In Eq. (199), all coefficients of the quadratic form are now of order unity, and the nonlinear term is a small parameter for $\bar{N} \to \infty$.

Landau theory follows from Eq. (199) via replacing $\phi(\vec{r})$ by its average $\langle\phi(\vec{r})\rangle$. For a lamellar structure we put

$$\phi(\vec{r}) = 2A\cos(q_0\vec{n}\cdot\vec{r}), \tag{200}$$

with $\vec{n}$ a unit vector oriented perpendicular to the lamellae. Now the amplitude A effectively serves as the order parameter of the transition: namely inserting Eq. (200) in Eq. (199) yields the standard Landau form for the free energy density $f_L(A)$,

$$f_L(A) = \frac{(\sqrt{6}R_g)^3}{k_BTV}\mathscr{H}\{\phi\} = \tau A^2 + \frac{u}{4}A^4. \tag{201}$$

Figure 46 reminds the reader how the transition from $\tau > 0 (\chi N < \chi_s N)$ to $\tau < 0$ $(\chi N > \chi_s N)$ yields a second order transition, with $A = \pm\sqrt{-2\tau/u}$ for $\tau < 0$.

Fredrickson and Helfand [58] used a selfconsistent Hartree approximation in Eq. (199) to study the Gaussian fluctuations around the solution, Eq. (200). While for ordinary second order transition the local function $\langle\phi^2(\vec{r})\rangle - \langle\phi\rangle^2$ stays small even at T_c, this local fluctuation diverges here as $T \to T_c$! The reason is that normally the "phase space" for critical fluctuations is only the vicinity of a point in reciprocal space (the surroundings of $\vec{q} = 0$ for a ferromagnet, the surroundings of a few discrete points $\vec{q}_B$ at the Brillouin zone boundary for antiferromagnets, etc.), while here it is the vicinity of a sphere ($|\vec{q}| = q^*$). Fluctuations lead here to a divergence of the mean square displacement of ϕ similar as it happens due to phonons in one-dimensional crystals.

Of course, local fluctuations must stay finite, and thus this result already shows that the fluctuations must lead to a "renormalization" of the parameters of the effective Hamiltonian. Working this out in the framework of the Hartree approximation yields [58]

$$\frac{1}{k_B T}\mathscr{H}_H(\bar{\phi}) = \int d^3\vec{r}\left\{\frac{1}{2}\bar{\phi}[\tau_R + e(\nabla^2 + q_0^2)]\bar{\phi} + (u_R/4!)\,\bar{\phi}^4 + \frac{w_R}{6!}\bar{\phi}^6\right\} \tag{202}$$

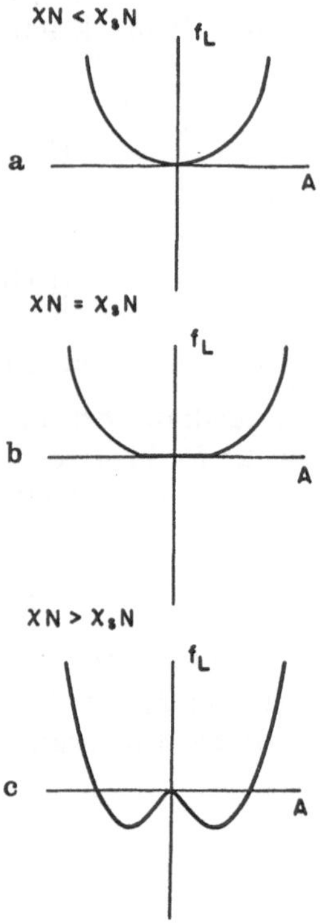

Fig. 46a–c. Free energy density $f_L(A)$ of a symmetrical diblock copolymer melt plotted vs the amplitude A of a concentration wave with $|\vec{q}| = q^*$. Above the critical point (**a**) only $A = 0$ is stable, while at the critical point (**b**) the curvature of the effective potential at $A = 0$ vanishes, and (**c**) below the critical point two symmetrical minima occur, corresponding to the stable lamellar phase. From Fredrickson and Binder [61].

instead of Eq. (199). The renormalized parameters of this "Hartree Hamiltonian" can be expressed as $\{d \equiv 3(q^*R_g)^2/2\pi\}$

$$\tau_R = \tau + du\tau_R{}^{-1/2},\ u_R = u[1 - \tfrac{1}{2}du\tau_R{}^{-3/2}]/[1 + \tfrac{1}{2}du\tau_R{}^{-1/2}]\,, \tag{203a}$$

$$w_R = 9du^3/[2\tau_R{}^{5/2}(1 + \tfrac{1}{2}du\tau_R{}^{-3/2})]^3\,. \tag{203b}$$

Thus u_R turns negative when the renormalized distance τ_R from the critical point gets small, i.e. for $\tau_R \propto u^{2/3} \propto (\bar{N})^{-3/2}$. Then τ differs from τ_R by a term of order $u\tau_R{}^{-1/2} = (\bar{N})^{-1/2}(\bar{N})^{+1/6} = (\bar{N})^{-1/3}$. Now $u_R < 0$ means a first order transition occurs, not for $\tau = 0$ but for a somewhat shifted value, namely for [58]

$$(\chi N)_H = 10.495 + 41.0(\bar{N})^{-1/3}\,. \tag{204}$$

This Brazovskii [327] mechanism for a fluctuation induced first order transition hence means that the strong increase of local fluctuations drives the fourth order coefficient of the Landau expansion negative, and thus a critical divergence of the local fluctuation is prevented.

Now the renormalized Hamiltonian can again be analysed in the spirit of the Landau theory, using once more Eq. (200). This yields the Hartree free energy density [61]

$$f_H = \frac{(\sqrt{6}R_g)^3}{k_BT\,V}\mathscr{H}_H(\bar{\Phi}) = \tau_R A^2 + \frac{u_R}{4}A^4 + \frac{w_R}{36}A^6. \tag{205}$$

Figure 47 shows the qualitative behavior of this free energy density. A crucial feature is that the renormalized distance τ_R corresponds still to the inverse scattering intensity $S^{-1}(q)$ at $q = q^*$. Since $\tau \propto \chi \propto 1/T$ in simple polymers, the nonlinear relation between τ and τ_R then implies a nonlinear relation between τ_R and $1/T$. Thus while Leibler's theory [43] predicts a linear variation of $S^{-1}(q^*)$ with $1/T$ (near the temperature where $S^{-1}(q^*)$ should vanish for $f = 1/2$), the fluctuation effects of Helfand and Fredrickson [58] imply a curved variation of $S^{-1}(q^*)$ with $1/T$. Such a curved variation indeed is found both in experimental data [317–323] and simulations [325, 328], see Figs. 43b, 48. Of course, due to finite size problems in the simulation one cannot as yet detect the small jump singularity that signals the mesophase separation transition in the experiment (Fig. 48).

The original version of the theory of Fredrickson and Helfand [58] did not yield any renormalization of the parameter q_0 in Eq. (199) or (202), respectively, and hence predict no shift of q^* with temperature, in disagreement with some experimental data (e.g. Fig. 48) and simulation (Fig. 45). In a more recent version such a shift was obtained taking the q-dependence of higher order terms in the Landau expansion into account when making the Hartree approximation [329]. Although this treatment yields an improved agreement with experiment [319], we feel that it does not incorporate fully the gradual stretching of the chains that occurs already in the disordered phase (see Sect. 5.4) and is partially responsible for the decrease of q^* when the transition is approached [325, 326, 328]. In

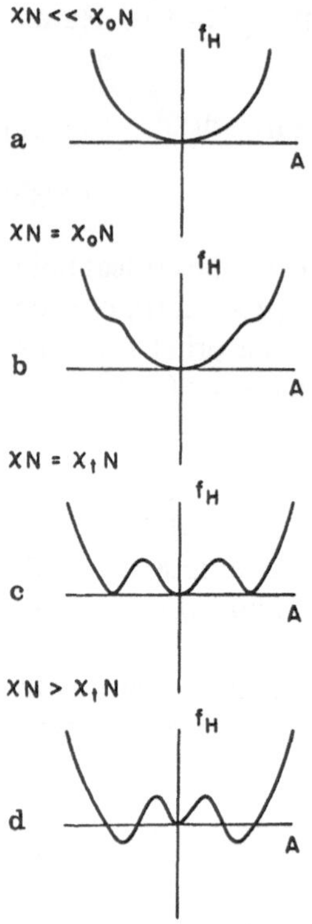

Fig. 47a–d. Hartree approximation for the free energy density $f_H(A)$ of a symmetrical diblock copolymer melt plotted vs the amplitude A of a concentration wave with $|\vec{q}| = q^*$. For high temperatures ($\chi < \chi_0$) only the disordered phase ($A = 0$) exists(**a**). Atχ_0 N the limit of metastability of the lamellar ordering in the disordered phase appears (**b**), two metastable minima at $A \neq 0$ develop, which become stable for $\chi = \chi_t$. (**c**). For$\chi > \chi_t$ the disordered state($A = 0$) is only metastable (**d**), the lamellar phase being in stable thermal equilibrium. From Fredrickson and Binder [61]

principle, the change of coil configurations due to the thermal interactions can be calculated systematically [330, 331] applying the "Edwards Hamiltonian" [332] method. At the time of writing, a full treatment of blockcopolymers along such lines seems lacking.

We now briefly mention some other extensions of Leibler's theory. Tang and Freed [333] use the Ohta-Kawasaki [57] formulation of the strong segregation theory of block copolymer ordering for a scaling analysis of chain stretching and local segregation. The quadratic term of the free energy functional in their approach is written as $\mathscr{F}_2 N/k_B T \cong \frac{1}{2}\int d^3\vec{q}\{A(f)/(N\sigma^2 q^2) + B(f)\ N\sigma^2 q^2 - 2[\chi N - C(f)]\}\psi(\vec{q})\ \psi(-\vec{q})$, where the coefficients A(f), B(f) and C(f) are given as $A(f) = 9/[f^2(1-f^2)]$, $B(f) = 1/[12f(1-f)]$ and $C(f) = s(f)/[4f^2(1-f^2)]$, s(f) being a function weakly dependent of f, with $s(1/2) = 0.9$. Higher order contributions are assumed [333, 57] to be local functionals of $\psi(\vec{r})$, e.g. $(g/4)\int d^3\vec{r}[\psi(\vec{r}) - \frac{1}{2}]^4$. Tang and Freed [333] now assume microdomains of scale λ^* and estimate that the domain interior yields a contribution $A(f)\ \lambda^*/N\sigma^2$, while the interfacial free energy per unit area becomes [57] $f_{int} = \frac{1}{9\sqrt{2}}\left[\left(\chi - \frac{C(f)}{N}\right)/B(f)\right]^{1/2}$. This yields a total free energy density $F/V \propto A(f)\ \frac{\lambda^{*2}}{N^2\sigma^2} + \frac{\sigma/\lambda^*}{3\sqrt{2B(f)}}[\chi - C(f)/N]^{1/2}$, which is minimized by

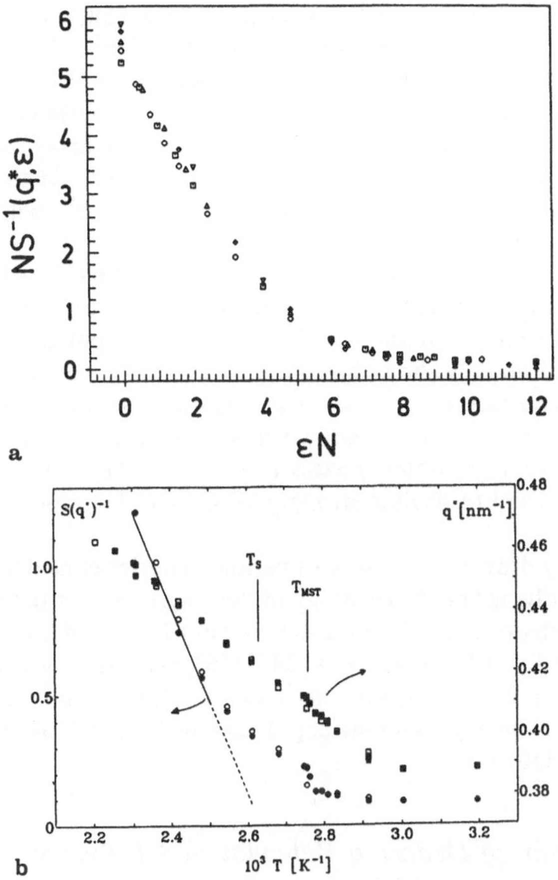

Fig. 48a. Normalized inverse scattering intensity $NS^{-1}(q^*,\varepsilon)$ observed in the Monte Carlo simulation of a block copolymer model on the simple cubic lattice (see Fig. 44) plotted vs the normalized inverse temperature εN. **b** Reciprocal structure factor $\{S(q^*)\}$-1(*circles*, left scale) and q^*(*squares*, right scale) plotted vs temperature for a nearly symmetric diblock copolymer of polystyrene/poly (*cis* – 1, 4) isoprene ($M_W = 15\,700$). *Filled symbols* refer to cooling, *open symbols* to heating runs. The *straight line* indicates the extrapolation to a spinodal temperature (T_s) that occurs above the actual transition temperature (T_{MST}), where the data show a jump. From Stühn et al. [323].

$\lambda^* \propto [\chi - C(f)/N]^{1/6} N^{2/3} \sigma$. In the strong segregation limit $N \to \infty$, i.e. C(f) /N can be omitted. However, in the regime where χ and C(f)/N are comparable, an "effective exponent" in the relation $\ln\lambda^* \propto \ln N$ applies which is even larger than the asymptotic value (2/3) of the strong segregation regime, consistent with some experiments [319, 334] and simulations [325, 326]. This theory thus elaborates the scaling analysis of Sect. 5.1.

Another interesting treatment [335] connecting Leibler's [43] mean field theory of the weak segregation limit (where $\lambda^* \propto N^{1/2}$) and the strong segregation limit (where $\lambda^* \propto N^{2/3}$) is achieved in terms of a density functional theory. As is well-known, density functional theories are very successful in describing fluid-solid transitions in general [336, 337], and hence this approach is most promising for the related transition of mesophase ordering using the information on the disordered melt. Unlike the Landau-type expansion, Eq. (184), high-order nonlinear terms are included selfconsistently.

Another approach [64, 338] combines the Hartree fluctuation corrections of the Fredrickson-Helfand theory [58] with contributions from multiple harmonics in the concentration expansion, chosen compatible with the considered symmetry of the ordered phase. This approach also predicts a shift of the characteristic wavevector q^*R_g with the parameter χN, unlike Leibler's theory [43]. Perhaps the most interesting prediction of this theory is that for $N < 10^9$ the lamellar phase is unstable, and rather ordering occurs in a particular hexagonal structure, the "lamellar cantenoid" where a hexagonal array of cylinders connects lamellar layers perpendicular to the cylindrical axes. At the time of writing, it seems unclear whether this prediction is compatible with experiment.

Holyst and Schick [339] study the phase diagram and scattering of AB symmetric diblock copolymers diluted with A and B homopolymers (in equal concentrations) having the same chain length $N_A = N_B = N$ as the copolymers. Constructing a Landau expansion, they show that the wave vector q^* vanishes at a critical copolymer concentration $\phi_c = 2/3$ and identify the ordering transition there as that of a Lifshitz tricritical point, where the disordered phase, lamellar phase, A-rich and B-rich separated phases can coexist. The critical behavior near this point is expected to deviate strongly from mean field theory [339].

We conclude this section by drawing attention to various theories considering the dynamics of block copolymer melts: rheology of these systems has been considered [340–342], single chain dynamics and selfdiffusion [343, 344], nucleation of the ordered phase [61], ordering kinetics [345, 346], and dynamics of concentration fluctuations [347]. These topics are not under consideration here, just as other extensions of the theory: random copolymer melts [348, 349], multiblock copolymer melts [350] etc.

5.4 Monte Carlo Simulation of Ordering Behavior and Polymer Structure in Block Copolymer Melts

As explained in the last section, the ordering behavior of block copolymer melts poses many challenging questions. Although corresponding experiments have contributed a great deal towards the understanding of these problems, the information that comes from experiment is still quite limited – the only static quantity that is measured is the collective structure factor S(q), no direct information on chain structure is available although dramatic changes in chain conformation are expected, from $R_g \propto N^{1/2}$ ($T \gg T_{MST}$) to $R_g \propto N^{2/3}$ ($T \ll T_{MST}$). Thus, it is desirable to obtain complementary information from Monte Carlo computer simulations. Though such simulations have their own limitations (only rather short chain lengths are accessible, there are finite system size effects and equilibration problems at low temperature), the advantages are that the models are perfectly well-characterized and ideal (e.g., the chains are perfectly symmetric, perfectly monodisperse, etc.) and information is accessible in as microscopic detail as desired.

The simulations by Minchau et al. [328] and by Fried and Binder [325, 326] treat the block copolymer analog of the lattice model of Sariban and Binder

[107] for polymer blends (Fig. 48a). Polymers are modelled as self- and mutually avoiding chains on the simple cubic lattice, using the slithering snake algorithm [289, 299] with chain lengths N = 16 to N = 60, for f = 1/2, so each chain has N/2 A-monomers and N/2 B-monomers. Pairwise energies between monomers (ε_{AA}, ε_{BB}, ε_{AB}) are restricted to nearest neighbors (Fig. 49a), and in order to avoid a clustering of vacancies as much as possible (which occurs for $\varepsilon_{AA} = \varepsilon_{BB} < 0$ and low temperatures [155]), we put $\varepsilon_{AA} = \varepsilon_{BB} = 0$, working with a purely repulsive model $\varepsilon_{AB} > 0$. In order to allow a rather fast approach

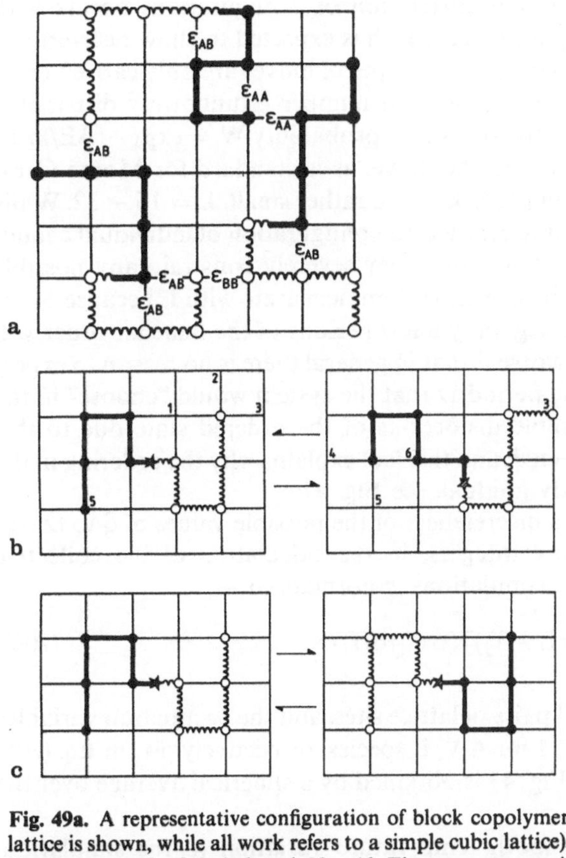

Fig. 49a. A representative configuration of block copolymers on the lattice (For clarity a square lattice is shown, while all work refers to a simple cubic lattice). Three symmetric diblock copolymers are shown, each of chain length N = 10. The two monomeric species are labeled A-type (*full dots*) and B-type (*open dots*). The vacancies are not shown explicitly, but are assumed to reside on each lattice site left unoccupied by either of the two species of monomer. A volume fraction of $\phi_v = 0.2$ is used, since experience with blends [107] has shown that such a system behaves like a very dense melt. The energy contributions ε_{AA}, ε_{BB} and ε_{AB} are shown. **b** Examples of typical "slithering-snake" [298, 299] motion: monomer situated at point labelled by 5 is removed, and one of sites 1, 2, 3 is randomly chosen for occupation. Note that unlike Refs. [298, 299] also the junction point needs to be displaced accordingly, as shown in the figure. For the reverse process, monomer at 3 is removed and the sites 4, 5, 6 are considered for attachment (of course, a move to site 6 is rejected due to excluded volume constraints). **c** Interchange of A-Block and B-Block of a diblock copolymer chain. From Fried and Binder [325].

to thermal equilibrium (and also for the sake of having data for binary blend models at the same vacancy concentration ϕ_V available [107], $\phi_V = 0.2$ has been chosen [325, 326]. At this value of ϕ_V, excluded volume interactions are screened off down to distances of a few lattice spacings, and the experience with blends where ϕ_V may be varied over a wide range [101] suggests that data at $\phi_V = 0.2$ are already representative of the behavior in the limit $\phi_V \to 0$.

Figure 49b, c illustrates the choice of copolymer motions used in the dynamic Monte Carlo algorithm of Fried and Binder [325, 326]: "slithering snake" moves [298, 299] were generalized to adapt them to the blockcopolymer architecture, and also A-B block exchanges. The motivation for the latter move is that one expects that a faster decorrelation of configurations occurs with respect to the lamellar order parameter, which is expected to show near-critical slowing down at the MST. Of course, both types of moves are only carried out if either they lower the energy or if a random number ξ (uniformly distributed between zero and one) exceeds the transition probability $W = \exp(-\Delta E/k_B T)$, ΔE being the energy change due to the move, as is standard for Monte Carlo simulation. Lattice sizes used in [325, 326] are rather small, $L = 16 - 32$. While these sizes are large enough not to restrict the configuration of individual chains, even if they are somewhat stretched out, they severely constrain any possible long range order, which must be necessary commensurate with L because of the periodic boundary conditions: e.g. only a few periods of the lamellar order will "fit" into the box, and what is worse is that in general there is no reason to expect that L will be a multiple of the period D that the system would "choose" in the limit $L \to \infty$. Hence considerable distortions of the ordered state due to this finite-size caused misfit is expected, and this fact explains why the ordering in the simulation appears particularly gradual, see Fig. 47a.

The finite size also causes a discreteness of the possible values of $\vec{q}$ to $(2\pi/L)$ (v_x, v_y, v_z), where (v_x, v_y, v_z), are integers, in the calculation of the collective structure factors, which in the simulations is normalized as

$$S(\vec{q}) = L^{-3} \sum_{i,\delta} \exp[iq \cdot (\vec{r}_i - \vec{r}_j)] \langle \sigma(\vec{r}_i) \sigma(\vec{r}_j) \rangle \,, \tag{206}$$

where the sum is taken over all pairs of lattice sites, and the occupation variables have the values $\sigma(\vec{r}) = -1, 0, 1$ for A,V, B species, respectively. From Eq. (206) the structure factor shown in Fig. 43 is obtained by a spherical average over the direction of $\vec{q}$.

Since $S(q^*)$ has such a smooth temperature variation in the simulation, Fig. 47a, Fried and Binder [325] rely on dynamic criteria for locating the MST, motivated by the experimental finding that the MST shows up most clearly in the dynamic response of the blockcopolymer melt (e.g. by an abrupt change in the frequency dependence of viscoelastic response functions at T_{MST} [317–323]. The idea is that in the lamellar phase there will be a spontaneous symmetry breaking, $S(\vec{q})$ depends on the direction of $\vec{q}$ for $T < T_{MST}$, since the orientation perpendicular to the lamellar is singled out, while for $T > T_{MST}$ the symmetry of the disordered phase requires that $S(\vec{q})$ is spherically symmetric. However, this spherical symmetry holds as an average property only, at each instant of time there will be a preferred orientation, reflecting the lamellar short range order in the system, but for $T \gg T_{MST}$ this orientation will rapidly change with time. As $T \to T_{MST}$, the lamellar short range order gets more and more pronounced, and we expect a slowing down in the change of the preferred orientation with time. If the MST were a second order transition, we would in fact expect critical slowing down and the spontaneous symmetry breaking in the "time averaging" done by the Monte Carlo method would appear as an "ergodicity breaking", since the relaxation time for this reorientation of the preferred direction diverges. This concept can be made precise by interpreting the instantaneous observations of $S_t(\vec{q})$ at time t, as a fictitious "mass" distribution, and associating principle moments of inertia $I_1(t)$, $I_2(t)$, $I_3(t)$ to it. We can introduce also associated principle directions (if $I_1 \leqq I_2 \leqq I_3$, the greatest mass is concentrated around the direction $\vec{e}_1(t)$). One now defines a time correlation function and associated relaxation time

$$\phi_s(t) = \langle |\vec{e}_1(t_1) \cdot \vec{e}_2(t_1 + t)| \rangle - \tfrac{1}{2}, \quad \tau_s = \frac{2}{\tau_{obs}} \int_0^{T_{obs}} \phi_s(t)\,dt\,. \tag{207}$$

Note that τ_s in Eq. (207) is measured in units of the *total* observation time (which was typically chosen as [325] $\tau_{obs} = 5000$ MCS/monomer), so τ_s is normalized between zero and unity. For $T \gg T_{MST}$, we expect $\int_0^{T_{obs}} \phi_s(t)\,dt = \int_0^{\infty} \phi_s(t)\,dt$ to be a time much smaller than τ_{obs}, and then $\tau_S \approx 0$. On the other hand, for $T < T_{MST}$ when a lamellar orientation is essentially frozen-in over the time interval τ_{obs},

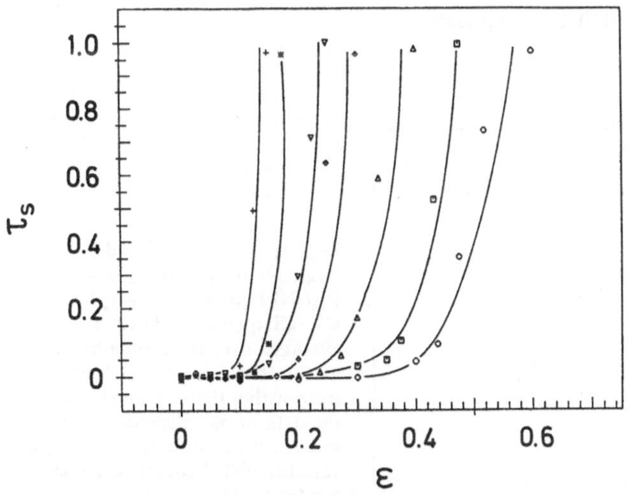

Fig. 50. The lamellar relaxation time $\tau_s(\varepsilon, N)$ plotted against ε for N = 16 (*circles*), 20 (*squares*), 24 (*triangles*), 32 (*diamonds*), 40 (*triangles on top*), 48 (*asterisks*), and N = 60 (*crosses*). Lines are guides for the eye only. From Fried and Binder [325].

$\phi_s(t) \approx 1/2$ for $0 < t < \tau_{obs}$, and hence $\tau_s = 1$. Near T_{MST} one expects that τ_s rises sharply from small values to order unity, and this is indeed seen (Fig. 50). In this way, the simulations allow us (at least roughly) to estimate where the MST occurs in this model, namely at [325]

$$\varepsilon(T_{MST})N \approx 7.5\text{–}9 . \tag{208}$$

The huge advantage of the simulations as compared to experiment is, of course, that arbitrary geometric characteristics of the chains are easily accessible. So one may consider the average mean square distance $\langle \vec{R}^2_{AB} \rangle$ between the center of gravity of the A-block ($\vec{R}^A_{cg}$) and of the B-block ($\vec{R}^B_{cg}$), $\vec{R}_{AB} \equiv (\vec{R}^A_{cg} - \vec{R}^B_{cg})$. For gaussian chains one easily derives $(\vec{R}^2_{AB}) = 2R^2_g$, R_g being the gyration radius. Figure 51 shows that upon cooling of the block copolymer melt $R_{AB}(\varepsilon^N) \equiv (\langle R^2_{AB} \rangle)^{1/2}$ increases much more rapidly than R_g. Thus the question arises: can it be that the shift of q^* with εN, Fig. 45, merely reflects the chain stretching that one sees in the increase of $R_{AB}(\varepsilon, N)$? This is tested in Fig. 52 by comparing $x^* = q^*(\varepsilon, N) R_g(\varepsilon, N)$ with the analoguous quantity $z^* = q^*(\varepsilon, N) R_{AB}(\varepsilon, N)/\sqrt{2}$. It is seen that z^* still decreases with increasing ε, N, i.e. the collective length scale $\lambda^*(\varepsilon, N) = 2\pi/q^*(\varepsilon, N)$ increases faster than any single-chain property. Thus λ^* reflects the characterized wavelength of lamellar "clusters" formed by many block copolymers, it is not simply a geometrical characteristics of a single chain. All these properties seen in Figs. 51, 52 represent pronounced deviations from the simple form of RPA, used in Leibler's theory (43), where chain configurations are gaussian independent of ε. Note also that this breakdown of RPA occurs completely gradually, there is no sharp transition where the coil stretching sets in, unlike the suggestions of Ref. 319. The latter suggestion was based on a log-log plot of q^* vs N at fixed temperature, Fig. 53, where one could distinguish two regimes.

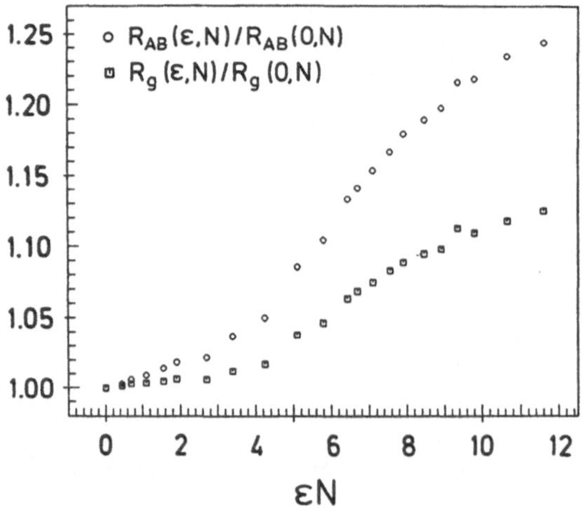

Fig. 51. Normalized radii $R_{AB}(\varepsilon,N)/(0,N)$ (*circles*) and $R_g(\varepsilon,N)/N_g(0,N)$ (*squares*) plotted vs εN (all quantities have been averaged over the results for the various chain lengths at fixed εN, since within the statistical errors the data do not depend on ε,N separately but only on one scaling variable εN). From Fried and Binder [325].

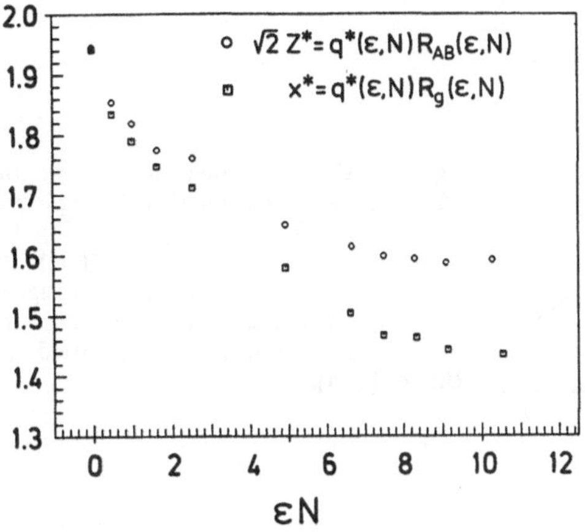

Fig. 52. Normalized characteristic ratios $x^* = q^*(\varepsilon,N)\,R_g(\varepsilon,N)$ (*squares*) and $z^* = q^*(\varepsilon,N)\,R_{AB}(\varepsilon,N)/\sqrt{2}$ (*circles*) plotted vs εN. (All quantities have been averaged over the results for the various chain lengths at fixed εN as in Fig. 51. Note that for $\varepsilon N = 0$ the Leibler value [43] $x^* = z^* \cong 1.95$ is recovered, as expected. From Fried and Binder [325].

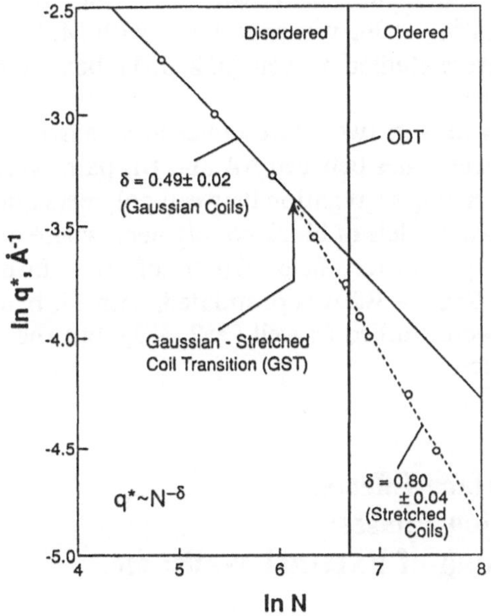

Fig. 53. Log-log plot of q^* vs molecular weight, for poly(ethylenepropylene)-polyethylethylene) (PEP-PEE) diblock copolymers containing 55% by volume PEP. From Almdal et al. [319].

As Fig. 54 shows, the replotting of the smooth variation of Fig. 52 [326] in the log-log form is suggestive of a rather sharp onset of the stretching, and thus there is no disagreement between the experimental observations [319] and corresponding simulations [325, 326].

Also the ordering of block copolymers in two space dimensions has been simulated [351]. In this case attractive energies between AA and BB pairs were used and in addition to the ordering a strong segregation between polymers and vacancies was observed. Coarse-grained models of block copolymers, where ad

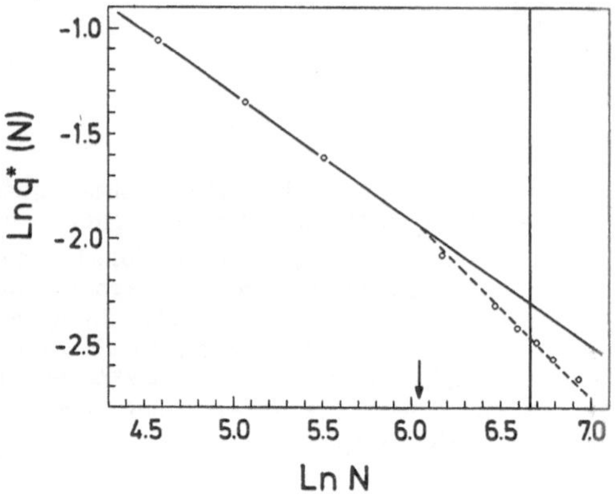

Fig. 54. Data obtained for q^* in the Monte Carlo simulation (Fig. 52) replotted in the same form as Fig. 53, using the scaling property of $x^* = x^*(\varepsilon N)$ to calculate q^* at fixed ε over a wide range of N. *Vertical straight line* indicates the order-disorder transition. From Fried and Binder [326].

hoc an equation for the order parameter field $\psi(r,t)$ of the form $(\partial/\partial t)\psi(r,t) = M\nabla^2[-b\phi + u\phi^3 - K\nabla^2\phi] - B\phi$ was postulated, with M, b, u, K, B phenomenological parameters, were studied as well [352, 353], but these models are outside of consideration here.

Also the ordering of block copolymers in two space dimensions has been simulated [351]. In this case attractive energies between AA and BB pairs were used and in addition to the ordering a strong segregation between polymers and vacancies was observed. Coarse-grained models of block copolymers, where ad hoc an equation for the order parameter field $\psi(r,t)$ of the form $(\partial/\partial t)\psi(r,t) = M\nabla^2[-b\phi + u\phi^3 - K\nabla^2\phi] - B\phi$ was postulated, with M, b, u, K, B phenomenological parameters, were studied as well [352, 353], but these models are outside of consideration here.

6 Outlook on Surface and Interface Effects: Interface Structure in Unmixing Phases; Surface Enrichment and Wetting of External Walls; etc.

Both in compatible and in incompatible polymer blends, the dynamics of chains at interfaces and the static interfacial structure are of very great theoretical and practical interest [354–356]: adhesion of polymer layers to walls, mechanical properties of inhomogeneous blends etc. may affect the application of polymeric materials, and at the same time fundamental questions are involved. This field of research is very active, and a complete coverage of the ongoing research in this area is not at all intended; rather we indicate only a few topics that are closely related to problems treated in previous sections of the present review.

In Sect. 2.4, we discussed the interdiffusion of compatible polymers where an interface is prepared as an initial condition of the experiment, by bringing an

A-rich layer on top of a B-rich layer and studying the interfacial broadening [211, 287–293, 355, 356], in an attempt to measure interdiffusion constants. Since most polymer mixtures exhibit a miscibility gap, it is also of interest to consider the situation of interfacial broadening for these systems, where the final equilibrium state is heterogeneous, two layers with compositions given by the coexistence curve being separated by an equilibrium interface (we have considered the structure of such interfaces in Sect. 2.3). The dynamics of interfacial broadening in such a situation has been studied both experimentally [357] and theoretically [358–360] recently. While in compatible mixtures the width W(t) of the interface in such interfacial broadening experiments follows $W(t) \propto t^{1/2}$ law, as is common for interdiffusion problems [211, 355], a slower growth law $W(t) \propto t^{a}$ with $a \approx 1/3$ was observed for the case of partial miscibility [357]. Theoretically exponents $a = 1/4$ have been suggested [359, 360]. None of these theories, however, considers complications due to entanglement formation (or tube renewal, respectively) which one should expect on the basis of reptation theory [116]. Such reptation effects have been identified in some studies of the initial stages of interdiffusion of polymeric layers, e.g. [293]. The fact that A-chains crossing an interface and making an excursion into the B-rich phase are entangled with the B-chains there, is, of course, very important for the viscoelastic flow behavior of such heterophase systems as for the mechanical properties of amorphous materials produced from cooling down such blends [355]. Such phenomena are outside the scope of the present review, however, as well as the modifications of interfacial structure that are possible when a third species is added that may get enriched at the A-B interface (small molecule surfactants, block copolymers etc., see e.g. [361–364].

Here we rather focus on effects of external surfaces (e.g. hard walls) on polymer blends. In general, one expects that the forces between the wall and monomers of type A will differ from those between the wall and monomers of type B, as it generally occurs at the surfaces of small-molecule mixtures as well [365]. For polymer mixtures that are partially compatible, the interactions in the bulk (as described by the Flory-Huggins χ-parameter) must be relatively small, however, since the entropy of mixing is down by a factor of N (for simplicity, the following discussion is restricted to a symmetric mixture, $N_A = N_B = N$). However, there is no reason that the difference of wall-A and wall-B forces is similarly small [125]. Thus one may expect rather pronounced "surface enrichment" effects in polymer mixtures [125]. Indeed some experimental evidence for this prediction has been found [37, 38, 126, 127].

The first theories for this phenomenon used the simple forms for the Flory-Huggins free energy functional discussed in Sect. 2.3, and augmented them by a local boundary condition at the surface [124, 125]. Denoting A as the surface area of the wall, the free energy functional per unit area then is, cf. Eq. (47)

$$\frac{\Delta \mathscr{F}}{A k_B T} = \int dz \left[f_{FH}(\phi) + \frac{a^2}{36\phi(1-\phi)} \left(\frac{d\phi}{dz} \right)^2 \right] - \mu_1 \phi_1 - \frac{1}{2} g \phi_1^2. \tag{209}$$

Here the "bare" surface free energy $f^{(b)} = -\mu_1\phi_1 - \frac{1}{2}g\phi_1^2$ is assumed a quadratic function of the local surface concentration ϕ_1, μ_1 and g are phenomenological coefficients. This ansatz can be justified as a Taylor expansion in the case where $\phi_1 \ll 1$ or where $1 - \phi_1 \ll 1$, respectively [125], or alternatively by alluding to Ising-models of small molecule mixtures [366, 367], where μ_1 corresponds to the difference in wall-A, wall-B forces alluded to above, and g results from effects due to "missing neighbors", changes in the pairwise interaction between A and B near the wall, etc. If one accepts Eq. (209), the further treatment is very simple: minimizing the surface excess free energy with respect to the concentration profile $\phi(z)$ {z being the distance from the wall that is situated at $z = 0$, while at $z \to +\infty$ one requires that $\phi(z)$ approaches the bulk value $\phi(\infty) = \phi_\infty$ of the concentration} one must satisfy the boundary condition at the surface

$$\frac{a^2}{18\phi_1(1-\phi_1)}\left.\frac{d\phi}{dz}\right|_{z=0} = -\mu_1 - g\phi_1\,, \tag{210}$$

whereas the concentration profile is described by [125]

$$\frac{a^2}{36\phi(1-\phi)}\left.\left(\frac{d\phi}{dz}\right)^2\right|_{z'=0}^{z'=z} = f_{FH}(\phi(z)) - f_{FH}(\phi_1)\,. \tag{211}$$

Equation (211) is simply integrated as

$$\frac{6z}{a} = \int_{\phi_1}^{\phi(z)} d\phi/\{\phi(1-\phi)[f_{FH}(\phi) - f_{FH}(\phi_\infty)]\}^{1/2}\,, \tag{212}$$

whereas Eq. (210) with the help of Eq. (211) becomes

$$\pm\left[\frac{a}{3}\,\frac{f_{FH}(\phi_1) - f_{FH}(\phi_\infty)}{\phi_1(1-\phi_1)}\right]^{1/2} = -\mu_1 - g\phi_1\,, \tag{213}$$

and the sign has to be taken such that the solution is actually the minimum of the surface excess free energy. It turns out that the solution of Eqs. (212), (213) may give rise to wetting phenomena [124, 125]. In the nonwet state of the surface, the enhancement of the volume fraction $\phi(z)$ of the preferred species near the wall decays to its bulk value ϕ_∞ at a distance of the order of the gyration radius of the chain (or smaller, if χ is not too small). If the system in the bulk is in an unmixed state, however, such that $\phi_\infty = \phi_{coex}^{(1)} < \phi_{coex}^{(2)}$ is the smaller concentration of the two branches of the coexistence curve, the surface may be in a wet state: This means the profile decays from its surface value $\phi_1(\phi_1 > \phi_{coex}^{(2)}$ then) first to $\phi_{coex}^{(2)}$, and at a large distance from the wall there is then another interface, separating the two coexisting phases. One thus has a (macroscopically thick) wetting layer of composition $\phi_{coex}^{(2)}$ on the surface of a mixture of composition $\phi_{coex}^{(2)}$, stabilized by the wall. This phenomenon is well known in small molecule mixtures [368] but has been observed in polymer mixtures only recently [127].

It turns out [125] that the wetting transition separating the nonwet state of the surface is always first order when $g > 0$, while it may be second order for $g < 0$. Figure 55 shows the phase diagram [125] resulting from this simple mean field theory of wetting in polymer mixtures. A second order wetting transition occurs when the solution of Eq. (213) reaches the value $\phi_1^{crit} = \phi_{coex}^{(2)}$. The surface response function $\kappa_{11} = (\partial\phi_1/\partial\mu_1)_T$ then exhibits a jump singularity, whereas it diverges at the wetting tricritical point where the order of the wetting transition changes from second to first order. One finds [125] that such a tricritical transition occurs for $\phi_\infty^t = -(a/g)/\sqrt{18N}$, $\mu_1^t = -g + a/\sqrt{18N}$. Thus for large N the model mostly predicts second order wetting transitions rather than first order transitions, if the coefficients g, μ_1 are of order unity (while $\chi \ll 1$).

Of course, this theory in many aspects is too simple: the gradient-square term in Eq. (1) is useful only for the description of large-scale concentration variations on length scales exceeding the chain gyration radius, as discussed in Sect. 2.3. An extension of the theory which does not suffer from this restriction was presented by Carmesin and Noolandi [369]. Also it is probably not a good approximation that the range of the forces between the wall and the monomers is so short – eq. (209) implies a range of the order of the lattice spacing a, i.e. the size of effective monomers – there are reasons [368] to expect wall forces that

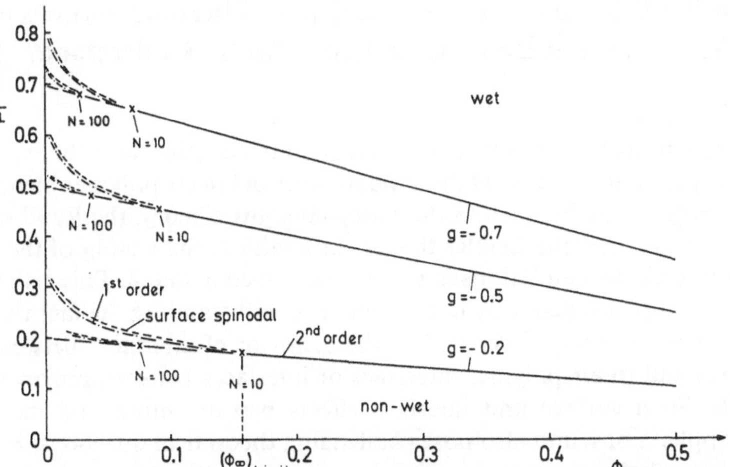

Fig. 55. Surface phase diagram in the plane of variables μ_1 and ϕ_∞ for three values of g. The region where the surface is non-wet (at small μ_1) is separated from the wet region by a phase boundary which describes the wetting transition. For $\phi_\infty > \phi_\infty^t$ (second-order wetting) this is just the straight line $\mu_1^{crit} = -g(1-\phi_\infty)$. The region of first order wetting is shown for symmetrical mixtures with $N_A = N_B = N = 10$ and $N = 100$, respectively, and the first-order transitions are shown by dash-dotted curves. In this regime metastable wet and non wet phases are possible up to the stability limits ("surface spinodals") denoted by dashed curves. Assuming that μ_1 and g are essentially independent of temperature T, variation of T essentially means variation of ϕ_∞. From Schmidt and Binder [125].

decay with the distance from the wall according to a power law. Chen et al. [370] have extended the theory to this situation. But even for the short range problem, Monte Carlo simulations [255] have called the phase diagram of Fig. 55 into question. Again this work [255, 371, 372] draws attention to the fact that the RPA, on which Eq. (209) is based, does not hold because chains attached to walls strongly deviate from gaussian coils. A rather fundamental objection to Eq. (209) was also recently raised by Fredrickson and Donley [373] who showed in the presence of a wall one finds a contribution to the free energy that is linear in the gradient of the concentration $\nabla\phi(z)$ and with a range that extends from the surface a distance of order of the radius of gyration. This term derives from conformational restrictions on the polymers in the surface layer. We shall not discuss in depth all these recent developments here – clearly this is a rapidly evolving field and much work still remains to be done.

The same fact is true for surface effects on the dynamics of spinodal decomposition in polymer blends [374, 375], dynamics of surface enrichment in blends [367, 375, 376], and, last but not least, for surface effects on block copolymers: there one may have surface – induced ordering [377, 379] and interesting competition effects between the lamellar ordering (of wavelength λ) and film thickness D in thin block copolymer films [380–388]. These phenomena are outside of our consideration here.

7 Discussion: To What Extent is the Statistical Thermodynamics of Polymer Mixtures and Block Copolymer Melts Understood?

In this article, emphasis has been laid on a tutorial introduction into the basic theoretical concepts on the statistical thermodynamics of binary polymer blends and of block copolymer melts, such as the Flory-Huggins-Theory, the Random Phase Approximation, and the Leibler theory. In addition, the testing of these concepts by Monte Carlo simulations have been described in detail. This article has also emphasized phase behavior in the bulk, not withstanding the fact that much recent research is being developed to the behavior of interfaces between coexisting phases and to air-polymer interfaces or interfaces between polymers and solid walls. Such surface and interface effects become more and more important for applications, and also pose challenging theoretical questions, but have only occasionally been mentioned, first because the author lacks expertise, secondly in order to keep the size of this article manageable. The same reasons prevented us from treating dynamic properties (other than interdiffusion and early stages of spinodal decomposition), although there are many challenging questions: e.g., how do the tube diameters of entangled polymers in a binary polymer blend depend on composition? How is selfdiffusion of polymers and their viscoelastic response affected by strong local segregation in unmixed polymer blends or block copolymer melts? Clearly, such questions will be a dominating issue of research for years to come, as well as other topics that have hardly been touched upon: ternary blends (polymer A-polymer B-polymer C), see Ref. [389] for a review, effects of solvent or plasticizers added to binary blends, role of surfactants in incompatible binary blends, etc. However, it is our feeling that for a profound theoretical understanding, progress in the theoretical description of the "simple" binary systems will be extremely crucial.

According to the author's personal point of view, the theoretical understanding of the thermodynamics of polymer blends is still rather limited: simple concepts like Flory-Huggins theory and RPA etc. are qualitatively correct in many circumstances, but in many aspects not quantitatively reliable. For example, while the RPA (in its simplest version) assumes gaussian coil configurations independent of the thermodynamic state of the blend (as described by the effective Flory-Huggins parameter χ, volume fractions ϕ_A, ϕ_B and density of the melt), there is quite compelling evidence (from computer simulations) that such an approximation is accurate only for $\chi N_A \lesssim 1$, $\chi N_B \gtrsim 1$, but the size of the coil does depend on χ and the volume fractions, when these inequalities are not fulfilled. Since the estimation of χ from small angle scattering data is largely based on the RPA, it is clear that this problem must induce spurious temperature – and composition dependence, and this in turn hampers the molecular interpretation of χ for many systems.

Of course, these problems are widely appreciated now and many very interesting and challenging theoretical approaches that go far beyond the simple theories have been proposed by a variety of authors. These recent theories certainly have not been adequately covered in this review: first of all, the author again lacks expertise, and secondly, this is an extremely active area of current research and even if one could give a "snapshot picture" of the current state of these theories it would be rapidly outdated.

In conclusion, it is hoped that the present review will interest additional researchers in this fascinating field, and that it gives an understandable introduction to the newcomer, as well as a useful guide to the more specialized literature for the experts.

Acknowledgements. The author's knowledge of this topic has strongly benefitted from interactions with numerous colleagues: F. S. Bates, E. W. Fischer, G. H. Fredrickson, H. L. Frisch, P. G. deGennes, C. C. Han, T. Hashimoto, E. Helfand, K. Kawasaki, J. Klein, E. J. Kramer, L. Leibler, L. Monnerie, A. Onuki, P. Pincus, D. Schwahn, K. Schweizer, T. Springer, G. Strobl.

In addition it is a pleasure for me to thank numerous coworkers who have collaborated fruitfully with the author on various related research topics, in particular H. P. Deutsch, H. Fried, B. Minchau, A. Sariban, and I. Schmidt (Carmesin).

References

1. Flory PJ (1953) Principles of polymer chemistry. Cornell University Press, Ithaca
2. DeGennes PG (1979) Scaling concepts in polymer physics. Cornell University Press, Ithaca
3. Klempner D, Frisch K (eds) (1977) Polymer Alloys. Plenum, New York

4. Olabasi M (ed) (1979) Polymer-polymer miscibility. Academic, New York
5. Walsh DS, Higgins JS, Maconnachie A (eds) (1985) Polymer blends and mixtures. Martinus Nijhoff Dordrecht
6. Koningsveld R (1968) Adv Colloid Interface Sci 2: 151
7. Koningsveld R, Kleinjens LA, Schoffelers HM (1974) Pure Appl Chem 39: 1
8. Koningsveld R, Kleinjens LA, Nies E (1987) Croat Chim Acta 60: 53
9. Binder K (1987) Colloid & Polymer Sci 265: 273
10. Hashimoto T (1987) In: Ottenbrite RM, Utracki LM, Inoue S (eds) Current topics in polymer science, vol 2. Hanser, Munich, p 199
11. Nose T (1987) Phase Transitions 8: 245
12. Hashimoto T (1988) Phase Transitions 12: 47
13. Flory PJ (1941) J Chem Phys 9: 660
14. Flory PJ (1942) J Chem Phys 10: 51
15. Huggins ML (1941) J Chem Phys 9: 440
16. Huggins ML (1942) J Phys Chem 46: 151
17. Huggins ML (1942) J Am Chem Soc 64: 1712
18. Staverman AJ (1941) Recl Trav Chim 60: 640
19. Staverman AJ, Van Santen JH (1941) Recl Trav Chim 60: 76
20. Guggenheim EA (1945) Proc Roy Soc (London) A183: 203
21. Guggenheim EA (1945) Proc Roy Soc (London) A183: 231
22. Scott RL (1949) J Chem Phys 17: 279
23. Cahn RW, Haasen P (1983) Physical metallurgy. North Holland, Amsterdam
24. Stocks LM, Gonis A (eds) (1989) Alloy phase stability. Kluwer Dordrecht
25. Haasen P (ed) (1991) Phase transformations in materials. VCH Verlagsges., Weinheim
26. Flory PJ (1969) Statistical mechanics of chain molecules. Interscience, New York
27. Snyder HL, Meakin P, Reich S (1983) Macromolecules 16: 757
28. Hashimoto T, Izumitani T (1985) J Chem Phys 83: 3694
29. Okada N, Han CC (1986) J Chem Phys 85: 5317
30. Meier H, Strobl GR (1987) Macromolecules 20: 649
31. Higgins JS, Fruitwala H, Tomlins PE (1989) Macromolecules 22: 3674
32. Schwahn D, Yee-Madeira H (1987) Colloid Polym Sci 265: 867; Schwahn D, Springer T, Mortensen K, Yee-Madeira H (1988) In: Komura S, Furukawa H (eds) Dynamics of ordering processes in condensed matter. Plenum, New York, p 445
33. Schwahn D, Springer T, Yee-Madeira H, Mortensen K (1988) In: Richter D, Springer T (eds) Polymer motion in dense systems. Springer Proceedings in Physics, vol 29. Springer, Berlin Heidelberg New York p 296
34. Sato T, Han CC (1988) J Chem Phys 88: 2057
35. Wiltzius P, Bates FS, Heffner WR (1988) Phys Rev Lett 60: 1538
36. Bates FS, Wiltzius P (1989) J Chem Phys 91: 3258
37. Jones RAL, Kramer EJ, Rafailovich MH, Sokolov J, Schwarz SA (1989) Phys Rev Lett 62: 280
38. Sokolov J, Rafailovich MH, Jones RAL, Kramer EJ (1989) Appl Phys Lett 54: 590
39. Bates FS, Fredrickson GH (1990) Ann Rev Phys Chem 41: 525
40. Brown RA, Masters AJ, Price C, Yuan XF (1989) In: Allen G (ed) Comprehensive Polymer Science. Pergamon, Oxford, p 155
41. Helfand E, Wasserman ZR (1976) Macromolecules 9: 879
42. Helfand E, Wasserman ZR (1978) Macromolecules 11: 960
43. Leibler L (1980) Macromolecules 13: 1602
44. Schweizer KS, Curro JG (1988) Phys Rev Lett 60: 809
45. Schweizer KS, Curro JG (1988) J Chem Phys 88: 7242
46. Schweizer KS, Curro JG (1989) J Chem Phys 91: 5059
47. Schweizer KS, Curro JG (1990) Chem Phys 149: 105
48. Curro JG, Schweizer KS (1990) Macromolecules 23: 1402
49. Schweizer KS, Curro JG (1991) J Chem Phys 94: 3986
50. Curro JG, Schweizer KS (1991) Macromolecules 24: 6736
51. Dudowicz J, Freed KF, Madden WG (1990) Macromolecules 23: 4803
52. Lipson JEG (1991) Macromolecules 24: 1334
53. Hong KM, Noolandi J (1981) Macromolecules 14: 727
54. Noolandi J, Hong KM (1982) Macromolecules 15: 482
55. Hong KM, Noolandi J (1983) Macromolecules 17: 1443

56. Semenov AN (1985) Soviet Phys JETP 61: 733
57. Ohta T, Kawasaki K (1986) Macromolecules 19: 2621
58. Fredrickson GH, Helfand E (1987) J Chem Phys 87: 697
59. Kawasaki K, Ohta T, Kohrogu M (1988) Macromolecules 21: 2972
60. Noolandi J, Kawassalis TA (1988) In: Nagasawa M (ed) Molecular conformation and dynamics of macromolecules in condensed systems. Elsevier, Amsterdam, p 285
61. Fredrickson GH, Binder K (1989) J Chem Phys 91: 7265
62. Burger C, Ruland W, Semenov AN (1990) Macromolecules 23: 3339
63. Muthukumar M (1986) J Chem Phys 85: 4722
64. Olvera de la Cruz M (1991) Phys Rev Lett 67: 85
65. Khoklov AR (1991) In: Ciferri A (ed) Liquid Crystallinity in Polymers: Principles and Fundamental Properties, p 97 VCH Publ., Weinheim
66. Sperling LH (1981) Interpenetrating Polymer Networks and Related Materials. Plenum, New York
67. Binder K, Frisch HL (1984) J Chem Phys 81: 2126; Schulz M (1992) J Chem Phys, in press; Schulz M, Binder K (1993) J Chem Phys 98: 655
68. Bates FS, Muthukumar M, Wignall GD, Fetters LJ (1988) J Chem Phys 89: 535
69. Bates FS, Rosedale JH, Stepanek P, Lodge TP, Wiltzius P, Fredrickson GH, Hjelm PP, Jr (1990) Phys Rev Lett 65: 1893
70. Cumming A, Wiltzius P, Bates SF (1990) Phys Rev Lett 65: 863
71. Stepanek P, Lodge TP, Kedrowski C, Bates SF (1991) J Chem Phys 94: 8289
72. Koningsveld R, Kleintjens LA (1973) In: Macromolecular chemistry, vol 1, Butterworths, London, p 97
73. Stanley HE (1971) An introduction to phase transitions and critical phenomena. Oxford University Press, Oxford
74. Fisher ME (1974) Rev Mod Phys 46: 597
75. Privman V, Aharony A, Hohenberg PC (1991) In: Domb C, Lebowitz JL (eds) Phase transitions and critical phenomena, vol 14, Academic, New York
76. DeGennes PG (1977) J Phys Lett (Paris) 38: L44
77. Joanny JF (1987) J Phys A11: L117
78. Binder K (1983) J Chem Phys 79: 6387
79. Binder K (1984) Phys Rev A29: 341
80. Dudowicz J, Freed KF (1991) Macromolecules 24: 5076, 5112
81. Dudowicz J, Freed MS, Freed KF (1991) Macromolecules 24: 5096
82. Freed KF, Bawendi MG (1989) J Phys Chem 93: 2194
83. Bawendi MG, Freed KF (1988) J Chem Phys 88: 2741
84. Carmesin I, Kremer K (1988) Macromolecules 21: 2819
85. Carmesin I, Kremer K (1990) J Phys (Paris) 51: 915
86. Wittmann HP, Kremer K (1990) Computer Phys Commun 61: 309
87. Deutsch HP, Dickman R (1990) J Chem Phys 93: 8983
88. Deutsch HP, Binder K (1991) J Chem Phys 94: 2294
89. Paul W, Binder K, Heermann DW, Kremer K (1991) J Phys (Paris) II1: 37
90. Jilge W, Carmesin I, Kremer K, Binder K (1990) Macromolecules 23: 5001
91. Deutsch HP (1992) J Stat Phys 67: 1039
92. Deutsch HP, Binder K (1992) Macromolecules (in press)
93. Deutsch HP, Binder K (1992) Macromol Chem, Macromol Symp (in press)
94. Paul W, Binder K, Kremer K, Heermann DW (1991) Macromolecules 24: 6332
95. Binder K (1991) Macromol Chem, Macromol Symp 50: 1
96. Wittmann HP, Kremer K, Binder K (1992) J Chem Phys 96: 6291
97. Lopez-Rodriguez A, Wittmann HP, Binder K (1990) Macromolecules 23: 4327
98. Wittmer J, Paul W, Binder K (1992) Macromolecules 25: 7211
99. Deutsch HP, Binder K (1991) Europhys Lett 17: 697
100. Landau LD, Lifshitz EM (1958) Statistical Physics. Pergamon, Oxford
101. Sariban A, Binder K (1988) Macromolecules 21: 711
102. Sariban A, Binder K (1988) Macromol Chem 189: 2357
103. Sariban A, Binder K (1988) Colloid Polym Sci 266: 389
104. Kremer K, Binder K (1988) Computer Phys Rep 7: 259
105. Binder K, Heermann DW (1988) Monte Carlo simulation in statistical physics. An introduction. Springer, Berlin Heidelberg New York

106. Binder K (1992) In: Bicerano J (ed) Computer modelling of polymers. Marcel Dekker, New York p. 221
107. Sariban A, Binder K (1987) J Chem Phys 86: 5853
108. Sanchez IC, Lacombe RH (1976) J Phys Chem 80: 2352
109. Lacombe RH, Sanchez IC (1976) J Phys Chem 80: 2568
110. Sanchez IC, Lacombe RH (1978) Macromolecules 11: 1145
111. Sanchez IC, Balasz AC (1989) Macromolecules 22: 2325
112. Panayiotou C, Vera JH (1982) Polym J 14: 681
113. Chen SH, Chu B, Nossal R (eds) (1981) Scattering techniques applied to supramolecular and nonequilibrium systems. Plenum, New York
114. Jannink G, De Gennes PG (1968) J Chem Phys 48: 2260
115. A detailed pedagogic account can be found in: Kehr K (1991) In: 22.IFF Ferienkurs, Physik der Polymere. Forschungszentrum Jülich, Jülich, Germany, Chapter 5
116. Doi M, Edwards SF (1986) The theory of polymer dynamics. Clarendon Oxford
117. Benoit H, Benmouna M (1984) Macromolecules 17: 535
118. Brout R (1965) Phase transitions. Benjamin, New York
119. Joanny JF (1978) Comptes Rendu Acad Sci (Paris) 286B: 89
120. Binder K (1991) In: Materials science and technology, vol 5, Phase transformations in materials. VCH Verlagsges., Weinheim, Germany, chap 7
121. Komura S, Furukawa H (eds) (1988) Dynamics of ordering processes in condensed matter. Plenum, New York
122. Zettlemoyer AC (1969) Nucleation. Marcel Dekker, New York
123. Binder K (1987) Rep Progr Phys 50: 783
124. Nakanishi H, Pincus P (1983) J Chem Phys 79: 997
125. Schmidt I, Binder K (1985) J Physique (Paris) 46: 1631
126. Jones RAL, Norton LJ, Kramer EJ, Composto RJ, Stein RS, Russell TP, Mansour A, Karim A, Felcher GP, Rafailovich MH, Sokolov J, Zhao X, Schwarz SA (1990) Europhysics Lett 12: 41
127. Steiner U, Eiser E, Klein J, Budkowski A, Fetters LJ (1992) Science 258: 1126
128. Binder K (1991) In: Haasen P (ed) Materials science and technology, vol 5, Phase transitions in materials. VCH Verlagsges., Weinheim, Germany, chap 3
129. De Gennes PG (1980) J Chem Phys 72: 4756
130. Shibayama M, Yang H, Stein RS (1985) Macromolecules 18: 2179
131. Akcasu AZ, Sanchez IC (1988) J Chem Phys 88: 7847
132. Helfand E, Tagami Y (1971) J Chem Phys 56: 3592
133. Joanny JF, Leibler L (1978) J Physique (Paris) 39: 951
134. Binder K, Frisch HL (1984) Macromolecules 17: 2928
135. Roe RJ (1986) Macromolecules 19: 728
136. Tang H, Freed KF (1991) J Chem Phys 94: 1572, 6307
137. Zeng XC, Oxtoby DW, Tang H, Freed KF (1992) J Chem Phys 96: 4816
138. DeGroot SR, Mazur P (1962) Nonequilibrium thermodynamics. North-Holland, Amsterdam
139. Rouse PE (1957) J Chem Phys 21: 1273
140. Doi M, Edwards SF (1986) The theory of polymer dynamics. Claredon, Oxford
141. De Gennes PG (1971) J Chem Phys 55: 572
142. Pincus P (1981) J Chem Phys 75: 1996
143. Cook HE (1970) Acta metall 18: 297
144. Hohenberg PC, Halperin BI (1977) Rev Mod Phys 49: 435
145. Fredrickson GH, Bates FS (1986) J Chem Phys 85: 633
146. Kawasaki K (1970) Ann Phys 61: 1
147. Fredrickson GH (1986) J Chem Phys 85: 3556
148. Ginzburg VL (1960) Sov Phys Solid State 1: 1824
149. Cahn JW (1961) Acta Metall 9: 795
150. Cahn JW, Hilliard JE (1958) J Chem Phys 28: 258
151. Cahn JW (1968) Trans Metall Soc AIME 242: 166
152. Gunton JD, San Miguel M, Sahni PS (1983) In: Domb C, Lebowitz JL (eds) Phase transitions and critical phenomena, Academic, London, vol 8, p 267
153. Hashimoto T (1992) In: Kramer EJ (ed) Materials science and technology, vol 12, Structure and properties of polymers. VCH Verlag, Weinheim
154. Strobl GR (1985) Macromolecules 18: 558
155. Sariban A, Binder K (1991) Macromolecules 24: 578

156. Hill RG, Tomlins PE, Higgins JS (1985) Polymer 26: 1708
157. Hill RG, Tomlins PE, Higgins JS (1985) Macromolecules 18: 2555
158. Hashimoto T, Itakura M, Shimizu N (1986) J Chem Phys 85: 6773
159. Takenata M, Izumitani T, Hashimoto T (1987) Macromolecules 20: 2257
160. Onuki A, Hashimoto T (1989) Macromolecules 22: 879
161. Binder K (1989) In: Stocks LM, Gonis A (eds): Alloy phase stability, Kluwer, Dordrecht, p 233
162. Carmesin HO, Heermann DW, Binder K (1986) Z Physik B65: 89
163. Binder K, Frisch HL, Jäckle J (1986) J Chem Phys 85: 1505
164. Jäckle J, Pieroth M (1988) Z Phys B72: 25
165. Higgins JS, Fruitwala HA, Tomlins PE (1989) British Polymer J 21: 247
166. Schwahn D, Janßen S, Springer T (1992) J Chem Phys 97: 8775
167. Siggia ED (1979) Phys Rev A20: 595
168. Furukawa H (1985) Adv Phys 34: 705
169. Guenoun P, Gastaud R, Perrot F, Beysen D (1987) Phys Rev A36: 4876
170. Hashimoto T (1988) In: Komura S, Furukawa H (eds) Dynamics of ordering processes in condensed matter, Plenum, New York, p 421
171. Onuki A (1986) J Chem Phys 85: 1122
172. Izumitani T, Hashimoto T (1992) preprint, quoted in Ref 153
173. Hashimoto T, Takenaka M, Izumitani T (1992) J Chem Phys 97: 679
174. Hayward S, Heermann DW, Binder K (1987) J Stat Phys 49: 1053
175. Lironis G, Heermann DW, Binder K (1990) J Phys A23: L329
176. Kawasaki K, Sekimoto K (1989) Macromolecules 22: 3063
177. Lifshitz IM, Slyozov W (1961) J Phys Chem Solids 19: 35
178. Binder K, Stauffer D (1974) Phys Rev Lett 33: 1006
179. Binder K, Stauffer D (1976) Z Phys B24: 407
180. Takenaka M, Hashimoto T (1992) J Chem Phys 96: 6177
181. Chou YC, Goldburg WI (1980) Phys Rev A23: 858
182. Porod G (1951) Koll Z 124: 83
183. Porod G (1951) Koll Z 125: 51, 108
184. Yeung CY (1988) Phys Rev Lett 61: 1135
185. Furukawa H (1989) Phys Rev B40: 2341
186. Furukawa H (1989) J Phys Soc Jpn 58: 216
187. Ohta T, Nozaki H (1989) In: Tanaka F, Doi M, Ohta T (eds) Space-time organization in macromolecular fluids, Springer, Berlin Heidelberg New York, p 51
188. Chakrabarti A, Toral A, Gunton JD, Muthukumar M (1990) J Chem Phys 92: 6899
189. Shinozaki A, Oono Y (1991) Phys Rev Lett 66: 173; Akcasu AZ, Erman B, Bahar I (1992) Makromol Chem, Macromol Symp 62: 43; Akcasu AZ, Klein R (1992) Macromolecules, in press
190. Koga T, Kawasaki K (1991) Phys Rev A44: 817
191. Brochard F, Jouffroy J, Levison P (1983) Macromolecules 16: 1638
192. Akcasu AZ, Benmouna M, Benoit H (1986) Polymer 27: 1935
193. Kramer EJ, Green PF, Palmstrom CJ (1984) Polymer 25: 473
194. Sillescu H (1984) Makromol Chem, Rapid Commun 5: 519
195. Sillescu H (1987) Makromol Chem, Rapid Commun 8: 393
196. Schichtel T, Binder K (1987) Macromolecules 20: 1671
197. Kehr KW, Binder K, Reulein SM (1989) Phys Rev B39: 4891
198. Brochard F, De Gennes PG (1983) Physica 118A: 289
199. Brochard F, De Gennes PG (1986) Europhys Lett 1: 221
200. Brochard-Wyart F (1987) C R Acad Sci Sec II (France) 305: 657
201. Hess W, Frisch HL (1987) J Polym Sci Part C24: 269
202. Hess W, Akcasu AZ (1988) J Phys (Paris) 49: 1261
203. Akcasu AZ (1989) Macromolecules 22: 3682; Akcasu A Z, Nägele C, Klein R (1991) Macromolecules 24: 4408, Akcasu AZ, Tombakogcu M (1990) Macromolecules 23: 607; Akcasu AZ (1991) Macromolecules 24: 2109
204. Onuki A (1989) Phys Rev Lett 62: 2472
205. Helfand E, Fredrickson GH (1989) Phys Rev Lett 62: 2468
206. Onuki A (1990) J Phys Soc Jpn 59: 3423, 3427
207. Milner ST (1991) Phys Rev Lett 66: 1477
208. Doi M, Onuki A (1992) J Phys II (France) 2: 1631

209. Wittmann H-P, Fredrickson GH (1993) preprint
210. Brochard F (1988) In: Nagasawa M (ed) Molecular conformation and dynamics of macromolecules in condensed systems, Elsevier, Amsterdam, p 249
211. Binder K, Sillescu H (1989) In: Encyclopedia of polymer science and engineering, Supplement Volume, 2nd edn, J Wiley, New York, p 297
212. Cahn JW, Hilliard JE (1959) J Chem Phys 31: 668
213. Klein W, Unger C (1983) Phys Rev B28: 445
214. Heermann DW, Klein W (1983) Phys Rev B27: 1732
215. Schwahn D, Mortensen K, Yee-Madeira Y (1987) Phys Rev Lett 58: 1544
216. Janßen S, Schwahn D, Springer T (1992) Phys Rev Lett 68: 3180
217. Fischer EW, Meier G, Momper B, (1992) Macromol Chem; Macromol Symp 62: 120
218. Binder K (1988) In: Landau DP, Mon KK, Schüttler HB (eds) Computer simulation studies in condensed matter physics. Springer Proceedings in Physics, vol 33, Berlin Heidelberg New York, p 84
219. Pesci AI, Freed KF (1989) J Chem Phys 90: 2017
220. Freed KF, Dudowicz J (1992) Theor Chim Acta 82: 357
221. Dudowicz J, Freed KF (1992) J Chem Phys 96: 1644, 9147
222. Freed KF, Dudowicz J (1992) J Chem Phys 97: 2105
223. Han CC, Bauer BJ, Clark BJ, Muroya Y, Okads M, Tran-Cong Z, Sanchez IC (1988) Polymer 29: 2002
224. Yethiraj A, Schweizer KS (1992) J Chem Phys 97: 5927
225. Lipson JEG (1992) J Chem Phys 96: 1418
226. Hansen JP, McDonald IR (1986) Theory of simple liquids. Academic, London
227. Born M, Green HS (1946) Proc Royal Soc London, Ser A188: 10
228. Yvon Y (1935) Actual Scientifiques et Industriels. Herman et Cie, Paris
229. Kirkwood JG (1935) J Chem Phys 3: 300
230. Chandler D, Anderson HC (1972) J Chem Phys 57: 1930
231. Schweizer KS, Curro JG (1987) Phys Rev Lett 58: 256
232. Curro JG, Schweizer KS (1987) Macromolecules 20: 1928
233. Curro JG, Schweizer KS (1987) J Chem Phys 87: 1842
234. Curro JG, Schweizer KS, Grest GS, Kremer K (1989) J Chem Phys 91: 1357
235. Gehlsen MD, Rosedale JH, Bates FS, Wignall GD, Almdal K (1992) Phys Rev Lett 68: 2452
236. Honnell KG, Curro JG, Schweizer KS (1990) Macromolecules 23: 3496
237. Honnell KG, McCoy JD, Curro JG, Schweizer KS, Narten AH, Habenschuss A (1991) J Chem Phys 94: 4659
238. Joanny JF, Leibler L, Ball R (1984) J Chem Phys 81: 4640
239. Schäfer L, Kappeler Ch (1985) J Phys (Paris) 46: 1853
240. Broseta D, Leibler L, Joanny JF (1987) Macromolecules 20: 1935
241. Onuki A, Hashimoto T (1989) Macromolecules 22: 879
242. Fukuda T, Nagata M, Inagaki H (1984) Macromolecules 17: 548
243. Brereton MG, Vilgis TA (1989) J Phys (Paris) 50: 245
244. Onuki A, Kawasaki CK (1979) Annals of Physics (N.Y.) 121: 456
245. Imaeda T, Onuki A, Kawasaki K (1984) Progr Theor Phys 71: 16
246. Pistoor N, Binder K (1988) Colloid & Polymer Sci 266: 132
247. Pistoor N, Binder K (1988) In: Richter D, Springer T (eds) Polymer motion in dense systems, Springer, Berlin Heidelberg New York, p 285
248. Hashimoto T, Takebe T, Suehiro S (1988) J Chem Phys 88: 5874
249. Takebe T, Sawaoka R, Hashimoto T (1989) J Chem Phys 91: 4369
250. Hashimoto T, Takebe T, Fujioka K (1990) In: Onuki A, Kawasaki K (eds) Dynamics and patterns in complex fluids. Springer, Berlin Heidelberg New York
251. Binder K (1992) (ed) The Monte Carlo method in condensed matter physics Springer, Berlin Heidelberg New York
252. Allen MP, Tildesley DJ (1987) Computer simulation of liquids. Clarendon, Oxford
253. Ciccotti G, Frenkel D, McDonald IR (1987) (eds) Simulations of liquids and solids. North-Holland, Amsterdam
254. Olaj OF, Wimmer M, Zifferer G (1990) Makromol Chem Rapid Commun 11: 451
255. Wang JS, Binder K (1991) J Chem Phys 95: 8537
256. Rosenbluth MN, Rosenbluth AW (1955) J Chem Phys 23: 256
257. Deutsch HP, Binder K (1993) J Phys (France) II (in press)
258. Binder K (1988) Colloid & Polymer Sci 266: 871

259. Salsburg ZW, Jacobsen DJ, Fickett W, Wood WW (1959) J Chem Phys 30: 64
260. Ferrenberg AM, Swendsen RH (1988) Phys Rev Lett 61: 2635
261. Ferrenberg AM, Swendsen RH (1988) Phys Rev Lett 63: 1195
262. Binder K (1987) Ferroelectrics 73: 43
263. Privman V (1990) (ed) Finite size scaling and numerical simulation of statistical systems. World Scientific, Singapore
264. Binder K (1992) Ann Rev Phys Chem 43: 33
265. Sariban A, Binder K (1989) Colloid & Polymer Sci 267: 469
266. Deutsch HP (1993) J Chem Phys (in press)
267. Gauger A, Pakula T (1993) J Chem Phys 98: 3548
268. Binder K (1981) Z Physik B43: 119
269. Rovere M, Heermann DW, Binder K (1988) Europhys Lett 6: 585
270. Rovere M, Heermann DW, Binder K (1990) J Phys: Condensed Matter 2: 7009
271. Rovere M, Nielaba P, Binder K (1993) Z Phys B 90: 215
272. Sariban A, Binder K (1993) (To be published)
273. Binder K, Nauenberg M, Privman V, Young AP (1985) Phys Rev B31: 1498
274. LeGuillou JC, Zinn-Justin J (1980) Phys Rev B21: 3976
275. Binder K, Deutsch HP (1992) Europhys Lett 18: 667
276. Sariban A, Binder K, Heermann DW (1987) Phys Rev B35: 6873
277. Sariban A, Binder K, Heermann DW (1987) Colloid & Polymer Sci 265: 424
278. Meier G, Momper B, Fischer EW (1992) J Chem Phys (in press); Momper B, Meier G, Fischer EW (1991) J Noncryst Solids 130: 624
279. Budkowski A, Steiner U, Klein J, Schatz G (1992) Europhys Lett 18: 705
280. Chu B, Linliu K, Ying Q, Nose T, Okada M (1992) Phys Rev Lett 68: 3184
281. Olvera de la Cruz M, Edwards SF, Sanchez IC (1988) J Chem Phys 89: 1704
282. Freed KF (1988) J Chem Phys 88: 5871
283. Kumar SK (1992) J Chem Phys 97: 3550
284. Kumar SK, Szleifer I, Panagiotopoulos AZ (1991) Phys Rev Lett 66: 2935
285. Kumar SK, Szleifer I, Panagiotopoulos AZ (1992) Phys Rev Lett 68: 3456
286. Kumar SK (1992) J Chem Phys 96: 1490
287. Garbella RW, Wendorff JH (1986) Makromol Chem Rapid Comm 7: 591
288. Jones RAL, Klein J, Donald AM (1986) Nature 321: 16
289. Composto RJ, Mayer JW, Kramer EJ, White D (1986) Phys Rev Lett 57: 1312
290. Jordan AE, Ball RC, Donald AM, Fetters LJ, Jones RAL, Klein J (1988) Macromolecules 21: 235
291. Brautmeier D, Stamm M, Lindner P (1991) J Appl Cryst 24: 665
292. Bartels CR, Crist B, Fetters LJ, Graessley WW (1986) Macromolecules 19: 785
293. a. Stamm M, Hüttenbach S, Reiter G, Springer T (1991) Europhys Lett 14: 451
b. Stamm M (1992) Adv Polymer Sci 100: 357
294. Hair DW, Hobbie EK, Douglas J, Han CC (1992) Phys Rev Lett 68: 2476
295. Sariban A, Binder K (1989) Polymer Commun 30: 205
296. Baumgärtner A, Heermann DW (1986) Polymer 27: 1777
297. Forrest BM, Heermann DW (1991) J Phys (France) II1: 909
298. Kron AK (1965) Polymer Sci USSR 7: 1361
299. Wall FT, Mandel F (1975) J Chem Phys 63: 4592
300. Chakrabarti A, Toral R, Gunton JD, Muthukumar M (1989) Phys Rev Lett 63: 2072
301. Chakrabarti A, Toral R, Gunton JD, Muthukumar M (1990) J Chem Phys 92: 6899
302. Brown G, Chakrabarti A (1993) J Chem Phys 98: 2451
303. Thomas EL, Anderson DM, Henke CS, Hoffmann D (1988) Nature 334: 598
304. Ruland WJ (1971) Appl Crystallogr 4: 70
305. Helfand E, Tagami Y (1971) J Polym Sci B9: 741
306. Helfand E, Tagami Y (1972) J Chem Phys 57: 1812
307. Helfand E (1975) Acc Chem Res 8: 295
308. Helfand E (1975) J Chem Phys 62: 999
309. Helfand E (1975) J Chem Phys 63: 2192
310. Helfand E, Sapse AM (1975) J Chem Phys 62: 1327
311. Helfand E, Weber TA (1976) Macromoleculels 9: 311
312. Helfand E (1976) Macromolecules 9: 307
313. Helfand E, Wasserman ZR (1982) In: Goodman I (ed) Developments in block copolymers I. Applied Science, New York, p 99

314. Meier DJ (1969) J Polymer Sci C26: 81
315. Leary DF, Williams MC (1970) J Polymer Sci B8: 335
316. Anderson DM, Thomas EL (1988) Macromolecules 21: 3221
317. Bates FS, Rosedale JH, Fredrickson GH, Glinka CJ (1988) Phys Rev Lett 61: 2229
318. Bates FS, Rosedale JH, Fredrickson GH (1990) J Chem Phys 92: 6255
319. Almdal K, Rosedale JH, Bates SF, Wignall GD, Fredrickson GH (1990) Phys Rev Lett 65: 1112
320. Almdal K, Koppi K, Bates SF, Mortensen K (1992) preprint
321. Gerharz B, Fischer EW, Fytas G (1991) Polymer Commun 32: 469; Gerharz B, Vogt S, Fytas G, Fischer EW (1992) Coll Polym Sci (in press)
322. Holzer B, Lehmann A, Strobl G, Stühn B, Kowalski M (1991) Polymer 32: 1935
323. Stühn B, Mutter R, Albrecht T (1992) Europhys Lett 18: 427
324. Toledano JC, Toledano P (1987) The Landau theory of phase transitions. World Scientific, Singapore
325. Fried H, Binder K (1991) J Chem Phys 94: 8349
326. Fried H, Binder K (1991) Europhys Lett 16: 237
327. Brazovskii SA (1975) Soviet Phys JETP 41: 85
328. Minchau B, Dünweg B, Binder K (1990) Polymer Comm 31: 348
329. Barrat JL, Fredrickson GH (1992) J Chem Phys 95: 1281
330. Vilgis T, Borsali R (1990) Macromolecules 23: 3172
331. Vilgis T, Borsali R (1991) Phys Rev A43: 6857
332. Edwards SF (1965) Proc Phys Soc London 85: 613
333. Tang H, Freed KF (1992) J Chem Phys 96: 8621
334. Hadziioannou G, Skoulios A (1982) Macromolecules 15: 258
335. Melenkovitz J, Muthukumar M (1991) Macromolecules 24: 4199
336. Ramakrishnan TV, Youssouf M (1979) Phys Rev B19: 2775
337. Oxtoby D W (1990) In: Hansen JP, Levesque D, Zinn-Justin J (ed) Liquids, freezing and the glass transition. North-Holland, Amsterdam
338. Olvera de la Cruz M, Mayes AM, Swift BW (1992) Macromolecules 25: 944
339. Holyst R, Schick M (1992) J Chem Phys 96: 7728
340. Fredrickson GH (1986) J Chem Phys 85: 5306
341. Onuki A (1987) J Chem Phys 87: 3692
342. Fredrickson GH, Helfand E (1988) J Chem Phys 89: 5890
343. Fredrickson GH, Milner ST (1990) In: Safinya CR, Safran PA (eds) MRS Symposium Proceedings, vol 177, Macromolecular liquids, p 169 Material Research Society, Boston
344. Vilgis TA, Benmouna M (1992) Macromol Chem Theory and Simulations 1: 25
345. Hashimoto T (1987) Macromolecules 20: 465
346. Harkless CR, Singh MA, Nagler SE, Stephensen GB, Jordan-Sweet JL (1990) Phys Rev Lett 64: 2285
347. Kawasaki K, Sekimoto K (1989) Macromolecules 22: 3063
348. Fredrickson GH, Leibler L (1990) Macromolecules 23: 531
349. Fredrickson GH, Milner ST (1991) Phys Rev Lett 67: 835
350. Mayers AM, Olvera de la Cruz M (1989) J Chem Phys 91: 7228
351. Chakrabarti A, Toral R, Gunton JD (1989) Phys Rev Lett 63: 2661
352. Oono Y, Bahiana M (1988) Phys Rev Lett 61: 1109
353. Chakrabarti A, Toral R, Gunton JD (1991) Phys Rev A44: 6503
354. Wu S (1982) Polymer interfaces and adhesion. Dekker, New York
355. Kausch HH, Tirell M (1989) Ann Rev Mater Sci 19: 347
356. Klein J (1990) Science 250: 640
357. Steiner U, Krausch G, Schatz G, Klein J (1990) Phys Rev Lett 64: 1119
358. DeGennes PG (1989) C R Hebd Seances Acad Sci Ser B308: 13
359. Harden JL (1990) J Phys (Paris) 51: 1777
360. Puri S, Binder K (1991) Phys Rev B44: 9735
361. Leibler L (1988) Makromol Chem, Macromol Symp 16: 1; Semenov AN (1992) Macromolecules 25: 4967
362. Broseta D, Fredrickson GH, Helfand E, Leibler L (1990) Macromolecules 23: 123
363. Carton JP, Leibler L (1990) J Phys (Paris) 51: 1683
364. Leibler L (1991) Physica A172: 258
365. Blakely JM (1979) In: Vanselow R (ed) Chemistry and physics of solid surfaces vol 2, CRC Press, Boca Raton, Florida, p 1
366. Schmidt I, Binder K (1987) Z Physik B67: 369

367. Binder K, Frisch HL (1991) Z Physik B84: 403
368. Dietrich S (1988) In: Domb C, Lebowitz L (eds) Phase transitions and critical phenomena, vol 12, Academic, London, p 1
369. Carmesin I, Noolandi J (1989) Macromolecules 22: 1689
370. Chen YZ, Noolandi J, Izzo D (1991) Phys Rev Lett 66: 727
371. Cifra P, Karasz FE, MacKnight WJ (1992) Macromolecules 25: 4895
372. Wang JS, Binder K (1992) Makromol Chem: Theory and Simulation 1: 49
373. Fredrickson GH, Donley JP (1992) J Chem Phys 97: 8941
374. Jones RAL, Norton LJ, Kramer EJ, Bates FS, Wiltzius P (1991) Phys Rev Lett 66: 1326
375. Puri S, Binder K (1992) Phys Rev A46: R4487; Ball RC, Essery RLH (1990) J Phys: Condensed Matter 2: 10303
376. Jones RAL, Kramer EJ (1990) Phil Mag B62: 129
377. Fredrickson GH (1992) In: Sanchez IC (ed) Physics of polymer surfaces and interfaces. Butterworths, Boston
378. Fredrickson GH (1987) Macromolecules 20: 2535
379. Tang H, Freed KF (1992) J Chem Phys 97: 4496
380. Thomas HR, O'Malley J (1979) Macromolecules 12: 323
381. Hasegawa H, Hashimoto T (1985) Macromolecules 18: 589
382. Coulon G, Russell TP, Deline VR, Green PF (1989) Macromolecules 22: 2581
383. Russell TP, Coulon G, Deline VR, Miller DC (1989) Macromolecules 22: 4600
384. Anastasiadis SH, Russell TP, Satija SK, Majkrzak CF (1989) Phys Rev Lett 62: 1852
385. Shull KR (1992) Macromolecules 25: 2122
386. Kikuchi M, Binder K (1993) Europhysics Letter 21: 427
387. Koizuni S, Hasegawa H, Hashimoto T (1992) Macromol Chem, Macromol Symp 62: 75
388. Russell TP, Menelle A, Anastasiadis SH, Satija SK, Majkrzak CF (1992) Macromol Chem, Macromol Symp 62: 92
389. Vilgis TA, Benmouna M, Benoit H (1991) Macromolecules 24: 4481

Editor: Prof. H.-H. Kausch
Received November 16, 1992

Author Index Volume 101-112

Author Index Vols. 1-100 see Vol. 100

Subject Index

Springer-Verlag and the Environment

We at Springer-Verlag firmly believe that an international science publisher has a special obligation to the environment, and our corporate policies consistently reflect this conviction.

We also expect our business partners – paper mills, printers, packaging manufacturers, etc. – to commit themselves to using environmentally friendly materials and production processes.

The paper in this book is made from low- or no-chlorine pulp and is acid free, in conformance with international standards for paper permanency.